FROM PHYSICS TO CONTROL THROUGH AN EMERGENT VIEW

WORLD SCIENTIFIC SERIES ON NONLINEAR SCIENCE

Editor: Leon O. Chua
University of California, Berkeley

Series B. SPECIAL THEME ISSUES AND PROCEEDINGS

Volume 1: Chua's Circuit: A Paradigm for Chaos
Edited by *R. N. Madan*

Volume 2: Complexity and Chaos
Edited by *N. B. Abraham, A. M. Albano, A. Passamante, P. E. Rapp, and R. Gilmore*

Volume 3: New Trends in Pattern Formation in Active Nonlinear Media
Edited by *V. Perez-Villar, V. Perez-Munuzuri, C. Perez Garcia, and V. I. Krinsky*

Volume 4: Chaos and Nonlinear Mechanics
Edited by *T. Kapitaniak and J. Brindley*

Volume 5: Fluid Physics — Lecture Notes of Summer Schools
Edited by *M. G. Velarde and C. I. Christov*

Volume 6: Dynamics of Nonlinear and Disordered Systems
Edited by *G. Martínez-Mekler and T. H. Seligman*

Volume 7: Chaos in Mesoscopic Systems
Edited by *H. A. Cerdeira and G. Casati*

Volume 8: Thirty Years After Sharkovskiĭ's Theorem: New Perspectives
Edited by *L. Alsedà, F. Balibrea, J. Llibre, and M. Misiurewicz*

Volume 9: Discretely-Coupled Dynamical Systems
Edited by *V. Pérez-Muñuzuri, V. Pérez-Villar, L. O. Chua, and M. Markus*

Volume 10: Nonlinear Dynamics & Chaos
Edited by *S. Kim, R. P. Behringer, H.-T. Moon, and Y. Kuramoto*

Volume 11: Chaos in Circuits and Systems
Edited by *G. Chen and T. Ueta*

Volume 12: Dynamics and Bifurcation of Patterns in Dissipative Systems
Edited by *G. Dangelmayr and I. Oprea*

Volume 13: Modeling and Computations in Dynamical Systems: In Commemoration of the 100th Anniversary of the Birth of John von Neumann
Edited by *E. J. Doedel, G. Domokos & I. G. Kevrekidis*

Volume 14: Dynamics and Control of Hybrid Mechanical Systems
Edited by *G. Leonov, H. Nijmeijer, A. Pogromsky & A. Fradkov*

WORLD SCIENTIFIC SERIES ON NONLINEAR SCIENCE Series B Vol. 15

Series Editor: Leon O. Chua

FROM PHYSICS TO CONTROL THROUGH AN EMERGENT VIEW

edited by

Luigi Fortuna
University of Catania, Italy

Alexander Fradkov
Russian Academy of Sciences, Russia

Mattia Frasca
University of Catania, Italy

World Scientific

NEW JERSEY • LONDON • SINGAPORE • BEIJING • SHANGHAI • HONG KONG • TAIPEI • CHENNAI

Published by

World Scientific Publishing Co. Pte. Ltd.

5 Toh Tuck Link, Singapore 596224

USA office: 27 Warren Street, Suite 401-402, Hackensack, NJ 07601

UK office: 57 Shelton Street, Covent Garden, London WC2H 9HE

British Library Cataloguing-in-Publication Data
A catalogue record for this book is available from the British Library.

FROM PHYSICS TO CONTROL THROUGH AN EMERGENT VIEW
World Scientific Series on Nonlinear Science, Series B — Vol. 15

ISBN-13 978-981-4313-14-8
ISBN-10 981-4313-14-9

Printed by FuIsland Offset Printing (S) Pte Ltd. Singapore

PREFACE

This thematic book contains selected papers from the 4th International Conference on Physics and Control (PhysCon2009), held at the University of Catania, Engineering Faculty, 1-4 September 2009 and organized by the University of Catania and the International Physics and Control Society (IPACS). In particular, the idea was to focus on including the contributions presented during the special sessions and minisymposia of the Conference. The plenary talks and some invited contributions are also taken into account.

We thank all the authors of the papers presented at PhysCon2009, their contributions made the conference more scientifically appealing. All the papers of PhysCon2009 have been included in a special issue of the IPACS electronic library.

The limited number of papers included in this book gives a complete overview of the activities that covered the area of Physics and Control of PhysCon2009. The interdisciplinarity of both the areas has made this book the first volume covering a wide range of topics on Physics and Control.

This approach leads to a better understanding of the environment in which we live, also allowing us to face new problems that have appeared in control engineering disciplines. We tried to find links among the infinitely many interesting issues that characterize the existing areas of Physics and Control.

In the practice of Physics and Control Engineering, considerable interest has been devoted to the question of whether it is possible to control physical systems, and vice versa, whether it is possible by using intrinsic physical principles to control general systems. The genesis of Physics and Control connection is very old. Furthermore, over the recent decades exploring complexity, the discovery of fundamental new properties of matter, far from equilibrium conditions, the central discovery of prevalence instability in physics, economics and the social world have led to a crucially important perspective to integrate phenomenological evidences in embedded information based devices and circuits devoted to advanced control strategies for large scale systems. Both experimental and analytical methods are combined to emphasize the two paradigms of investigations in both Physics and Control Engineering for the real view of the problems and their practical understanding, looking into the possibility of generalizations by using mathematical modeling strategies.

In this view, the understanding of innovative control strategies arising from physics methods is considered appealing. Besides, physical theories and applications are nonlinear as more real control applications are nonlinear.

A common principle in this book is to emphasize the intrinsic nonlinearity and the growing complexity in control systems. This publication hopes to stimulate new research routes, and create new research networks by encouraging emergent research strategies. Here, we remark that multidisciplinarity is strategic for research in this century.

From the topics included in the book the following key methods, items and reference terms have been identified.

Key methods:

- advanced matrix linear algebra;
- quantum computation;
- multistability;
- absolute stability;
- harmonic balancing;
- set membership description;
- nonlinear stochastic systems;
- delayed feedback control;
- network theory;
- chaos-based computing;
- synchronization.

Key items:

- biology;
- neuroscience;
- nanotechnology;
- lasers;
- semiconductors;
- microfluidics;
- electromechanical systems;
- manufacturing systems;
- geophysics;
- plasma;
- sensors;
- embedded systems;

- electronic circuits.

Selected terms:

- uncertain destination dynamics;
- chaos computing;
- set membership description;
- virtual sensors;
- institutionalization.

The book is divided into 11 parts. Part A collects the plenary talks of invited speakers and the contributions of the speakers partially supported by the European Grants' Committee of the European Physical Society (EPS). The other chapters are each devoted to one special session. In particular, part B consists of five chapters selected from the contributions on the special session "Modelling and control of coupled stochastic oscillators" organized by Prof. G. Rigatos. Part C collects four contributions from the special session "Multistability in natural and laboratory-scale nonlinear systems" organized by Prof. B. K. Goswami and Prof. U. Feudel. Part D consists of three contributions from the special session "Linear and matritial algebra, open problems related to control theory" organized by Prof. M. Isabel García-Planas. Part E consists of three selected papers from the special sessions on "Localization of oscillations in dynamical systems" (organized by Prof. G. A. Leonov) and "Control of oscillatory delayed-coupled networks" (organized by Prof. C. Masoller and Prof. J. Garcia-Ojalvo). Part F collects four papers from the special session "Microfluidics: Theory, methods and applications" organized by Prof. M. Bucolo. Part G consists of eight chapters selected from the contributions to the special session "Mathematical modelling of dynamic systems for volcano physics" organized by Prof. C. Del Negro. Part H consists of six papers from the special session "Geometric control for quantum and classical models" organized by Prof. A. Sarychev and Prof. U. Boscain. Part I consists of eight chapters from the special session "Control problems for dynamical systems under uncertainty and conflict" organized by Prof. T. F. Filippova. Part J collects seven papers from the special session "Physics and control in fusion plasma devices" organized by Prof. G. Vagliasindi, Prof. G. Mazzitelli and Prof. A. Murari. Part K consists of four contributions from the special session on "Modeling and optimization of beam and plasma dynamics" organized by A. Ovsyannikov. The editors of the book would like to thank all the organizers of the special session and Prof. J. Pereira for having organized the mini-symposium on "EU-funded projects in the area of control and large-scale systems".

The editors of the book would also like to thank all the sponsors and institutions who have supported and contributed to the success of the conference, and, in particular, the University of Catania, IPACS (International Physics and Control Society), the EPS European Physical Society, the IEEE Institute of Electrical and Electronics Engineers, the Scuola Superiore di Catania, the INFN Istituto Nazionale di Fisica Nucleare, the City University of Hong Kong, the INGV — Istituto Nazionale di Geofisica e Vulcanologia, the Schneider Electric, the COST Action B27 ENOC, the ST Microelectronics and the Azienda Agricola Avola. The Conference has been held under the patronage of the "Accademia Gioenia di Catania".

L. Fortuna
Catania, Italy
A. Fradkov
St. Petersburg, Russia
M. Frasca
Catania, Italy

INTERNATIONAL PROGRAMME COMMITTEE

CHAIR

L. Fortuna

CO-CHAIR

M. Frasca

MEMBERS

Y. Astrov	M. Berz	S. Boccaletti	A. Bulsara
C. Celikovsky	M. Di Bernardo	R. Genesio	B. Gluckman
C. Grebogi	M. Hasler	D.J. Hill	A. Kovaleva
V. Latora	M. Lattuada	G. Leonov	G. Mazzitelli
R. Meucci	I. Mezic	J.L. Moiola	M. Ogorzalek
D. Ovsyannikov	P. Parmananda	U. Parlitz	M. Pavon
L. Pecora	A. Pikovsky	K. Pyragas	A. Rapisarda
G. Rizzotto	T. Roska	E. Scholl	S. Sieniutycz
Y.P. Tian	L. Viola	N. Wessel	X. Yu

HONORARY MEMBERS

F.T. Arecchi	I. Blekhman	G. Chen	F. Chernousko
L. O. Chua	A. Fradkov	H. Haken	H. Kimura
J. Kurths	H. Rabitz	A. Sharkovsky	

LOCAL ORGANIZING COMMITTEE

B. Andò	P. Arena	S. Baglio	A. Basile
M. Bucolo	A. Buscarino	C. Camerano	R. Caponetto
C. Del Negro	F. Di Grazia	G. Dongola	A. Gallo
G. Ganci	S. Graziani	M. La Rosa	G. Muscato
D. Nicolosi	G. Nunnari	A. Rizzo	G. Sciuto
G. Sicurella	G. Tomarchio	G. Vagliasindi	M.G. Xibilia

CONTENTS

PART A

Distinguished talks (plenary talks and EPS talks)

HIDDEN OSCILLATION IN DYNAMICAL SYSTEMS

G. A. LEONOV

Faculty of Mathematics and Mechanics
St.Petersburg State University
Russia
E-mail: leonov@math.spbu.ru

Effective methods of searching periodic oscillations of dynamical systems are described. Their applications to Hilbert's 16-th problem for quadratic systems and Aizerman's problem are considered. The synthesis of the method of harmonic linearization, the applied theory of bifurcations, and the numerical methods of computation of periodic oscillations are described.

Keywords: Hidden oscillation, Hilbert problem, Aizerman problem.

In the making and original development of the theory of nonlinear oscillations in the first half of twentieth century, particular attention has been given to analysis and synthesis of oscillating systems, for which a solution of the problem of existence of oscillating modes was not too difficult. The structure itself of many mechanical, electromechanical, and electronic systems was such that they have oscillating modes of operation, the existence of which was "almost obvious". Therefore the main attention of researchers was given to the analysis of forms and properties of these oscillations ("almost" harmonic, relaxation, synchronous, circular, orbitally stable, and so on).

In the 50-s of last century the attention of many scholar was concentrated on two famous problems: Hilbert's 16-th problem and Aizerman's problem, for which the proof of existence of periodic solutions was a nontrivial problem. And in studying these problems it was made much progress. It turns out that they have many similar features: while Hilbert has formulated, at first, the problem of searching periodic solutions for two-dimensional polynomial systems, in studying the Aizerman problem it was stated that differential equations of systems of automatic control, satisfying the generalized Routh–Hurwitz conditions, also can have periodic solutions. In addition for further investigations in this direction it also becomes actual a problem of searching periodic solutions of such differential equations.

These problems stimulate a great flow of investigations in the second half of twentieth century. While Hilbert's 16-th problem greatly encouraged a development of the theory of bifurcations and the theory of normal forms, the Aizerman problem encouraged a development of the theory of absolute stability.

Arnold writes [1]: "To estimate the number of limit cycles of quadratic vector fields on plane, A.N. Kolmogorov had distributed a few hundreds of such fields (with randomly chosen coefficients of quadratic expressions) among a few hundreds of students of Mechanics and Mathematics Faculty of MGU for a further mathematical practice. Each student had to find the number of limit cycles of his/her field. The result of this experiment was absolutely unexpected: not a single field had a limit cycle! It is known that a limit cycle persists under a small change of field coefficients. Therefore, the systems with one, two, three (and even, as has become known later, four) limit cycles form an open set in the space of coefficients and in this case for a random choice of polynomial coefficients, the probability of hitting in it is positive. The fact that this did not occur permits one to suggest that the above-mentioned probabilities are, evidently, small."

The result of this experiment shows also the necessity of development of goal-oriented methods (as analytic, as numerical ones) for the search of periodic oscillations, which would use all horsepower of modern computational technique. The present survey is devoted to description of the certain of such methods.

We shall consider the problem of Kolmogorov and clear up if there exist two-dimensional quadratic dynamical systems, for which the students taking part in the above-described practical work could find limit cycles. For this purpose we reduce an arbitrary quadratic system to special Lienard equation.

Consider a quadratic system

$$\begin{aligned} \dot{x} &= a_1x^2 + b_1xy + c_1y^2 + \alpha_1x + \beta_1y \\ \dot{y} &= a_2x^2 + b_2xy + c_2y^2 + \alpha_2x + \beta_2y, \end{aligned} \tag{1}$$

where $a_i, b_i, c_i, \alpha_i, \beta_i$ are real numbers.

Proposition 1. Without loss of generality, one can assume that $c_1 = 0$.

Proposition 2. In studying limit cycles, for $b_1 \neq 0$ system (1) with $c_1 = 0$ can be reduced to the Lienard equation

$$\dot{x} = y, \quad \dot{y} = -f(x)y - g(x), \tag{2}$$

where

$$f(x) = R(x)e^{p(x)} = R(x)|\beta_1 + b_1x|^q,$$

$$g(x) = P(x)e^{2p(x)} = P(x)|\beta_1 + b_1 x|^{2q}; \quad q = -c_2/b_1$$

$$R(x) = -\frac{(b_1 b_2 - 2a_1 c_2 + a_1 b_1)x^2}{(\beta_1 + b_1 x)^2} + \\ + \frac{(b_2\beta_1 + b_1\beta_2 - 2\alpha_1 c_2 + 2a_1\beta_1)x + \alpha_1\beta_1 + \beta_1\beta_2}{(\beta_1 + b_1 x)^2}$$

$$P(x) = -\left(\frac{a_2 x^2 + \alpha_2 x}{\beta_1 + b_1 x} - \frac{(b_2 x + \beta_2)(a_1 x^2 + \alpha_1 x)}{(\beta_1 + b_1 x)^2} + \frac{c_2(a_1 x^2 + \alpha_1 x)^2}{(\beta_1 + b_1 x)^3}\right)$$

and for $b_1 = 0$ to equation (2) with the functions

$$f(x) = R(x)e^{qx}, \quad g(x) = P(x)e^{2qx}; \quad q = -c_2/\beta_1$$

For equation (2) with the smooth functions f and g it is well known the theorem of Lienard on the existence of limit cycle [2,3]. This theorem will be extended here to the case of discontinuous functions f and g and be applied then to the problem of Kolmogorov. We suppose that the functions $f(x)$ and $g(x)$ are differentiable on the interval $(a, +\infty)$ and for certain numbers $a < \nu_1 \leq x_0 \leq \nu_2$ the following conditions

1) $g(x) < 0, \ \forall x \in (a, x_0); \ g(x) > 0, \ \forall x \in (x_0, +\infty)$

$$\lim_{x \to a} G(x) = \lim_{x \to +\infty} G(x) = +\infty; \; G(x) = \int_{x_0}^{x} g(z)dz$$

2) $f(x) > 0, \ \forall x \in (a, \nu_1) \cup (\nu_2, +\infty); \ F_1(\nu_2) \geq 0; F_k(x) =$

$$= \int_{\nu_k}^{x} f(z)dz, \; k = 1, 2$$

are satisfied.

Theorem 1. [4] Let Conditions 1 and 2 be satisfied and the point $x = x_0, y = 0$ is unstable Lyapunov focal equilibrium. Then system (2) has a limit cycle.

Theorem 1 provides Kolmogorov's problem solving. The parameters of system (1), discriminated by Theorem 1, correspond to the limit cycles, which can easily be computed by using modern packets. Thus, for example, for $a_1 = b_1 = \beta_1 = 1$, $c_1 = \alpha_1 = 0$, $b_2 = 0$, $c_2 = 3/4$, $\beta_2 = 1$, $\alpha_2 = -2$ by Theorem 1 the limit cycle of system (1) exists and can be found. It is shown in Fig. 1

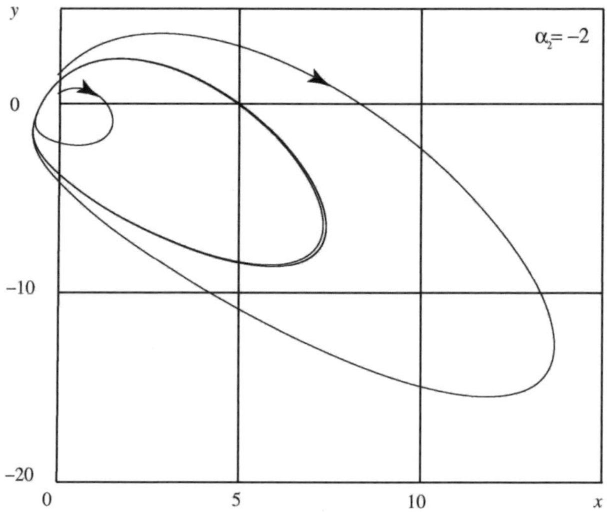

Fig. 1.

Consider a system

$$\begin{aligned}\dot{x}_1 &= -\omega_0 x_2 + b_1 \varphi_0(x_1 + c^* x_3)\\ \dot{x}_2 &= \omega_0 x_1 + b_2 \varphi_0(x_1 + c^* x_3)\\ \dot{x}_3 &= A x_3 + b \varphi_0(x_1 + c^* x_3).\end{aligned} \tag{3}$$

Here A is a constant $((n-2) \times (n-2))$-matrix, all eigenvalues of which have negative real arts, b and c are constant $(n-2)$-vectors, b_1 and b_2 are certain numbers:

$$\varphi_0(\sigma) = \begin{cases} \mu\sigma, & \forall \sigma \in (-\varepsilon, \varepsilon) \\ M\varepsilon^3, & \forall \sigma > \varepsilon \\ -M\varepsilon^3, & \forall \sigma < -\varepsilon \end{cases},$$

where μ, M are certain positive numbers, ε is a small positive parameter.

Let us write now a transfer function of system (3):

$$W(p) = \frac{\alpha p + \beta}{p^2 + \omega_0^2} + c^*(A - pI)^{-1} b.$$

Here $\alpha = -b_1, \beta = b_2 \omega_0$.

Theorem 2. [5] If the inequalities $\alpha > 0$ and $\mu \beta r^* q > \alpha \omega_0^2$ are satisfied, then system (3) has periodic solutions such that

$$x_1(0) = O(\varepsilon^2), \quad x_3(0) = O(\varepsilon^2), \quad x_2(0) = -\sqrt{\frac{\mu(\mu\beta r^* q - \alpha\omega_0^2)}{3\omega_0 M\alpha}} + O(\varepsilon).$$

In a number of cases this theorem provides Aizerman's problem solving.

References

1. V.I. Arnold, *Experimental mathematics*, (Fazis, M, 2005).
2. L. Cesari, *Asymptotic Behavior and Stability Problems in Ordinary Differential Equations*, (Berlin– Springer, 1959).
3. S. Lefschertz, *Differential Equations: Geometric Theory*, (N. Y.: L.: Interscience, 1957).
4. G.A. Leonov, *Intern. J. Bifurcation and Chaos.* **18**, (2008).
5. G.A. Leonov, *Automation and remote control.* **7**, (2009).

ANHARMONICITY AND SOLITON-MEDIATED ELECTRIC TRANSPORT. IS A KIND OF SUPERCONDUCTION POSSIBLE AT ROOM TEMPERATURE?

M. G. VELARDE*, W. EBELING and A. P. CHETVERIKOV

Instituto Pluridisciplinar, Universidad Complutense de Madrid
Paseo Juan XXIII 1, Madrid-28040, Spain
*E-mail: *mgvelarde@pluri.ucm.es*
www.ucm.es/info/fluidos

We discuss here models for soliton-mediated electron transport that may underlie the possibility of superconductivity.

Keywords: Morse interaction, anharmonicity, lattice solitons, solectrons, electron transfer, high T superconductivity.

1. Introduction

It is customary to denote by *solitary* waves certain localized (single-event) nonlinear waves of translation, i.e., waves that cause a net displacement of e.g. the liquid in the direction of the wave motion like tsunami in the ocean or bores in rivers. This denomination may also apply to nonlinear periodic waves or wave trains. Surfing on a river bore or on a huge wave approaching the sea-shore is a form of wave-mediated transport. Some of those single waves or wave peaks may exhibit particle-like behavior upon collision among themselves or reflection at walls as already noted long ago by the pioneer Russell. Their particle-like behavior led Zabusky and Kruskal to introduce the concept of soliton (bores or hydraulic jumps or even kinks are also called "topological" solitons, whereas waves of "elevation" or "depression" are denoted as non-topological solitons -aka "bright" and "dark" solitons, respectively- in condensed matter physics). We also wish to highlight the work done by Toda on a lattice (he invented) with an exponential interaction (with repulsion akin to Morse and Lennard-Jones potentials) that due to its integrability permitted obtaining exact explicit analytical solutions. The Toda interaction yields the hard rod impulsive force in one limit (the fluid-like or "molten" phase) while in another limit it becomes a harmonic oscillator (the lattice crystal-like solid phase). Worth also mentioning is that Schrieffer, Heeger, and collaborators have used (topological) solitons to explain the electric conductivity of polymers though in this case solitons come from the degeneracy of the ground state and not from an originally underlying lattice anharmonicity. The present authors have shown that lattice solitons like in a Toda lattice can trap electrons thus forming dynamic bound states called solectrons. The latter may act as electric carriers thus generalizing the polaron concept and quasiparticle introduced by Landau and Pekar.

2. One-dimensional (1D) models

We consider a 1D nonlinear lattice which is treated classically, augmented with an added excess electron that will be considered within the quantum tight binding approximation (TBA). The lattice interactions are assumed to be of Morse type, hence allowing for phonon -and soliton- longitudinal vibrations with compressions governed by the repulsive part of the potential. Thus we consider the Hamiltonian $H = H_{lattice} + H_{electron}$, with

$$H_{lattice} = \sum_n \left\{ \frac{p_n^2}{2M} + D\left(1 - exp\left[-B\left(q_n - q_{n-1}\right)\right]\right)^2 \right\}, \tag{1}$$

and

$$H_{electron} = E_n\left(q_k\right) c_n^* c_n - \sum_n V_{nn-1}\left(q_k\right)\left(c_n^* c_{n-1} + c_n c_{n-1}^*\right), \tag{2}$$

with n denoting the lattice site where the electron is (in probability density) "placed"; the complex quantities c_n give the n-th component of the electron wave function, and $p_n = |c_n|^2$ gives the probability of finding the electron at site n. The state energy at site n may depend on the particle displacements of the neighbors. We can use the linear ansatz

$$E_n = E_n^0 + \chi_0 q_n + \chi_1\left(q_{n+1} - q_{n-1}\right), \tag{3}$$

for rather low values of χ_0 and χ_1 as we want to neglect effects owing to energy shifts relative to hopping effects controlled by V_{nn-1}, the transfer matrix elements along the chain (considering only nearest neighbors). A reasonable choice for V_{nn-1} is

$$V_{nn-1} = V_0 exp\left[-\alpha\left(q_n - q_{n-1}\right)\right], \tag{4}$$

where α accounts for the strength of the electron-lattice coupling. We shall measure all energies in units 2D except the energy levels which are scaled with $\hbar\omega_0$ (quantum of the oscillations around the minimum of the Morse potential). Further we consider the system in a "thermal bath" characterized by a Gaussian white noise, ξ_j, of zero mean and time delta correlated.

We take ω_0^{-1} and B^{-1} as unit of time and displacements, respectively. Then $\tau = V_0/\hbar\omega_0$ gives the ratio of the two time scales involved in the dynamics. The system (1)-(2) permits both phonon and soliton (solectron) assisted hopping. An important role plays the temperature T. Indeed, for solitons to be sustained moving unaltered along the lattice the temperature must be high enough. According to the specific heat characteristics of the lattice the solitons are expected beyond the Dulong-Petit plateau (for biomolecules this corresponds to room temperature). When the lattice is heated both phonons (infinitessimal, linear disturbances) and solitons (finite amplitude, nonlinear disturbances) may be excited. However, we do not have just one or two solitons, but many of them with a finite life-time up to a few picoseconds. The solectron picture is now that the electron probability density is concentrated in local "hot spots" created by the local soliton thermal excitations.

Using the corresponding classical and quantum evolution equations obtained from (1) and (2), respectively, in a series of computer experiments we released an electron into à lattice already appropriately heated by means of the friction and noise sources. These sources were switched-off at $t = 0$. Then the electron was "placed" at a lattice site. The result is shown in Fig. 1 in which we represent the evolution of the electron probability density in a lattice heated to $T = 0.2$ (in our dimensionless units). The electron probability density evolves in time confined in a kind of cone. Clearly the electron probability density splits into many small spots bound at the thermally excited solitons. These "hot spots" may comprise up to 10 lattice sites. The overall process is time-dependent as the "hot spots" are erratically created and annihilated in the thermal process. Recall that the spots denote only probability density. We have done computer simulations also at low and at high temperatures. It occurs that at $T = 0.2$ there is some kind of optimum for the creation of solectron spots. We understand that such process is not fully diffusion-controlled. There are other tunneling contributions. "Surfing" of the electrons on thermal solitons provides not a fast transport mechanism, due to the quick and erratic changes of thermal soliton directions. However this ride is for free since thermal solitons are always present in real systems. Another important finding is that solectrons may form bosons with the peculiarity that electron pairing occurs both in momentum space and in real space due to the above mentioned solectron bound states.

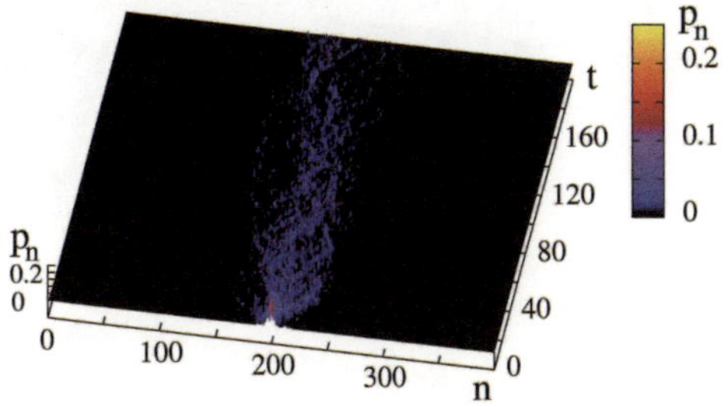

Fig. 1. Heated Morse lattice. An electron is placed (in probability density) at site 200 of a heated lattice ($T = 0.2$, $\tau = 10$, $\alpha = 1.5$, $V = 0.4$). The probability density tends to be concentrated at places of local soliton excitations (embracing up to 10 lattice sites) and survives there for a finite time (a few picoseconds). Subsequently, it moves to another "hot spot" and so on.

3. Two-dimensional (2D) models

Due to the difficulty of defining (stable) solitons and their evolution in 2D lattices and the fact that specifying lattice symmetry may jeopardize the "universality"of our predictions we prefer now to explore an alternative path. Using again the Morse interaction but not in the frame of a "lattice-kind" model but in the frame on an "ensemble-kind" of 2D system by considering the evolution of the n-th particle with coordinates (x_n, y_n) on the complex plane $Z = x + iy$, as a result of interaction with the other $(N-1)$ particles of the ensemble we can write

$$\ddot{Z}_n = \sum_k F_{nk}\left(|Z_{nk}|\right) z_{nk} + \left[-\gamma \dot{Z}_n + \sqrt{2D_v}\left(\xi_{nx} + i\xi_{ny}\right)\right], \qquad (5)$$

with $Z_{nk} = Z_n - Z_k$, $z_{nk} = \frac{Z_n - Z_k}{|Z_n - Z_k|}$ and $F_{nk}\left(|Z_{nk}|\right) = -\frac{dU^M(r)}{dr}|r = |Z_{nk}|$.

Based on trajectories of particles $Z_n(t)$ derived as a result of computer simulations of (5) we restrict the interaction of atoms with electrons within a "polarization" radius r_0

$$U(Z) = \sum_n U_{el}\left(|Z - Z_n|\right), \qquad (6)$$

where $U_{el}(r) = -U_e\left[r_0^2/\left(r^2 + r_0^2\right)\right]^2$. Further we use the Boltzmann approximation for the lattice particles distribution

$$\rho_{el}(Z) = exp\left[-U(Z)/k_B T\right]. \qquad (7)$$

In Fig. 2 we show a typical illustration of the evolution of the lattice particles density in the 2D system.

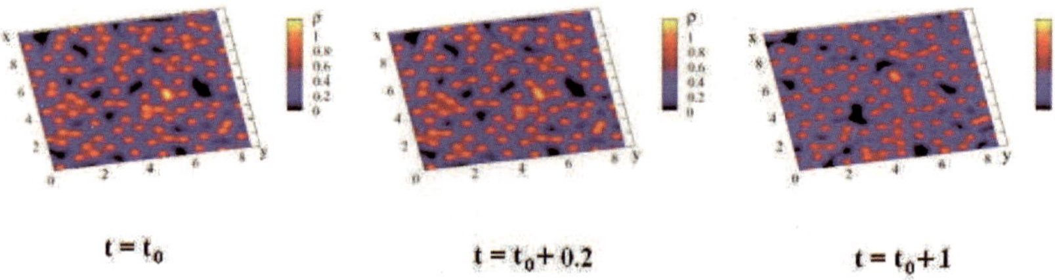

Fig. 2. Time evolution of density landscape of Morse lattice particles for a 2D system illustrating solectron islands that may lead to percolation and hence to a form of second-order phase transition and thus a solectron (soliton-mediated) kind of superconduction (at least supersonic).

4. Possibility of superconducting (SC) states. What is to be done

As it is known and earlier noted, handling solitons and their evolution in 2D systems is a hard task. The exploration sketched in the previous Section points as way out to the problem using a phenomenological approach to the dynamics (and thermodynamics). Following Landau one assumes that the Gibbs free energy of the system is a function $G(T, p, \psi)$ [where ψ is the order parameter]:

$$G(T, p, \psi) = G_0(T, P) + A(p)(T - T_c)\psi^2 + C\psi^4 + \dots \tag{8}$$

Then following Ginzburg and Landau (GL) ψ is considered as a kind of macroscopic wave function. Then the free energy density is

$$g(T, p, \psi(r)) = g_0(T, P, r) + A(T)|\psi(r)|^2 + \frac{1}{2}|\psi(r)|^4 + \frac{\hbar^2}{2m^*}|\nabla\psi|^2 + \dots \tag{9}$$

By minimization one can obtain from (9) a *nonlinear* Schroedinger equation for the wave function ψ which describes the SC phase. Could such an approach work with our solectrons?: (i) solectrons are the new charge carriers (quasiparticles) stable up to high temperatures (room temperatures for biomolecules). They permit forming bosons with appropriate electron pairing occurring both in momentum space and in real space! (ii) in 1D the solectrons form charged islands and there is no chance to unify them to a highly conducting path. However, in 2D we may obtain connected conducting regions leading to percolation and hence a new conducting phase, which could be taken as a candidate for SC phase; (iii) the solectron equations may be derived from a variational principle for two fields, the soliton density $\Phi(r)\psi$ and the charge density $\rho\ (r)$, thus offering the possibility of a *nonlinear* Schroedinger equation permitting a kind of soliton-bearing equation for the two fields $\Phi(r)$ and $\rho(r)$. Thus the phenomenological GL approach offers a sound path to understand solectron SC.

Details about some of the results so far obtained can be found in Refs. [1–13].

Acknowledgments

This research has been sponsored by the European Union under Grant SPARK-II-FP7-ICT-216227 and by the Spanish Ministerio de Ciencia e Innovacion (MICINN) under Grant EXPLORA (2009).

References

1. M. G. Velarde, W. Ebeling and A. P. Chetverikov, *Int. J. Bifurcation Chaos* **15**, 245 (2005).
2. M. G. Velarde, W. Ebeling, D. Hennig and C. Neissner, *Int. J. Bifurcation Chaos* **16**, 1035 (2006).
3. A. P. Chetverikov, W. Ebeling and M. G. Velarde, *Eur. Phys. J. B* **51**, 87 (2006).
4. D. Hennig, C. Neissner, M. G. Velarde and W. Ebeling, *Phys. Rev. B* **73**, 024306 (2006).
5. M. G. Velarde, W. Ebeling, A. P. Chetverikov and D. Hennig, *Int. J. Bifurcation Chaos* **18**, 521 (2008).
6. M. G. Velarde and C. Neissner, *Int. J. Bifurcation Chaos* **18**, 885 (2008).
7. M. G. Velarde, W. Ebeling and A. P. Chetverikov, *Int. J. Bifurcation Chaos* **18**, 3815 (2008).
8. D. Hennig, M. G. Velarde, W. Ebeling and A. P. Chetverikov, *Phys. Rev. E* **78**, 066606 (2008).
9. A. P. Chetverikov, W. Ebeling and M. G. Velarde, *Eur. Phys. J. B* **70**, 217 (2009).
10. A. P. Chetverikov, W. Ebeling and M. G. Velarde, *Contrib. Plasma Phys.* **49**, 529 (2009).
11. M. G. Velarde, *J. Comput. Applied Maths.* **233**, 1432 (2010).
12. A. P. Chetverikov, W. Ebeling and M. G. Velarde, *Int. J. Quantum Chem.* **110**, 46 (2010).
13. W. Ebeling, M. G. Velarde and A. P. Chetverikov, *Condensed Matter Phys.* (2010) to appear.

SOFT SENSORS AND ARTIFICIAL INTELLIGENCE WITH APPLICATIONS TO NUCLEAR FUSION EXPERIMENTS

ALESSANDRO RIZZO

Dipartimento di Elettrotecnica ed Elettronica, Politecnico di Bari
Via E. Orabona 4 - Bari, 70125, Italy
E-mail: rizzo@deemail.poliba.it
http://dee.poliba.it/rizzo

This paper contains the Author's notes concerning his Invited Lecture with the same title, given at PHYSCON09. The lecture is part of the IEEE Distinguished Lecturer Program of the IEEE Nuclear and Plasma Science Society.

Keywords: Soft sensors, inferential models, nuclear fusion, industrial processes, fault detection and diagnosis, monitoring.

1. Introduction

Industrial plants are being increasingly required to improve their production efficiency while respecting government laws that enforce tight limits on product specifications and on pollutant emissions, thus leading to more and more efficient measurement and control policies. In this context, the importance of monitoring a large set of process variables using adequate measuring devices is clear. However, key obstacles to the implementation of large-scale plant monitoring and control policies are posed by both the high cost of on-line measurement devices and the difficulty for operators to keep hundreds of measurements under control.

This need is also strongly felt in the experimental physics community. The huge number of measurements that has to be monitored during an experiment can lead to an ineffective management of the experimental campaigns. Moreover, in this field the interest on monitoring is manifold. In fact, different kinds of signals must be observed and processed for several purposes, such as safety, measurement validation, fault detection, experiment conduction, experiment evaluation.

Suitable mathematical models of processes, called *soft sensors, virtual sensors,* or *inferential models,* designed on the basis of experimental data via system identification procedures, can greatly help to reduce the need for measuring devices, monitor sets of significant measurements, and develop tight control policies. In this framework, the use of soft computing techniques and/or hybrid approaches can help in dealing with the intrinsic uncertainty of real world problem, exploiting efficiently both experimental data and human expertise.

In the invited lecture, design procedures for virtual sensors based on data-driven approaches are described, and relevant case studies referring to experimental physics are illustrated. In this paper some notes on the main points of the lecture are highlighted.

2. Soft Sensors

Measurement devices (hardware sensors) are obviously indispensable to the operational life of industrial plants. Nevertheless, there are many points of weakness connected with the presence of hardware sensors: beyond their cost and physical space requirements, they often have to work in hostile environments, they are subject to strict and costly maintenance protocols and, in spite of this, they are subjected to unexpected fault. Moreover, the limitations of measuring technology lead in some cases to delays in the measurement chain which can deteriorate the performance of control loops. Finally, it is difficult to interpret the sensors outcome as long as more and more measurement devices are installed.

Soft sensors can help to overcome or mitigate the effects of the drawbacks listed above [1]. As stated in the Introduction, they are mathematical models designed to estimate relevant process variables. They are usually built by either *modelling* or *identification* techniques and constitute a low-cost alternative to expensive hardware device. They are able to work in parallel with hardware sensors (e.g. for *fault detection* purposes), and are often implementable on existing hardware. Moreover, they can be designed to have real-time estimation capabilities, in order to overcome the time delays introduced by slow hardware sensors. The main applications of soft sensors can be listed as: back-up of measuring devices, reduction of measuring hardware requirements, real-time estimation for monitoring and control, sensor validation, fault detection and diagnosis, what-if analysis.

Even though several techniques are suitable to build effective soft sensors, their design procedure usually follows the steps depicted in Fig. 1. Feedback loops in the flowchart highlight the possibility of getting back to reconsider some aspects of past steps, if the current step leads to poor results, as the design of a soft sensor is very often done on a trial and error basis.

Due to space constraints, we are not able to go into details of each single design phase. These are illustrated in full detail in [1] and references therein.

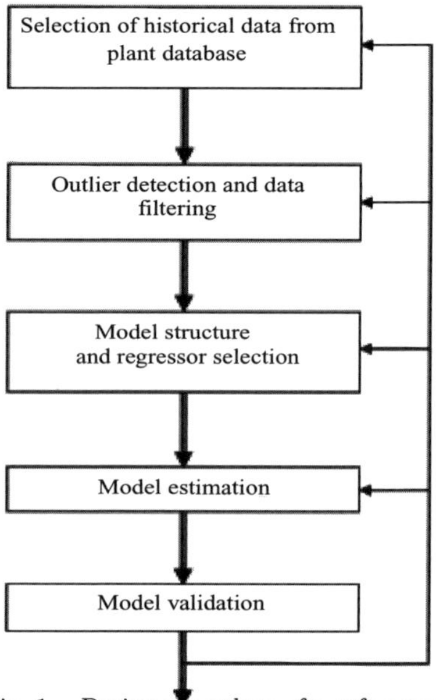

Fig. 1. Design procedure of a soft sensor.

We will spend here some few words on the third and fourth step, i.e. model structure and regressor selection, and model estimation. Several techniques can be adopted to tackle these tasks. An interesting categorization of these techniques is based on how much knowledge about the physics of the system is retained by the soft sensor designer. The modelling problem can then be dealt with in the *white box* framework, where both the system constitutive laws and its parameters are known, up to the *black box* approach, where only experimental data are available, and no or little knowledge of the system physics is available. Many intermediate situations can occur, all of them falling in the *grey box* approach, as illustrated in Fig. 2.

Complex industrial applications must often be tackled through a black box, data driven approach, as very often the knowledge of the phenomena involved can hardly be embedded in a first principle model. In this context, approaches based on soft computing techniques are very effective [2]. Moreover, these approaches allow the designer to embed into a model the empirical knowledge provided by the plant experts. Other data driven approaches adopted for soft sensor design in industrial applications are linear and nonlinear regressors, PCA and PLS based regressors (and their nonlinear extensions), support vector machines, multivariate statistics [3]. A main point in data driven soft sensor design is that a data driven model cannot

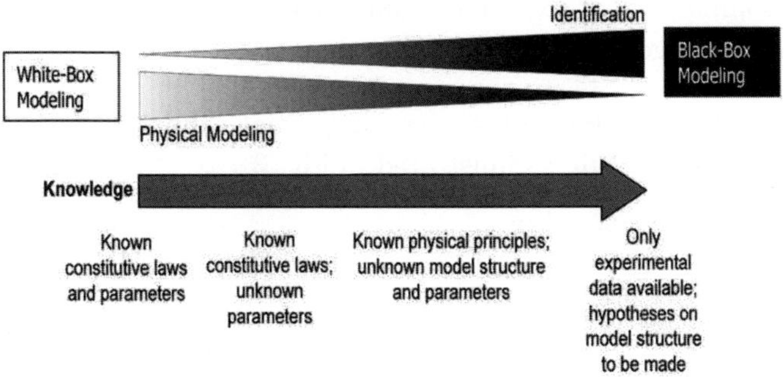

Fig. 2. Modelling techniques: white, grey, and black box.

provide more representative information than that used for its design. A main difficulty in the design of data driven soft sensors is that very often, in industrial applications, it is not possible to plan an experimental campaign to obtain data to be specifically used for a model design. Therefore, the steps of data selection from the available database, followed by outlier detection and data filtering can be crucial. There are even some cases in which the available dataset would be too small to obtain, in principle, an effective model. These cases can be coped with by using specific *small data set* techniques [4].

An interesting application of soft sensors is for *sensor validation* [1]. This is a particular case of *fault detection and diagnosis* [1,5], in which the aim is to monitor and evaluate the functionality of one, or a set of, sensors. Reliability of measurements is a fundamental requirement in industrial applications, in order to achieve better product quality, plant efficiency and availability, enhanced safety and environmental protection, better performance of feedback control loops. Very often, *hybrid approaches* trying to conjugate the benefits of signal analysis, knowledge engineering, approximate reasoning, nonlinear black-box modelling, and optimization tools led to satisfactory results in this specific field [1,6].

The design procedure for a soft sensor for sensor validation can be sketched as:

- Acquisition of the expert knowledge (causality relations, known faults, known causes, heuristic knowledge, etc.);
- Definition of symptoms (variables or information which reflect changhes from normal system behaviour);
- Design of detection tools to generate symptoms (this involves the whole design procedure illustrated in Fig. 1);

- Comparison of expected symptoms and actual ones and, consequently, generation of suitable alarms.

The problem can be broken down in partial validation tasks, followed by a decision stage [1,6], as illustrated in Fig. 3.

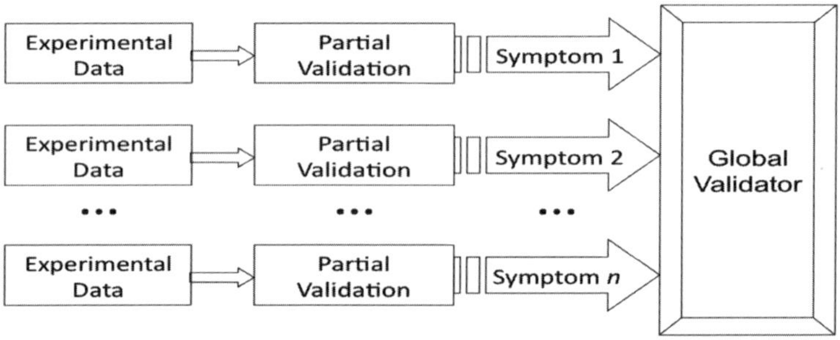

Fig. 3. The hybrid approach to sensor validation.

3. Soft Sensors for Nuclear Fusion Experiments

There are many nuclear fusion laboratories worldwide. Nuclear fusion is still in its experimental phase, and research is carried out in order to define the best operational conditions to achieve a stable, steady-state reaction with the final aim of building reactors able to provide a clean, safe, and low-cost energy production.

Sensor validation is particularly needed in the nuclear fusion community, as many thousands of sensors are installed in each lab to monitor the experiments, and measurement reliability is indispensable for experiment analysis. On the other hand, sensor validation is particularly challenging in this field, for at least three reasons. Firstly, the pulsating nature and the complexity of the phenomena often prevent to build first principle (i.e. white box) models. Secondly, the experimental nature of the plant often prevents to define normal behaviours, leading to the difficulty in distingushing sensor faults from "odd" behaviours (which may be of great interest for researchers). Finally, the same sensor set is often observed by many experts, each of whom carries a personal expertise on the phenomena and focuses his interest on particular aspects of the measurements. So, the measurement interpretation is often equivocal. Again, in this case, hybrid approaches, mostly based on soft computing techniques, can help in embedding all the expert's points of view and the information hidden in the past measurement database.

Details on applications of soft sensor to the nuclear fusion field can be retrieved in [1,6–9].

4. Conclusions

Soft sensors are a valuable, low-cost tool with many applications to engineering problems. Exploiting expert knowledge, experimental data and hybrid approaches can help in overcoming modelling difficulties. The need for soft sensors is strongly felt also in the experimental physics community, especially for data monitoring, sensor validation, improvement of control loops performance, what-if analysis and experiment diagnosis.

In our view, challenges for future research activity are: the automation of each step of the soft sensor design procedure, especially in nonlinear cases; the adaptive validation, maintenance and re-tuning of models; the small data set techniques; the improvement of prediction performance through the combination of the outcomes of different models.

References

1. L. Fortuna, S. Graziani, A. Rizzo and M. Xibilia, *Soft Sensors for Monitoring and Control of Industrial Processes* (Springer-Verlag, 2007).
2. L. Fortuna, A. Rizzo, M. Sinatra and M. Xibilia, *Control Engineering Practice* **11**, 1491 (2003).
3. P. Kadlec, B. Gabrys and S. Strandt, *Computers and Chemical Engineering* **33**, 795 (2009).
4. L. Fortuna, S. Graziani and M. Xibilia, *IEEE Transactions on Instrumentation and Measurement* **58**, 2444 (2009).
5. R. Isermann, *Fault-Diagnosis Systems - An Introduction from Fault Detection to Fault Tolerance* (Springer-Verlag, 2006).
6. G. Buceti, L. Fortuna, A. Rizzo and M. Xibilia, *Fusion Engineering and Design* **60**, 381 (2002).
7. B. Esposito, Y. Kaschuck, A. Rizzo, L. Bertalot and A. Pensa, *Nuclear Instrumentation and Measurement, Part A* **518**, 626 (2004).
8. A. Rizzo and M. Xibilia, *IEEE Transactions On Control Systems Technology* **10**, 421 (2002).
9. G. Buceti, C. Centioli, F. Iannone, M. Panella, A. Rizzo and V. Vitale, A rating system for post pulse data validation, in *SOFT2002, 22nd Symposium on Fusion Technology*, (Helsinki, Finland, 2002).

PREPROCESSING METHOD FOR IMPROVING ECG SIGNAL CLASSIFICATION AND COMPRESSION VALIDATION

L. GORAS

Gh. Asachi Technical University of Iasi
and
Institute for Computer Science, Romanian Academy
Iasi, Romania
E-mail: lgoras@etc.tuiasi.ro

M. FIRA

Institute for Computer Science, Romanian Academy
Iasi, Romania
E-mail: mfira@etc.tuiasi.ro

A method for improving electrocardiogram (ECG) signal classification in time domain is presented. The main idea is to preprocess the segmented waveforms in order to obtain an alignment of the ECG with respect to the maximum value of the R beat while keeping the information on its initial position as a feature. We propose two simple preprocessing methods basically consisting in the alignment of the R-waves either by translation or by a slight nonlinear time scaling. These techniques are also used to improve the validation of ECG compression.

Keywords: ECG signals; alignment; compression validation; classification.

1. Introduction

ECG signal classification is an important issue in clinical monitoring and has been thoroughly studied [1] [2]. The process involves extraction of attributes from ECG waveforms and subsequent comparisons with patterns characteristic to known diseases.

In all cases classification is related more or less to the concept of distance [3] [4]; distances between signals belonging to the same class are generally smaller than distances between signals belonging to different classes [5] [6].

The simplest way to segment ECG signals consists in taking samples between the middles of two successive RR periods so that the R wave will be placed somewhere about the middle of the signal support. The waveforms obtained in this way can be normalized to a constant number of samples to eliminate the influence of heart rate variability and to be easily compared using for instance the L2 norm, neural networks, etc. Such a technique has been used in [7] to validate the efficiency of a new compression method.

2. Materials and methods

We start from the simple observation that, even after normalization to the same number of samples, the position of the R wave peaks for various frames containing normal or pathological beats generally does coincide neither as position nor as amplitude. In a standard normal beat the R wave has a peak about three times higher than any other maxima in the heart-beat. Even though the duration of the R beat is about 5% of the waveform support, due to its high amplitude the weight in the distance of the R wave contribution is significantly important.

The main idea of this paper is based on the finally fully confirmed conjecture that aligning ECG waveforms with respect to the R wave will increase the time domain techniques classification rate.

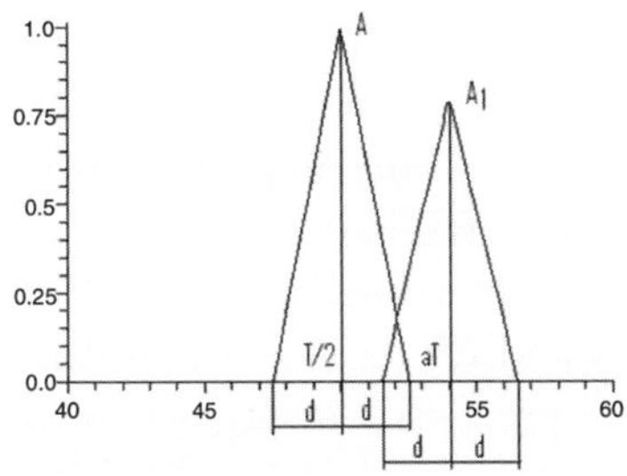

Fig. 1. Stylized R-beats with different amplitudes and positions with details on another temporal scale.

To make an image on the contribution in the distance between two beats coming from the un-alignment of the R waves we have considered as a 'toy' example two (highly) stylized R beats assimilated for simplicity with two isosceles triangles with the same basis width 2d and different heights A, A1 placed at different abscissas T/2 and aT respectively ($a > T/2$) as shown in Fig. 1.

Considering a value of $d = 2.5$ for a waveform with a normalized duration of 100 units (A=A1=1), the maximum L2 distance between the two waves which corresponds to the case when the support of the triangles bases are disjoint (which happens for values of a higher than 0.55 or less than 0.45) is where $1.825 = 1.291\sqrt{2}$, a significant value, is the norm ('length') of each triangle. We have studied two R wave alignment techniques.

2.1 First method (R-wave centering through shifting) consist in shifting the patterns to the left or to the right so that the R waves are placed just in the middle of the waveform interval. After shifting, the samples that leave out the interval at one end are discarded and the missing samples at the other end are replaced with the (same) rightmost/leftmost known value.

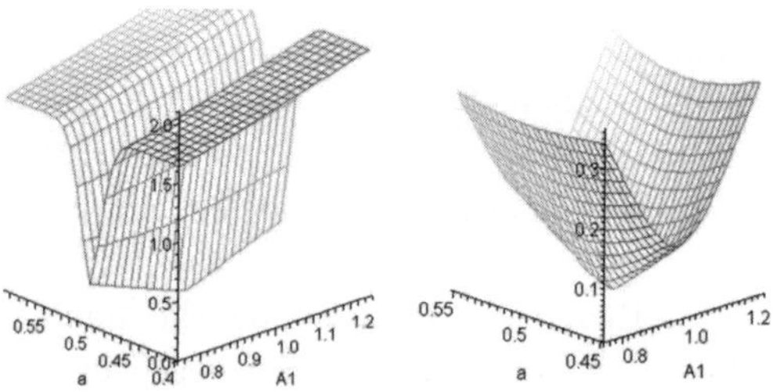

Fig. 2. Distance between the two triangles as a function of A1 and a ($A = 1$, $d = 2.5$)(left) Dependence on a and A1 of the distance between triangles after alignment(right).

To make an image on the contribution in the distance between two waveforms coming only from the un-alignment of the R waves we will consider first the 'stylized' R-waves described above. The distance between the two waves varies with a and A1 for $A = 1$ and constant $d = 2.5$ as shown in Fig. 2. Note that the previously mentioned value $1.825 = 1.291\sqrt{2}$ for the distance between two non-overlapping equal triangles can be found in (Fig. 2a) for $\text{A1} = A = 1$ and $a > 0.55$ or $a < 0.45$.

Coming back to real heartbeats, the distance between the two real R-waves selected between the two dotted lines in Fig. 1 is 1.676 while the distance between the whole waves is 1.737. A typical value for the distance between two beats with completely nonoverlapping R waves is 2.

2.2 Second method (R-wave centering through shrinking/ stretching) consists in using a nonlinear time scaling to make the R waves coincide. An intuitive way to describe the method is the following. Imagine that the waveform already normalized to a specified duration T = 100 is drawn on an elastic sheet with width T and fastened lateral margins. Suppose that the waveform has the R peak placed to the right of T/2. Let us further imagine a rigid vertical line passing through the R wave peak fastened with the sheet. Now move the line to the left until its abscissa reaches T/2 thus slightly shrinking/stretching the left/right part of the waveform. In this way any waveform can be processed such that the peak has an abscissa of T/2. Moreover, the waveforms can be vertically (linearly) scaled as well to a normalized value of the peak equal to unity.

We only mention that if the A1 triangle is translated and deformed until the abscissa of its peak reaches the value T/2, the abscissas of its bases become T/2-$d/(2a)$ and T/$d+d(2(1\text{-}a))$. In this case, the A1 triangle is no Iore isosceles: in our example its left part has been slightly shrunk while it left part slightly stretched. This is why the distance between the two triangles vanishes only in the trivial case when A1 $= A = 1$ and $a = 1/2$. For any other value of a, even with a vertical scaling, the distance between the triangles will be not exactly zero.

2.3 ECG preprocessing: The ECG signals used to illustrate the proposed method were taken from the MIT-BIH Arrhythmia database and consisted in ECG signals digitized through sampling at 360 samples/s. Since the segmented output patterns had different dimensions, each pattern was resampled to 100 samples. The simulations used three classes of waveforms. The first class, denoted by **I (initial)** consisted of the above described resampled patterns which were used for classification with no extra processing. The second class, denoted by **S (shifted)** was obtained from the first one by shifting the pattern until the R wave reached T/2, discarding samples that left out the interval at one end and replacing the missing samples at the other end with the last known value. The last category, denoted by **S/S (stretch/shrink)** was obtained from the first one through resampling the patterns with 50 samples on each side of the R wave which is equivalent to a linear stretching/shrinking of the two sides of the pattern when the R wave was not in the middle of the interval. Since the information about the period variability is surely a non-negligible aspect and should not be discarded, it can always be displayed in association to the processed waveforms.

3. Results

In order to verify the advantages of the proposed methods and to compare results, the two methods mentioned above have been used. The aim was to show the classification improvements for the above two methods. It is expected that the method would improve the results of other classification techniques as well, with even better recognition rates.

An MLP (a back-propagation algorithm with gradient descendent) with a hidden layer containing 50 neurons has been trained with heartbeats from 8 classes taken from 23 records of the MIT-BIH Arrhythmia database. From a total of 7500 patterns (60% training, 20% validation and 20% testing).

Table 1. Results obtained with the MLP and k-NN for original patterns with I, S and S/S. Features: * initial length of the pattern, ** initial lengths of the left and right sides of the pattern.

	MLP without features		**MLP with features**	
	No scaling	*vertical scaling*	*No scaling*	*vertical scaling*
(I)	90.9	91.4	90.4*	90.7*
(S)	91.7	92	93.23**	94.41**
(S / S)	92.8	93.1	94.9**	95.3**
	k-NN no vertical scaling		**k-NN vertical scaling**	
(I)	93.36		93.84	
(S)	94.48		93.68	
(S/S)	95.60		95.76	

From Table 1 can be observed that the classification results are better for the S and S/S classes and further improve when using features and vertical normalization.

It is also apparent that using features for comparing initial waveforms i.e., without centering and vertical scaling gives worse results. On the other side, the classification rate increased in the S/S case with vertical scaling with 4.4% compared to the un-processed situation.

Besides the ANN based classification, we have also used the k-NN algorithm for patterns, this time without features since the principle of the method is based solely on the concept of distance. The best results were obtained considering only the first neighbor and (S/S) with vertical scaling segmentation method (95.76%) and are presented in Table 1.

3.1 Compression validation: Taking into account the need for relevant errors measures for the biomedical signal compression, but also that these measures should be easily calculated, we suggest that the classification error of reconstructed signals can be used as a distortion measure of the compression method, i.e., a higher classification rate of the reconstructed signals reflects a higher quality compression.

Based on the above considerations on preprocessing, we have tested the validation of the compression method proposed in [7] by comparing the classification results of a MLP trained with original waveforms and then with reconstructed ones.

4. Conclusion

We have presented an improved method for electrocardiogram (ECG) signal classification. By preprocessing the waveforms in order to obtain an 'alignment' of the R beats it has been shown that the classification results

Table 2. Results obtained with the MLP and k-NN for reconstructed patterns with I, S and S/S.

	MLP without features		**MLP with features**	
	No scaling	*vertical scaling*	*No scaling*	*vertical scaling*
(I)	78.2	81	83.5*	85.6*
(S)	84.7	86.5	89.89**	89.29**
(S/S)	86.4	88.9	90.05**	90.54**

	k-NN no vertical scaling	**k-NN vertical scaling**
(I)	77.71	79.30
(S)	90.80	90.39
(S/S)	93.16	93.98

have been significantly improved for both tested methods (ANN and k-NN) and are expected to improve for other classification techniques as well.

A major advantage of the proposed method based on cardiac beat alignment consists in a reduced computational complexity and robustness. Last but not least, considering the results obtained for reconstructed signals after compression, the method can be used for compression validation as well.

References

1. Wei Jiang Kong and S. G. Peterson, ECG signal classification using block-based neural networks, in *Proc. IJCNN 2005*, 1p. 326, 2005.
2. Nyongesa H., Classification Of ECG By Auto-Regressive Modelling And Neural Networks, *IEEE AFRICON*, pp. 841-841, 2004.
3. Jekova I., Comparison of five algorithms for the detection of ventricular fibrillation from the surface ECG, *Physiol. Measur 2000*, 21, pp. 429-439, 2000.
4. Christov I., Jekova I. and Bortolan G., Premature ventricular contraction classification by the Kth nearest-neighbours rule, *Physiol. Meas*, 26, pp. 123-130, 2005.
5. P. De Chazal, M. O'Dwayer, R. B. Reilly, Automatic Classification of Heartbeats Using ECG Morphology and Heartbeat Interval Features, *IEEE Trans Biomed Eng*; 51, 7, pp 1196- 1206, 2004.
6. G. Krishna Prasad, J. S. Sahambi, Classification of ECG Arrhythmias using Multi Resolution Analysis and Neural Networks, *Proc TENCON*, 2003.
7. M. Fira, L. Goras, An ECG signals compression method and its validation using NN's, *IEEE Trans Biomed Eng*, 45, pp. 1319-1326, 2008.

CHAOS OF GENERALIZED TWO-SIDED SYMBOLIC DYNAMICAL SYSTEMS

GUANRONG CHEN[1], CHUANJUN TIAN[2] and SHENGLI XIE[3]

[1] *Department of Electronic Engineering, City University of Hong Kong Kowloon, Hong Kong SAR, P. R. China*

[2] *College of Information Engineering, Shenzhen University Shenzhen 518060, P. R. China*

[3] *College of Electronic Information, South China University of Technology Guangzhou 510641, P. R. China*

In this paper, a mathematical model of generalized two-sided symbol dynamical systems is proposed and analyzed, which is proved to be chaotic in the sense of Devaney.

Keywords: Chaos, generalized symbol dynamical system.

Recently, chaos in discrete spatiotemporal systems have been studied extensively [1,3,4,6,8]. In [13], a generalized (one-sided) symbol dynamical system is introduced as a special case of the general class of discrete spatiotemporal systems. Following this approach, generalized two-sided symbol dynamical systems is proposed and analyzed in this paper.

Throughout this paper, let I be a bounded subset of $R = (-\infty, \infty)$, $Z = \{\cdots, -1, 0, 1, \cdots\}$. For any $k \in Z$, denote $Z[k, \infty) = \{k, k+1, \cdots\}$, $I^\infty = \{x = (x_0, x_1, \cdots) = \{x_n\}_{n=0}^{\infty} \mid x_n \in I,\, n \in Z\}$ and $I_\infty^\infty = \{x = (\cdots, x_{-1}, x_0, x_1, \cdots) = \{x_n\}_{n=-\infty}^{\infty} \mid x_n \in I,\, n \in Z\}$. Define

$$d(\alpha, \beta) = \sum_{n=-\infty}^{\infty} \frac{|x_n - y_n|}{2^{|n|}}, \quad \forall \alpha = \{x_n\}_{n=-\infty}^{\infty}, \beta = \{y_n\}_{n=-\infty}^{\infty} \in I_\infty^\infty,$$

$$\overline{d}(\eta, \gamma) = \sum_{n=0}^{\infty} \frac{|x_n - y_n|}{2^n}, \quad \forall \eta = \{x_n\}_{n=0}^{\infty}, \gamma = \{y_n\}_{n=0}^{\infty} \in I^\infty. \tag{1}$$

Then, both (I_∞^∞, d) and $(I^\infty, \overline{d})$ are metric spaces.

Let $f : I_\infty^\infty \to I$ be a function and consider the following discrete spatiotemporal systems:

$$x_{m+1,n} = f(\cdots, x_{m,n-1}, x_{m,n}, x_{m,n+1}, \cdots), \quad m = 0, 1, 2, \cdots, \quad n \in Z, \tag{2}$$

where $x_{0,n} \in I$ for $n \in Z$. In particular, if $f(x) \equiv g(x_0, x_1, \cdots)$ for any $x = \{x_n\}_{n=-\infty}^{\infty} \in I_\infty^\infty$, where $g : I^\infty \to I$ is a function, then (2) becomes

$$x_{m+1,n} = g(x_{m,n}, x_{m,n+1}, \cdots), \quad m = 0, 1, 2, \cdots, \quad n \in Z. \tag{3}$$

Chaos of system (3) was studied in [8,11,13].

Let $x = \{x_{m,n}\}$ be a solution of system (2) and $x_m = \{x_{m,n}\}_{n=-\infty}^{\infty} \in I_\infty^\infty$, $m = 0, 1, 2, \cdots$. Then, system (2) is equivalent to the following infinite-dimensional discrete system: for $m = 0, 1, 2, \cdots$,

$$x_{m+1} = (\cdots, x_{m+1,-1}, x_{m+1,0}, x_{m+1,1}, \cdots) = F(x_m), \tag{4}$$

The function $F : I_\infty^\infty \to I_\infty^\infty$ defined by (4) is said to be induced by system (2) (or, by function f), and system (4) is also said to be induced by system (2).

Obviously, a sequence $x = \{x_{m,n} | \, n \in Z, m = 0, 1, 2, \cdots\}$ is a solution of (2) if and only if $x = \{x_m = \{x_{m,n}\}_{n=-\infty}^{\infty}\}_{m=0}^{\infty}$ is a solution of (4). Hence, it is possible to study the properties of some discrete spatiotemporal systems by using knowledge of the corresponding infinite-dimensional discrete systems.

In the following, some basic properties of two-sided symbolic dynamical systems are first reviewed [3–5].

Let $k \geq 2$ be an integer, $S = \{0, 1, \cdots, k-1\}$, and $S_\infty^\infty = \{\{s_n\}_{n=-\infty}^{\infty} \mid s_n \in S, \, n \in Z\}$. Then, (S_∞^∞, d) is a metric space, where d is defined by (1).

Define a shift map $\sigma : S_\infty^\infty \to S_\infty^\infty$ as follows:

$$\sigma(\alpha) = \sigma(\cdots, x_0, x_1, \cdots) = \beta = \{y_n = x_{n+1}\}_{n=-\infty}^{\infty}, \tag{5}$$

where $\alpha = (\cdots, x_0, x_1, \cdots)$ and $\beta = (\cdots, y_0, y_1, \cdots)$. From the map σ, for any $\alpha = \{x_n\}_{n=-\infty}^{\infty} \in S_\infty^\infty$, one can obtain a sequence $O(\alpha) = (\sigma^0(\alpha) = \alpha, \sigma(\alpha), \sigma^2(\alpha), \cdots)$. Obviously, $O(\alpha)$ is a solution of the following discrete system:

$$\alpha_{m+1} = \sigma(\alpha_m), \quad \alpha_0 = \alpha = \{x_n\}_{n=-\infty}^{\infty} \in S_\infty^\infty \quad m = 0, 1, \cdots. \tag{6}$$

It is easy to verify that system (6) is equivalent to an infinite-dimensional discrete system induced by the following discrete spatiotemporal system:

$$x_{m+1,n} = x_{m,n+1}, \quad x_{0,n} \in S, \quad m = 0, 1, 2, \cdots, \quad n \in Z. \tag{7}$$

Hence, two-sided symbolic dynamical systems are special cases of discrete spatiotemporal systems. In this paper, if $I = S = \{0, 1, \cdots, k-1\}$ for an integer $k \geq 2$, then system (2) is said to be a generalized two-sided symbolic dynamical system.

In [13], chaos in the sense of Devaney on $(S^\infty, \overline{d})$ for a generalized one-sided symbolic dynamical system is studied, where $\overline{d}$ is defined by (1). Based on this discussion, in this paper, chaos in the sense of Devaney on (S^∞_∞, d) for a generalized two-sided symbolic dynamical system is further studied and several new results of this system will be derived.

Throughout, let (X, d) be a metric space, $f : X \to X$ be a map, $f^0(x_0) = x_0 \in X$, $f^{n+1}(x_0) = f(f^n(x_0))$, $n = 0, 1, 2, \cdots$. Then, the sequence $O(x_0) = \{x_n = f(x_{n-1})\}_{n=0}^\infty$ is a solution of the following discrete system:

$$x_{n+1} = f(x_n), \quad x_0 \in X, \quad n = 0, 1, 2, \cdots . \tag{8}$$

Definition 1. [2] Let $f : X \to X$ be a map on a metric space (X, d). The map f or system (8) is said to be chaotic in the sense of Devaney if

1. f or system (8) is transitive;
2. the set of periodic points of f or system (8) is dense in X;
3. f or system (8) has sensitive dependence on initial conditions.

Definition 2. [8-12] Let I be a bounded subset of R, (I^∞_∞, d) be a metric space and $f : I^\infty_\infty \to I$ be a function. System (2) with the (system) function f is said to be chaotic (on (I^∞_∞, d)) if system (4) induced by system (2) is chaotic on (I^∞_∞, d). In particular, if system (4) induced by system (2) is chaotic in the sense of Devaney on (I^∞_∞, d), then system (2) is said to be chaotic in the sense of Devaney.

In the generalized symbolic dynamical system (2), assume that there exist two integers $r, q \geq 0$ with $r + q \neq 0$, such that its system function $f : S^\infty_\infty \to S$ satisfies

$$f(x) = g(x_{-r}, \cdots, x_q) = ax_{-r} + W(x_{-r+1}, \cdots, x_{q-1}) + bx_q \;\; mod(k), \tag{9}$$

where $S = \{0, 1, \cdots, k-1\}$, $x = \{x_n\}_{n=-\infty}^\infty$, a, b are two nonnegative integers, $k \geq 2$ is an integer, a and k, b and k are coprime (i.e., their greatest common divisor is 1), $mod(\cdot)$ denotes the modular operator, $W : S^{r+q-1} \to S$ is a function with $W \equiv 0$ if $r + q = 1$. Then, system (2) becomes

$$x_{m+1,n} = ax_{m,n-r} + W(x_{m,n-r+1}, \cdots, x_{m,n+q-1}) + bx_{m,n+q} \;\; mod(k), \tag{10}$$

where $x_{0,n} \in S$, $m \in Z[0, \infty)$ and $n \in Z$. Obviously, the usual two-sided generalized symbolic dynamical system (6) is a special case of system (10).

Assume that system (10) is equivalent to the following infinite-dimensional discrete system:

$$x_{m+1} = (\cdots, x_{m+1,-1}, x_{m+1,0}, x_{m+1,1}, \cdots) = F(x_m), \tag{11}$$

where $x_m = \{x_{m,n}\}_{n=-\infty}^{\infty}$ for any $m = 0, 1, \cdots$. Obviously, for any $\alpha = \{x_n\}_{n=-\infty}^{\infty} \in S_{\infty}^{\infty}$, $F(\alpha) = \beta = \{y_n = g(x_{n-r}, x_{n-r+1}, \cdots, x_{n+q})\}_{n=-\infty}^{\infty}$.

In the following, it is to prove that system (10) or system (11) is chaotic in the sense of Devaney on (S_{∞}^{∞}, d), where d is defined by (1) and $r, q \in Z[1, \infty)$.

First, it is to prove that system (11) is transitive.

Let U and V be any two nonempty open subsets of S_{∞}^{∞}. Then, for any $\alpha = \{u_n\}_{n=-\infty}^{\infty} \in U$ and $\beta = \{v_n\}_{n=-\infty}^{\infty} \in V$, there exists a constant $\theta \in (0,1)$ such that $B_\theta(\alpha) = \{\gamma \in S_{\infty}^{\infty} |\ d(\alpha, \gamma) < \theta\} \subseteq U$ and $B_\theta(\beta) \subseteq V$.

From the definition of the metric d in (1), there exists an integer $Q > 0$ such that

$$\{\{x_n\}_{n=-\infty}^{\infty} |\ x_i = u_i, x_j \in S, i = -Q+1, \cdots, Q-1; |j| \geq Q\} \subseteq B_\theta(\alpha).$$
$$\{y = \{y_n\}_{n=-\infty}^{\infty} |\ y_i = v_i, y_j \in S, |i| \leq Q-1; |j| \geq Q\} \subseteq B_\theta(\beta). \qquad (12)$$

For any $x = \{x_n\}_{n=-\infty}^{\infty} \in S_{\infty}^{\infty}$, denote

$$F(x) = (\cdots, g(x_{-r}, \cdots, x_q), g(x_{-r+1}, \cdots, x_{q+1}), \cdots) = \{x_i^{(1)}\}_{i=-\infty}^{\infty},$$
$$F^n(x) = (\cdots, g(x_{-r}^{(n-1)}, \cdots, x_q^{(n-1)}), g(x_{-r+1}^{(n-1)}, \cdots, x_{q+1}^{(n-1)}), \cdots) = \{x_i^{(n)}\}_{i=-\infty}^{\infty}$$

where $x_m^{(0)} = x_m$ and $x_m^{(n)} = g(x_{m-r}^{(n-1)}, \cdots, x_{m+q}^{(n-1)})$ for all $m \in Z$ and $n = 1, 2, \cdots$.

Obviously, for all $i = \cdots, -1, 0, 1, \cdots$, one can obtain

$$x_i^{(1)} = (ax_{i-r} + W(x_{i-r+1}, \cdots, x_{i+q-1}) + bx_{i+q}) \ mod(k),$$
$$x_i^{(2)} = (ax_{i-r}^{(1)} + W(x_{i-r+1}^{(1)}, \cdots, x_{i+q-1}^{(1)}) + bx_{i+q}^{(1)}) \ mod(k)$$
$$= (a^2 x_{i-2r} + W_1(x_{i-2r+1}, \cdots, x_{i+2q-1}) + b^2 x_{i+2q}) \ mod(k),$$

where $W_1 : S^{2r+2q-1} \to S$ is a nonnegative function irrelevant to $i \in Z$. In general, by induction, one has, for any integers $n > 0$ and $i \in Z$,

$$x_i^{(n)} = [ax_{i-r}^{(n-1)} + W(x_{i-r+1}^{(n-1)}, \cdots, x_{i+q-1}^{(n-1)}) + bx_{i+q}^{(n-1)}] \ mod(k)$$
$$= [a^n x_{i-nr} + W_{n-1}(x_{i-nr+1}, \cdots, x_{i+nq-1}) + b^n x_{i+nq}] \ mod(k),$$

where $W_{n-1} : S^{nr+nq-1} \to S$ is a nonnegative function irrelevant to i. Hence

$$x_i^{(n)} = \{a^n x_{i-nr} + W_{n-1}(x_{i-nr+1}, \cdots, x_{i+nq-1}) + b^n x_{i+nq}\} \ mod(k)$$
$$= \{[a^n x_{i-nr} + W_{n-1}(x_{i-nr+1}, \cdots, x_{i+nq-1})] \ mod(k) + b^n x_{i+nq}\} \ mod(k)$$
$$= \{a^n x_{i-nr} + [W_{n-1}(x_{i-nr+1}, \cdots, x_{i+nq-1}) + b^n x_{i+nq}] \ mod(k)\} \ mod(k).$$

It follows from the given conditions that for any given integers $n > 0$ and $m, \overline{m}, m_{-nr}, \cdots, m_{nq} \in S$, there exist two integers $v, w \in S$ such that

$$\{[a^n m_{-nr} + W_{n-1}(m_{-nr+1}, \cdots, m_{nq-1})] \ mod(k) + b^n v\} \ mod(k) = \overline{m},$$

$$\{a^n w + [W_{n-1}(m_{-nr+1}, \cdots, m_{nq-1}) + b^n m_{nq}] \; mod(k)\} \; mod(k) = m. \quad (13)$$

Otherwise, if for a certain integer $m \in S$ there does not exist an integer $w \in S$ such that (13) holds, then there are two different integers $y, z \in S$ such that

$$\{a^n y + [W_{n-1}(m_{-nr+1}, \cdots, m_{nq-1}) + b^n m_{nq}] \; mod(k)\} \; mod(k)$$
$$= \{a^n z + [W_{n-1}(m_{-nr+1}, \cdots, m_{nq-1}) + b^n m_{nq}] \; mod(k)\} \; mod(k).$$

Hence, there exists an integer $p \neq 0$ such that $a^n y = pk + a^n z$, i.e., $pk = a^n(y - z)$. Because $y, z \in \{0, 1, \cdots, k-1\}$, one can see that k and a^n are not coprime, which is a contradiction to the fact that k and a are coprime. The other equality of (13) can be proved similarly.

Let n be a sufficiently large positive integer and $Q = np$ with $p = min\{r, q\}$. Then, in view of (12) and (13), there exists a point $\gamma = (\cdots, c_{-1}, c_0, c_1, \cdots) \in B_\theta(\alpha)$ such that $c_i = u_i$ for all $i \in \{-nr, -nr+1, \cdots, -1, 0, 1, \cdots, nq-1\}$ and $c_j \in S$ for all $j \in \{\cdots, -nr-2, -nr-1\} \cup \{nq, nq+1, \cdots\}$, and that for all $s = \cdots, -2, -1, 0, 1, 2, \cdots$, one has

$$\{a^n c_{-nr+s} + W_{n-1}(c_{-nr+s+1}, \cdots, c_{nq+s-1}) + b^n c_{nq+s}\} \; mod(k) = v_s.$$

Hence, $F^n(\gamma) = \beta \in V$ and $\gamma \in U$, i.e., system (11) is transitive.

Second, it is to prove that system (11) has a dense set of periodic points.

Let $\alpha = \{u_n\}_{n=-\infty}^{\infty}$ be any point of S_∞^∞ and U be any neighborhood of α. Then, there exist an $\varepsilon > 0$ and a large integer $Q = n \times min\{r, q\} > 0$ such that $B_\varepsilon(\alpha) \subseteq U$ and

$$\{y = \{y_n\}_{n=-\infty}^{\infty} \mid y_i = u_i, y_j \in S, i = -Q+1, \cdots, Q-1; \; |j| \geq Q\} \subseteq B_\varepsilon(\alpha).$$

Similarly to the above proof of transitivity, there exists a point $\beta = \{v_n\}_{n=-\infty}^{\infty} \in B_\varepsilon(\alpha)$ such that $v_i = u_i$ for all $i \in \{-nr, -nr+1, \cdots, 0, \cdots, nq-1\}$ and $v_j \in S$ for all $j \in \{\cdots, -nr-2, -nr-1\} \cup \{nq, nq+1, \cdots\}$, and that

$$\{a^n v_{-nr+s} + W_{n-1}(v_{-nr+s+1}, \cdots, v_{nq+s-1}) + b^n v_{nq+s}\} \; mod(k) = v_s, \quad s \in Z,$$

where the function W_n is the same as above, for all $n \in Z[1, \infty)$. Hence, $\beta \in B_\varepsilon(\alpha) \subseteq U$ and $F^n(\beta) = \beta$. Therefore, system (11) has a dense set of periodic points.

Finally, it is to prove that system (11) has sensitive dependence on initial conditions.

Let $\delta = 0.25$, $\alpha = \{u_n\}_{n=-\infty}^{\infty}$ be any point of S_∞^∞, and U be any neighborhood of α. Then, there exist an $\varepsilon > 0$ and an integer $n > 0$ such that $Q = n \times min\{r, q\}$, $B_\varepsilon(\alpha) \subseteq U$ and $\{y = \{y_n\}_{n=-\infty}^{\infty} | \ y_i = u_i, y_j \in S, |i| \leq Q - 1; \ |j| \geq Q\} \subseteq B_\varepsilon(\alpha)$.

Let $\beta = \{v_n\}_{n=-\infty}^{\infty} \in B_\varepsilon(\alpha)$ such that $v_i = u_i$ for all $i \in \{-nr, -nr + 1, \cdots, nq - 1\}$ and $v_j \in S$ for all $j \in \{\cdots, -nr - 2, -nr - 1\} \cup \{nq, nq + 1, \cdots\}$, and $u_0^{(n)} \neq v_0^{(n)}$. Then, $\beta \in B_\varepsilon(\alpha) \subseteq U$ and $d(F^n(\alpha), F^n(\beta)) \geq |u_0^{(n)} - v_0^{(n)}| \geq \delta$. Hence, system (11) has sensitive dependence on initial conditions.

From the above proof, one can see that system (11) is chaotic in the sense of Devaney on (S_∞^∞, d). Consequently, system (10) is chaotic in the sense of Devaney on (S_∞^∞, d), where d is defined by (1).

As a special case, consider the following generalized symbolic dynamical systems:

$$x_{m+1,n} = (x_{m,n-1} + x_{m,n} + x_{m,n+1}) \ mod(2), \quad x_{0,n} \in S = \{0, 1\}, \tag{14}$$

$$x_{m+1,n} = (x_{m,n-1} + x_{m,n} \otimes x_{m,n+1} + x_{m,n} \oplus x_{m,n+1} + x_{m,n+2}) \ mod(2), \tag{15}$$

where $x_{0,n} \in S = \{0, 1\}$, $\otimes$ denotes the logical "AND" operator and $\oplus$ denotes the logical "OR" operator.

From the above main result, one can see that systems (14)–(15) are both chaotic in the sense of Devaney on (S_∞^∞, d), where d is defined by (1).

References

1. T. Li, Y. Yorke, *Amer. Math. Monthly*, **82**, 985 (1975).
2. R. L. Devaney, *An Introduction to Chaotic Dynamical Systems*, 2nd Ed. (Addision-Wesley, NY, 1989).
3. B. L. Hao, *Introduction to Chaotic Dynamics – Starting from the Parabola* (in Chinese) (Shanghai Scientific and Technological Education Publishing House, Shanghai, 1995).
4. Z. L. Zhou, *Symbolic Dynamics* (in Chinese) (Shanghai Scientific and Technological Education Publishing House, Shanghai, 1997).
5. W. M. Zhen, B. L. Hao, *Applied Symbolic Dynamics* (in Chinese) (Shanghai Scientific and Technological Education Publishing House, Shanghai, 1994).
6. M. Vellekoop, R. Berglund, *Amer. Math. Monthly*, **101**, 353 (1994).
7. S. N. Elaydi, *Discrete Chaos* (Chapman & Hall/CRC, New York, 2000).
8. G. R. Chen, C. J. Tian, Y. M. Shi, *Chaos, Solit. Fractals*, **25**, 637 (2005).
9. C. J. Tian, G. R. Chen, *Physica A*, **376**, 246 (2007).
10. C. J. Tian, G. R. Chen, *Chaos, Solit. Fractals*, **28**, 1067 (2006).
11. C. J. Tian and G. R. Chen, *J. Applied Math. Computing*, **26**, 503 (2008).
12. C. J. Tian amd G. R. Chen, *J. Math. Anal. Appl.*, **356**, 800 (2009).
13. C. J. Tian amd G. R. Chen, *ACTA Math. Appl. Sinica*, **31**, 440 (2008).

PART B

Modelling and control of coupled stochastic oscillators

FEEDBACK-DEPENDENT CONTROL OF STOCHASTIC SYNCHRONIZATION IN COUPLED NEURAL SYSTEMS

PHILIPP HÖVEL, SARANG A. SHAH, MARKUS A. DAHLEM and ECKEHARD SCHÖLL*

Institut für Theoretische Physik, Technische Universität Berlin
Hardenbergstraße 38, 10623 Berlin, Germany
*E-mail: *schoell@physik.tu-berlin.de*
http://www.itp.tu-berlin.de/schoell

We investigate the synchronization dynamics of two coupled noise-driven FitzHugh-Nagumo systems, representing two neural populations. For certain choices of the noise intensities and coupling strength, we find cooperative stochastic dynamics such as frequency synchronization and phase synchronization, where the degree of synchronization can be quantified by the ratio of the interspike interval of the two excitable neural populations and the phase synchronization index, respectively. The stochastic synchronization can be either enhanced or suppressed by local time-delayed feedback control, depending upon the delay time and the coupling strength. The control depends crucially upon the coupling scheme of the control force, i.e., whether the control force is generated from the activator or inhibitor signal, and applied to either component. For inhibitor self-coupling, synchronization is most strongly enhanced, whereas for activator self-coupling there exist distinct values of the delay time where the synchronization is strongly suppressed even in the strong synchronization regime. For cross-coupling strongly modulated behavior is found.

Keywords: Synchronization, noise, coupling, time-delayed feedback.

1. Introduction

The control of unstable or irregular states of nonlinear dynamic systems has many applications in different fields of physics, chemistry, biology, and medicine [1]. A particularly simple and efficient control scheme is time-delayed feedback [2] which occurs naturally in a number of biological systems including neural networks where both propagation delays and local neurovascular couplings lead to time delays [3–5]. Moreover, time-delayed feedback loops might be deliberately implemented to control neural disturbances, e.g., to suppress undesired synchrony of firing neurons in Parkinson's disease or epilepsy [6–8]. Here we study coupled neural systems subject to noise and time-delayed feedback [9–12]. In particular we focus upon the question how stochastic synchronization of noise-induced oscillations of two coupled neural populations can be controlled by time-delayed feedback, and how robust this is with respect to different coupling schemes of the control force.

Time-delayed feedback control of noise-induced oscillations was demonstrated in a single excitable system [13–16]. The simplest network configuration displaying features of neural interaction consists of two coupled excitable systems.

In order to grasp the complicated interaction between billions of neurons in large neural networks, those are often lumped into groups of neural populations each of which can be represented as an effective excitable element that is mutually coupled to the other elements [8,17]. In this sense the simplest model which may reveal features of interacting neurons consists of two coupled neural oscillators. Each of these will be represented by a simplified FitzHugh-Nagumo (FHN) system [18,19], which is often used as a generic model for neurons, or more generally, excitable systems [20].

This paper is organized as follow: We introduce the model equations and the feedback scheme in Sec. 2. Sec. 3 is devoted to two measures of the stochastic synchronization. These are investigated for different coupling schemes of the feedback in Sec. 4. Finally, we conlcude in Sec. 5.

2. Model equations

Neurons are excitable units which can emit spikes or bursts of electrical signals, i.e., the system rests in a stable steady state, but after it is excited beyond a threshold, it emits a pulse. In the following, we consider electrically coupled neurons modelled by the FitzHugh-Nagumo system in the excitable regime:

$$\varepsilon_1 \frac{du_1}{dt} = f(u_1, v_1) + C(u_2 - u_1) \tag{1a}$$

$$\frac{dv_1}{dt} = g(u_1, v_1) + D_1 \xi_1 \tag{1b}$$

$$\varepsilon_2 \frac{du_2}{dt} = f(u_2, v_2) + C(u_1 - u_2) \tag{2a}$$

$$\frac{dv_2}{dt} = g(u_2, v_2) + D_2 \xi_2 \tag{2b}$$

with $f(u_i, v_i) = u_i - u_i^3/3 - v_i$ and $g(u_i, v_i) = u_i + a$ $(i = 1, 2)$. The fast activator variables u_i $(i = 1, 2)$ refer to the transmembrane voltage, and the slow inhibitor variables v_i are related to the electrical conductance of the relevant ion currents. The parameter a is the excitability parameter. For the purposes of this paper, a is fixed at 1.05, such that there are no autonomous oscillations (excitable regime). C is the diffusive coupling strength between u_1 and u_2. To introduce different time scales for both systems, ε_1 is set to 0.005 and ε_2 is set to 0.1. Both systems, when uncoupled, are driven entirely by independent noise sources, which in the above equations are represented by ξ_i ($i = 1, 2$, Gaussian white noise with zero mean and unity variance). D_i is the noise intensity, and for the purposes of this paper, D_2 will be held fixed at 0.09 [9].

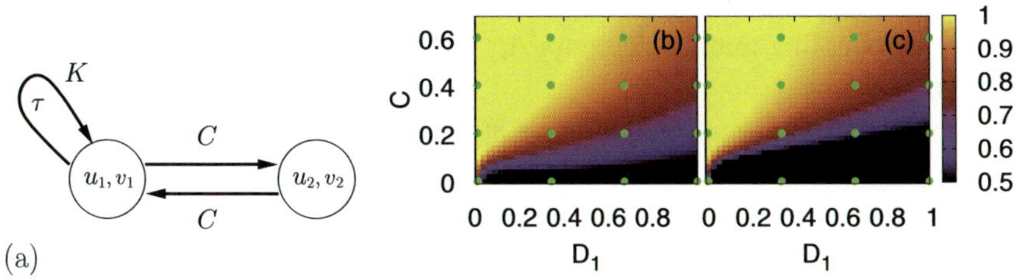

Fig. 1. Panel(a): Schematic diagram of two coupled FitzHugh-Nagumo systems with time-delayed feedback applied to the first subsystem. K and τ denote the feedback gain and time delay, respectively, and C is the coupling strength. Panels (b) and (c) show the ratio of interspike intervals $\langle T_1\rangle/\langle T_2\rangle$ and the phase synchronization index γ of the two subsystems as color code in dependence on the coupling strength C and noise intensity D_1, respectively. No control is applied to the system. The dots mark the parameter choice for different synchronization regimes used in the following. Other parameters: $\varepsilon_1 = 0.005$, $\varepsilon_2 = 0.1$, $a = 1.05$, and $D_2 = 0.09$.

The control force which we apply only to the first of the neural populations as schematically depicted in Fig. 1(a) is known as time-delay autosynchronization (TDAS) or time-delayed feedback control. This method was initially introduced by Pyragas [2] for controlling periodic orbits in chaotic systems. It has been effective in a variety of experimental applications at controlling oscillatory behavior and can be easily implemented in many analog devices [1]. TDAS constructs a feedback F from the difference between the current value of a control signal w and the value for that quantity at time $t - \tau$. The difference is then multiplied by the gain coefficient K

$$F(t) = K[w(t - \tau) - w(t)], \tag{3}$$

where w determines which components of the system enter the feedback as will be discussed in the following.

The variable w in the control force can be either the activator u_1 or the inhibitor v_1. Also, the control force can either be applied to the activator or the inhibitor differential equation. These possibilities lead to two self-coupling schemes (uu and vv) where either the activator is coupled to the activator equation or the inhibitor is coupled to the inhibitor equation, and two cross-coupling schemes (uv and vu). Thus, Eqs. (1) of the first subsystem can be rewritten including time-delayed feedback as

$$\begin{pmatrix} \varepsilon_1 \frac{du_1}{dt} \\ \frac{dv_1}{dt} \end{pmatrix} = \begin{pmatrix} f(u_1, v_1) + C(u_2 - u_1) \\ g(u_1, v_1) + D_1 \xi_1 \end{pmatrix} + K \begin{pmatrix} A_{uu} & A_{uv} \\ A_{vu} & A_{vv} \end{pmatrix} \begin{pmatrix} u_1(t-\tau) - u_1(t) \\ v_1(t-\tau) - v_1(t) \end{pmatrix}, \tag{4}$$

where the coupling matrix elements A_{ij} with $i, j \in \{u, v\}$ define the specific coupling scheme.

Next, we will discuss cooperative stochastic dynamics resulting in frequency synchronization and phase synchronization in the following Sections.

3. Measures of synchronization

A measure of frequency synchronization is the ratio of the interspike intervals (ISI) of the two neural populations [9,10]. The respective average ISI of each neural population is denoted by $\langle T_1 \rangle$ and $\langle T_2 \rangle$. The ratio $\langle T_1 \rangle / \langle T_2 \rangle$ compares the average time scales of both systems, where unity ratio describes two systems spiking at the same average frequency. It is for this reason that the ISI ratio is often considered as a measure of frequency synchronization. It does not contain information about the phase of synchronization, and a given ISI ratio can also result from different ISI distributions.

In order to account for the phase difference between two systems, one can define a phase [9,21,22]

$$\varphi(t) = 2\pi \frac{t - t_{i-1}}{t_i - t_{i-1}} + 2\pi(i-1) \tag{5}$$

where $i = 1, 2, \ldots$. t_i denotes the time of the ith spike. The phase difference between two consecutive spikes is 2π. The phase difference of 1:1 synchronization is

$$\Delta\varphi(t) = |\varphi_1(t) - \varphi_2(t)|, \tag{6}$$

where $\varphi_1(t)$ and $\varphi_2(t)$ are the phases of the first and second system, respectively. Two systems that are phase synchronized at a given time satisfy

$\Delta\varphi = const.$ Finally, the overall time-averaged phase synchronization of two systems can be quantified using the synchronization index

$$\gamma = \sqrt{\langle \cos \Delta\varphi(t) \rangle^2 + \langle \sin \Delta\varphi(t) \rangle^2}. \tag{7}$$

A value of 0 indicates no synchronization, while a value of unity indicates perfect synchronization.

Figure 1(b) and 1(c) depict both measures for stochastic synchronization in the (D_1, C) plane, both exhibiting very similar behavior. Panel (a) refers to the frequency synchronization characterized by the ratio of the average ISIs $\langle T_1 \rangle / \langle T_2 \rangle$ and panel (b) shows the phase synchronization index γ. The green dots mark parameter values used in Sec. 4. Note that both panels share the same color code. For a small value of D_1 and large coupling strength, the two subsystems display well synchronized behavior, $\langle T_1 \rangle / \langle T_2 \rangle \approx 1$ and $\gamma \approx 1$. The timescales in the interacting systems adjust themselves to $1:1$ synchronization. On average, they show the same number of spikes and the two subsystems are in-phase which is indicated by yellow color. The two subsystems are less synchronized in the dark blue and black regions.

In the following we show the ratio of the average interspike interval $\langle T_1 \rangle / \langle T_2 \rangle$ and the phase synchronization index γ which are color coded in the (τ, K) plane for fixed combinations of D_1 and C. For each coupling scheme of time-delayed feedback control (cross-coupling schemes uv and vu and self-coupling schemes uu and vv) we present a selection of (D_1, C) values. In all cases, only one element of the coupling matrix $\mathbf{A}$ is equal to unity and all other elements are zero.

4. Coupling schemes

After the introduction of the system and the coupling schemes, we will present results on frequency and phase synchronization in the following. We consider 16 different combinations of the noise intensity D_1 and the coupling strength C which are marked as green dots in Fig. 1(b) and 1(c). The ordering of panels in Figs. 2(a) to 3 is the following: The rows correspond to fixed coupling strength chosen as $C = 0.01, 0.21, 0.41$, and 0.61 from bottom to top. The columns in each figure are calculated for constant noise intensity $D_1 = 0.01, 0.34, 0.67$, and 1.0 from left to right.

4.1. *Frequency synchronization*

Figures 2(a) to 2(d) show frequency synchronization measured by the ratio of average interspike intervals $\langle T_1 \rangle / \langle T_2 \rangle$ calculated from the summarized

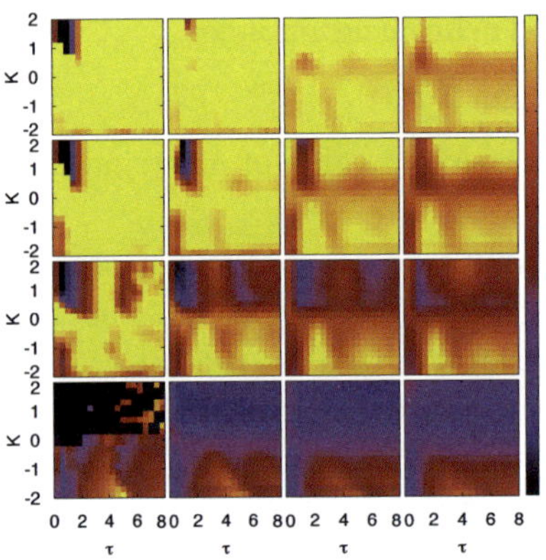

(a) uu-coupling

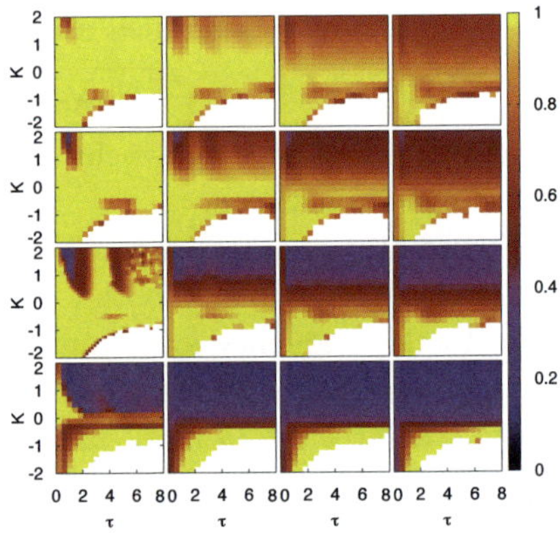

(b) uv-coupling

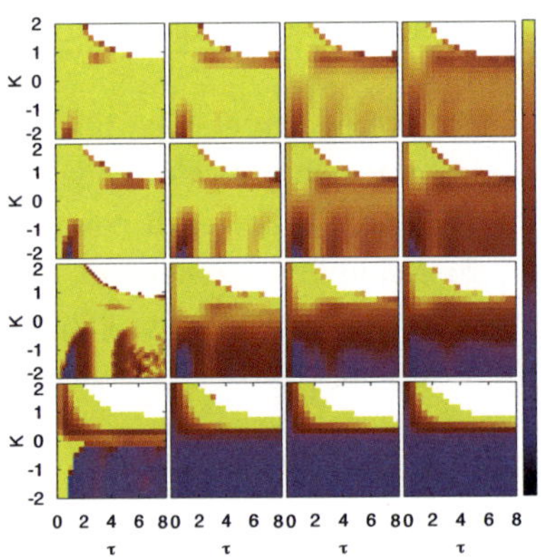

(c) vu-coupling

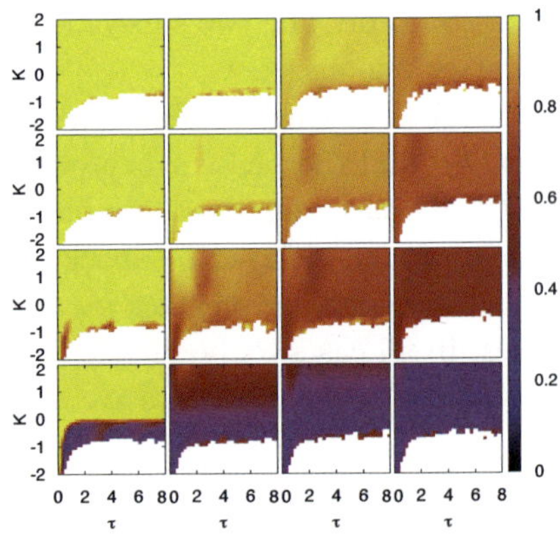

(d) vv-coupling

Fig. 2. Ratio of average interspike intervals $\langle T_1\rangle/\langle T_2\rangle$ as color code. Rows and columns correspond to constant coupling strength C and noise intensity D_1, respectively, as marked in Figs. 1(b) and 1(c) as green dots, and specified in the text. Other parameters are as in Fig. 1.

activator variable $u_\Sigma = u_1 + u_2$ as color code in dependence on the feedback gain K and the time delay τ. The system's parameters are fixed in each panel as described above. Figures 2(a) and 2(d) correspond to self-coupling (uu- and vv-coupling) and Figs. 2(b) and 2(c) depict the cross-coupling schemes (uv- and vu-coupling). The dynamics in the white regions is outside the excitable regime and does not show noise-induced spiking, but rather the system exhibits large-amplitude self-sustained oscillations.

One can see that appropriate tuning of the control parameters leads to enhanced or deteriorated synchronization displayed by bright yellow and dark blue areas, respectively. In each figure, all panels show qualitatively similar features like a modulation of the ratio $\langle T_1 \rangle / \langle T_2 \rangle$ whose range between maximum and minimum depends on D_1 and C. Comparing the rows, the systems are less (more strongly) synchronized for small (large) values of C indicated by dark blue (yellow) color. As the noise intensity D_1 increases, the dynamics of the coupled subsystems is more and more noise-dominated and the dependence on the time delay τ becomes less pronounced.

Note the symmetry in the cross-coupling schemes shown as Figs. 2(b) and 2(c) between K and its negative value $-K$ for the inverse cross-coupling. The reason is that enhancing the activator yields a similar effects on the dynamics as diminishing the inhibitor variable.

4.2. *Phase synchronization*

Figures 3(a) to 3 depict the phase synchronization index γ as color code depending on the control parameters K and τ for uu-, uv-, vu-, and vv-coupling, respectively. The noise intensity D_1 and coupling strength C are fixed for each panel as described in Sec. 4.1.

Comparing Figs. 3(a) to 3 with the respective plots for frequency synchronization, i.e., Figs. 2(a) to 2(d), one can see that both types of synchronization coincide qualitatively, but the phase synchronization index is more sensitive to the modulation features. Similar to the case of frequency synchronization, time delayed feedback can lead to either enhancement or suppression of phase synchronization depending on the specific choice of the feedback gain K and time delay τ indicated by yellow and dark blue regions. In general, these effects become less sensitive on the time delay as D_1 increases. For larger values of C, the two subsystems show enhanced phase synchronization.

5. Conclusion

In summary, we have shown that stochastic synchronization in two coupled neural populations can be tuned by local time-delayed feedback control of one population. Synchronization can be either enhanced or suppressed, depending upon the delay time and the coupling strength. The control

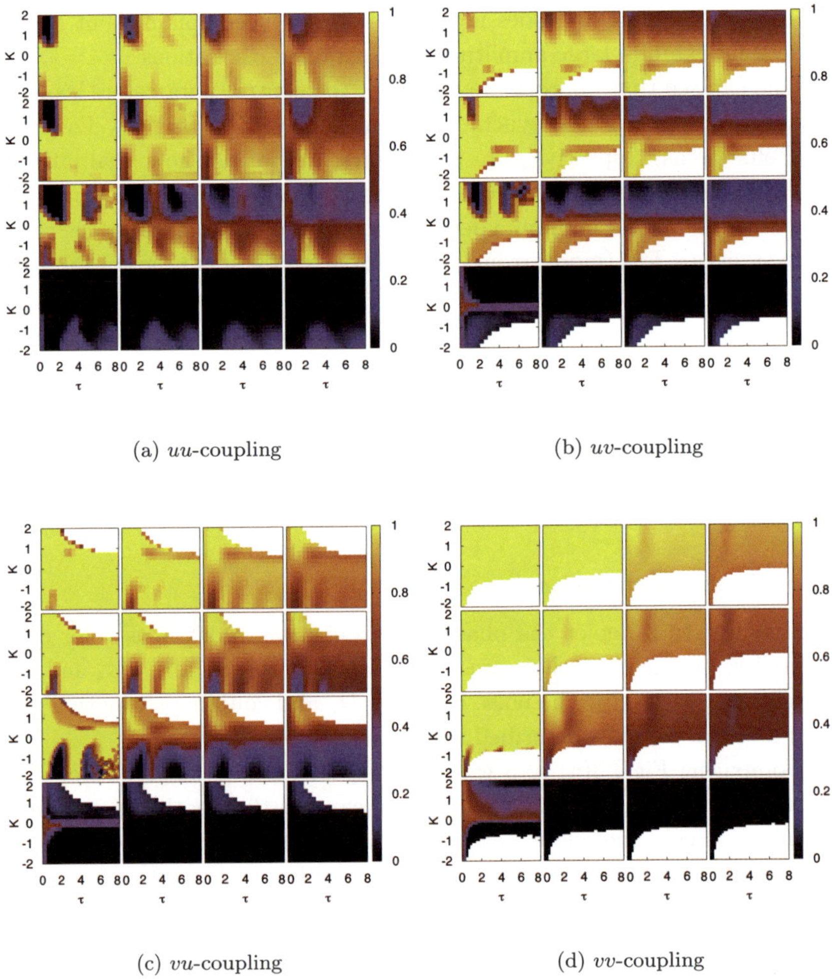

(a) uu-coupling

(b) uv-coupling

(c) vu-coupling

(d) vv-coupling

Fig. 3. Phase synchronization index γ. Noise intensity D_1 and coupling strength C chosen as described in Sec. 4.1. Other parameters as in Fig. 1.

dependents crucially upon the coupling scheme of the control force. For inhibitor self-coupling (vv) synchronization is most strongly enhanced, whereas for activator self-coupling (uu) there exist distinct values of τ where the synchronization is strongly suppressed even in the strong synchronization regime. For cross-coupling (uv, vu) there is mixed behavior, and both schemes exhibit a strong symmetry with respect to inverting the sign of

K. These observations might be important in the context of the deliberate application of control with the aim of suppressing synchronization, e.g. as therapeutic measures for Parkinson's disease.

Acknowledgments

This work was supported by DFG in the framework of Sfb 555 (Complex Nonlinear Processes). S. A. S. acknowledges support of the Deutsche Akademische Austauschdienst (DAAD) in the framework of the program Research Internships in Science and Engineering (RISE).

This paper is the result of two contributions to two different invited sessions at PhysCon 2009 on "Modelling and control of coupled stochastic oscillators" and on "Control of oscillatory delayed-coupled networks".

References

1. E. Schöll and H. G. Schuster (eds.), *Handbook of Chaos Control* (Wiley-VCH, Weinheim, 2008). Second completely revised and enlarged edition.
2. K. Pyragas, *Phys. Lett. A* **170**, 421 (1992).
3. H. Haken, *Brain Dynamics: Synchronization and Activity Patterns in Pulse-Coupled Neural Nets with Delays and Noise* (Springer Verlag, Berlin, 2006).
4. H. R. Wilson, *Spikes, Decisions, and Actions: The Dynamical Foundations of Neuroscience* (Oxford University Press, Oxford, 1999).
5. W. Gerstner and W. Kistler, *Spiking neuron models* (Cambridge University Press, Cambridge, 2002).
6. S. J. Schiff, K. Jerger, D. H. Duong, T. Chang, M. L. Spano and W. L. Ditto, *Nature (London)* **370**, 615 (1994).
7. M. G. Rosenblum and A. Pikovsky, *Phys. Rev. Lett.* **92**, 114102 (2004).
8. O. V. Popovych, C. Hauptmann and P. A. Tass, *Phys. Rev. Lett.* **94**, 164102 (2005).
9. B. Hauschildt, N. B. Janson, A. G. Balanov and E. Schöll, *Phys. Rev. E* **74**, 051906 (2006).
10. P. Hövel, M. A. Dahlem and E. Schöll, *Int. J. Bifur. Chaos (in print)* (2009), (arxiv:0809.0819v1).
11. E. Schöll, G. Hiller, P. Hövel and M. A. Dahlem, *Phil. Trans. R. Soc. A* **367**, 1079 (2009).
12. E. Schöll, P. Hövel, V. Flunkert and M. A. Dahlem, Time-delayed feedback control: from simple models to lasers and neural systems, in *Complex Time-Delay Systems*, ed. F. M. Atay (Springer, Berlin, 2009)
13. N. B. Janson, A. G. Balanov and E. Schöll, *Phys. Rev. Lett.* **93**, 010601 (2004).
14. A. G. Balanov, N. B. Janson and E. Schöll, *Physica D* **199**, 1 (2004).
15. T. Prager, H. P. Lerch, L. Schimansky-Geier and E. Schöll, *J. Phys. A* **40**, 11045 (2007).
16. A. Pototsky and N. B. Janson, *Phys. Rev. E* **77**, 031113 (2008).

17. M. G. Rosenblum and A. Pikovsky, *Phys. Rev. E* **70**, 041904 (2004).
18. R. FitzHugh, *J. Gen. Physiol.* **43**, 867 (1960).
19. J. Nagumo, S. Arimoto and S. Yoshizawa., *Proc. IRE* **50**, 2061 (1962).
20. B. Lindner, J. García-Ojalvo, A. Neiman and L. Schimansky-Geier, *Phys. Rep.* **392**, 321 (2004).
21. A. Pikovsky, M. G. Rosenblum and J. Kurths, *Europhys. Lett.* **34**, 165 (1996).
22. A. Pikovsky, M. G. Rosenblum and J. Kurths, *Synchronization, A Universal Concept in Nonlinear Sciences* (Cambridge University Press, Cambridge, 2001).

DYNAMICS OF CAVITY QED IN STOCHASTIC FIELD IN INTERACTING FOCK SPACE

P. K. DAS

Physics and Applied Mathematics Unit
Indian Statistical Institute
203, B. T. Road, Kolkata-700108
E-mail:daspk@isical.ac.in

The dynamical behaviour of Quantum Electrodynamics(QED) in Stochastic Medium is formulated by state equation in Interacting Fock Space which is a generalization of Langevin dynamics in Boson Fock Space.

Keywords: Quantum electrodynamics, interacting Fock space, Langevin dynamics.

1. Introduction

Quantum Control Theory is an Emerging Field with application to Modern Technology of Quantum Computer and Quantum Information Processing. The present work comes under the purview of field-field interaction of optical cavity and an external stochastic field and is concerned with the study of quantum electrodynamics in Interacting Fock Space and a modelling of quantum feedback systems of the optical QED cavities.

The dynamical behaviour and the state space representation in stochastic field of the quantum electrodynamic (QED) system provide the basis for developing quantum feedback QED control system.

In Sec. 2 the basic concept of the interacting Fock space and the general formulation of the interacting optical cavity with a noisy field is used to derive the dynamics of the QED cavity in the stochastic field [1,2,4,6]. The essential result is a master equation of the general state space model of the optical cavity QED system in interacting Fock space. The Langevin model of the QED bath in boson Fock space is also derived [8].

In Sec. 3 we consider the designing the optical QED system in feedback form in the quantum world. In Sec. 4 we present a brief note of the concluding discussion.

2. Dynamics of the optical QED in stochastic media

In this section we will present briefly the formulation of quantum stochastic process in which the computational relation are made and indeed become a powerful computational tool.

Let us begin with the definition of quantum stochastic process over the time interval $[t_0, t]$ by the operator

$$B_{in}(t, t_0) = \int_{t_0}^{t} b_{in}(s)ds \tag{1}$$

where the field $b_{in}(t)$ being the input to the cavity and represents also the field immediately before it interacts with the system.

Then the commutation relation can be formulated as[6]

$$[B_{in}(t, t_0), B_{in}^{+}(t, t_0)] = t - t_0 \tag{2}$$

Also

$$[dB_{in}(t), dB_{in}^{+}(t)] = [B_{in}(t + dt, t), B_{in}^{+}(t + dt, t)] = dt \tag{3}$$

The relation (3) leads to the natural definition of quantum stochastic process referred in [1].

We now consider the interaction of an interacting single-mode of quantized field confined in an optical cavity with a noisy external field. Let $\mathcal{H}_{\mathcal{A}}$ and $\mathcal{H}_{\mathcal{B}}$ be Hilbert spaces of the cavity and the external field respectively. The composite system is expressed by the tensor product space $\mathcal{H}_{\mathcal{A}} \otimes \mathcal{H}_{\mathcal{B}}$. The total Hamiltonian is given by

$$H_{total} = H_A + H_B + H_{int}, \tag{4}$$

H_A being described the Hamiltonian of the cavity mode and may be further subdivided into two parts H_{cav} and H. Here H is referred to as a *free Hamiltonian* determined by the optical medium in the cavity. H_B is the Hamiltonian of the external field.

After dropping the energy non-conserving terms in H_{int} corresponding to the rotating-wave approximation we obtain the simplified Hamiltonian

$$H_{int}(t) = i\sqrt{\gamma}[a(t)b^{+}(t) - a^{+}(t)b(t)] \tag{5}$$

with $[b(t), b^{+}(t')] = \delta(t - t')$, γ being a coupling constant. The operators a and b are respectively the annihilation operators of the cavity and the external field. The evolution of an arbitrary operator X is given by

$$X(t) = U^{+}(t)XU(t) \tag{6}$$

in which the unitary operator $U(t)$ is generated by the Hamiltonian in (4). The Hamiltonians H_{cav} and H_B drive the cavity and the external field respectively. We shall assume here H to be zero. The unitary operator of the system is then given by

$$U(dt) = e^{\sqrt{\gamma}(adB_{in}^{+} - a^{+}dB_{in})} \tag{7}$$

The increment of an arbitrary operator r of the system driven by the stochastic input b_{in} is given by

$$\begin{aligned} dr(t) = {} & \sqrt{\gamma}[a^{+}dB_{in} - adB_{in}^{+}, r(t)] + \tfrac{\gamma}{2}\{(N' + 1)(2a^{+}ra - a^{+}ar - ra^{+}a) \\ & + N'(2ara^{+} - aa^{+}r - raa^{+}) + M[a^{+}, [a^{+}, r]] + M^{*}[a, [a, r]]\}dt. \end{aligned} \tag{8}$$

We describe the behaviour of the optical cavity in the interacting Fock space $\Gamma(\mathbb{C})$ defined by

$$\Gamma(\mathbb{C}) = \bigoplus_{n=0}^{\infty} \mathbb{C}|n\rangle \tag{9}$$

where $\mathbb{C}|n\rangle$ is called the n-particle subspace. The different n- particle subspaces are orthogonal, that is, the sum in (9) is also orthogonal. The norm of the vector $|n\rangle$ is given by

$$\langle n|n\rangle = \lambda_n \tag{10}$$

where $\{\lambda_n\} > 0$. The norm introduced in (10) makes $\Gamma(\mathbb{C})$ a Hilbert space. on $\Gamma(\mathbb{C})$ creation and annihilation operators are defined as

$$a^{\dagger}|n\rangle = |n+1\rangle, \quad a|n+1\rangle = \tfrac{\lambda_{n+1}}{\lambda_n}|n\rangle. \tag{11}$$

The commutation relation of the operators then takes the form

$$[a, a^{\dagger}] = \frac{\lambda_{N+1}}{\lambda_N} - \frac{\lambda_N}{\lambda_{N-1}} \tag{12}$$

where N is the number operator defined by $N|n\rangle = n|n\rangle$.

The dynamical behaviour of the optical cavity in interacting Fock space is now described on replacing the general operator $r(t)$ in equation (8) by the operator $a(t)$ of the QED bath. Then using the commutation relations (3, 12) and the stochastic process given in [1] we get

$$\begin{aligned} da = a(t+dt) - a(t) = {} & \{-\tfrac{\gamma}{2}(\tfrac{\lambda_{N+1}}{\lambda_N} - \tfrac{\lambda_N}{\lambda_{N-1}})a \\ & -\sqrt{\gamma}(\tfrac{\lambda_{N+1}}{\lambda_N} - \tfrac{\lambda_N}{\lambda_{N-1}})b_{in}(t)\}dt. \end{aligned} \tag{13}$$

This implies

$$\dot{a}(t) = -\tfrac{\gamma}{2}(\tfrac{\lambda_{N+1}}{\lambda_N} - \tfrac{\lambda_N}{\lambda_{N-1}})a(t) - \sqrt{\gamma}(\tfrac{\lambda_{N+1}}{\lambda_N} - \tfrac{\lambda_N}{\lambda_{N-1}})b_{in}(t). \tag{14}$$

The state equation represented by (14) of the dynamics of the cavity A in the interacting Fock space is a generalization of the well known Langevin equation of the cavity in boson Fock space [1,2,8]. In the case of boson Fock space the commutator described in equation (12) becomes unity, that is,

$$[a, a^*] = \frac{\lambda_{N+1}}{\lambda_N} - \frac{\lambda_N}{\lambda_{N-1}} = \Lambda_N^{'} = 1.$$

Then the dynamics of the cavity reduces to the usual quantum Langevin form in boson Fock space

$$\dot{a}(t) = -\frac{\gamma}{2}a(t) - \sqrt{\gamma}b_{in}(t) \tag{15}$$

Due to interaction of the evolving incoming field with the optical cavity an outgoing field is produced and is given by

$$B_{out}(t, t_0) = \int_{t_0}^{t} b_{out}(s)ds \tag{16}$$

where

$$b_{out}(t) = U^+(dt)b_{in}(t)U(dt) \tag{17}$$

The input-output relation after the interaction at time t is given simply by the following derivation:

$$\begin{aligned} dB_{out}(t) &= U^+(dt)dB_{in}(t)U(dt) \\ &= dB_{in}(t) + \sqrt{\gamma}a[dB_{in}, dB_{in}^+] \end{aligned} \tag{18}$$

Now using (3) the above relation gives us

$$b_{out}(t)dt = b_{in}(t)dt + \sqrt{\gamma}adt. \tag{19}$$

we then have the required input-output relation of the cavity QED

$$b_{out}(t) = \sqrt{\gamma}a(t) + b_{in}(t). \tag{20}$$

We have seen that the cavity dynamics may be thought of as an operator equation in Hilbert space. The equations (14) and (20) give the state equation and the system output of a single cavity in different modes. The operator $b_{in}(t)$ is the input and the operator $b_{out}(t)$ is the output of the cavity. The state equation of the cavity dynamics along with the output equation can be represented in the general form of classical control system as

$$\dot{a}(t) = A^{'}a(t) + B^{'}b_{in}(t), \quad b_{out}(t) = C^{'}a(t) + D^{'}b_{in}(t) \tag{21}$$

where

$$\begin{aligned} A' &= -\tfrac{\gamma}{2}\left(\tfrac{\lambda_{N+1}}{\lambda_N} - \tfrac{\lambda_N}{\lambda_{N-1}}\right) \\ B' &= -\sqrt{\gamma}\left(\tfrac{\lambda_{N+1}}{\lambda_N} - \tfrac{\lambda_N}{\lambda_{N-1}}\right) \\ C' &= \sqrt{\gamma} \\ D' &= 1 \end{aligned} \tag{22}$$

The dynamics of the cavity in the interacting Fock space is a first order differential equation of the system operators with variable space parameter Λ'_N as coefficient. Applying Laplace transform in (21), assuming zero initial state of the QED bath, we get the transfer function representation of the optical QED system in interacting mode as,

$$b_{out}(s) = G(s) b_{in}(s), \tag{23}$$

$$G(s) = \frac{s - \frac{\gamma_A}{2}\Lambda'_N}{s + \frac{\gamma_A}{2}\Lambda'_N}. \tag{24}$$

We have seen that a cavity QED in interacting mode in some way closely analogous to the classical one in which the input and the output are described by operators in Hilbert space.

3. Quantum feedback control system of the optical cavity

The state space modelling of the optical QED bath in the interacting Fock Space is taken to be the basis of designing closed-loop feedback control system.

Utilizing the general procedure of classical feedback control theory, the mathematical model of the cavity QED system with unity feedback may be described using a quantum device, such as, beam splitter as shown in Fig. 1.

The two input signals b_{in} and b_1 to the beam splitter are related to the output signals b_0 and b_2 by

$$\begin{bmatrix} b_2 \\ b_0 \end{bmatrix} = \begin{bmatrix} \alpha & \beta \\ \beta & -\alpha \end{bmatrix} \begin{bmatrix} b_{in} \\ b_1 \end{bmatrix} \tag{25}$$

where α and β are real, positive and satisfy $\alpha^2 + \beta^2 = 1$. From the input-output relation (20) of the single cavity, we have

$$b_1 = \sqrt{\gamma_A} a_A + b_0 \tag{26}$$

Each signal in the single cavity feedback loop can now be written as

$$\begin{aligned} b_0 &= \tfrac{\beta}{1+\alpha} b_{in} - \tfrac{\alpha}{1+\alpha}\sqrt{\gamma_A} a_A \\ b_1 &= \tfrac{\beta}{1+\alpha} b_{in} + \tfrac{1}{1+\alpha}\sqrt{\gamma_A} a_A \\ b_2 &= b_{in} + \tfrac{\beta}{1+\alpha}\sqrt{\gamma_A} a_A \end{aligned} \tag{27}$$

The closed-loop transfer function of the feedback control system with unity feedback of a single optical cavity QED using beam splitter is described as

$$M_1(s) = \frac{\beta G(s)}{1 + \alpha G(s)} = \frac{\beta(s - \frac{\gamma_A}{2}\Lambda_N^{'})}{s(1+\alpha) + \frac{\gamma_A}{2}\Lambda_N^{'}(1-\alpha)} \tag{28}$$

The stability of the closed loop system can be easily analysed with the help of Nyquist stability criterion.

4. Conclusion

This study mainly explores various aspects of the experimental problems [4] of coupling a single photon with a single atom within an optical cavity QED, and of the problem of controlling a closed-loop feedback QED control systems.

References

1. Yanagisawa, M. and Kimura, H. *Transfer Function Approach to Quantum Control- Part I: Dynamics of Quantum Feedback Systems*, IEEE Transactions on Automatic Control, Vol. 48, no. 12, 2107,(2003).
2. Yanagisawa, M. and Kimura, H. *Transfer Function Approach to Quantum Control- Part II: Control Concepts and Applications*, IEEE Transactions on Automatic Control, Vol. 48, no. 12, 2121,(2003).
3. Accardi, L. and Bozejko, M. *Interacting Fock Space and Gaussianization of Probability Measures*, Infinite Dimensional Analysis, Quantum Probability and Related Topics, 1(4), 663,(1998).
4. Wiseman, H. M., and Milburn, G. J., *Quantum Theory of Optical Feedback via Homodyne Detection*, Phy. Rev. Letts, Vol. 70, no. 5, 548, (1993).
5. Das, P. K. *Coherent states and squeezed states in interacting Fock space.* International Journal of Theoretical Physics. vol. 41, no. 06, 1099-1106, (2002), MR. No. 2003e: 81091 (2003).
6. Das, P. K. and Roy, B. C., *State space modelling of quantum feedback control system in interacting Fock space*, International Journal of Control, vol. 79, no. 7, 729-738(2006).
7. Roy, B. C. and Das, P. K., *Optimal control of multilevel quantum system with energy cost functional*, International Journal of Control,vol. 80, no. 8, 1299-1306(2007).
8. Gardiner, C. W. and Zoller, P., *Quantum Noise*, Springer-verlag, Berlin(2000).

DELAYED FEEDBACK CONTROL IN STOCHASTIC EXCITABLE NETWORKS

NATALIA B. JANSON* and ANDREY POTOTSKY

School of Mathematics, Loughborough University
Ashby road, Loughborough, LE11 3TU, United Kingdom
*E-mail: *n.b.janson@lboro.ac.uk*
http://www-staff.lboro.ac.uk/~manbj/

SANDHYA PATIDAR

Actuarial Mathematics and Statistics
School of Mathematical and Computer Sciences, Heriot-Watt University
Edinburgh, EH14 4AS, Scotland, United Kingdom
S.Patidar@hw.ac.uk

A simplified model of a stochastic neural network is considered, being a system of a large number of identical excitable FitzHugh-Nagumo oscillators coupled via the mean field. The possibility to control the global dynamics of this network is investigated. The control tool being probed is Pyrgas delayed feedback constructed and applied through the mean field. It is shown that one can destroy or diminish stochastic synchronization in a partially synchronized network by a weak delayed feedback under the appropriate choice of delay.

Keywords: Network, neruon model, stochastic, delay, control.

1. Introduction

We consider the collective behavior of a network of excitable stochastic units coupled through the mean field. Each network element is represented by the FitzHugh-Nagumo system in the excitable regime under the influence of noise, which is a paradigmatic model of a single excitable unit [1]. This system serves as a rough model of a neural network. It has been earlier shown [2] that such a network is capable of demonstrating various kinds of collective behavior: from non-synchronized independently spiking units, through a few distinct stages when spiking of different units is synchronized only partially, to the perfectly synchronized network. The detection of different stages of synchronization is possible through the mean field, which demonstrates periodic or chaotic small oscillations around the only fixed point in the absence of synchronization, or periodic or aperiodic spiking. The effect of synchronization in a real neural network is two-fold. On the one hand, synchronization is believed to help better processing of information and is thus advantageous [3,4]. On the other hand, synchronization is suggested to be responsible for inducing a regular rhythmic activity in the brain, which is associated with Parkinson's disease, essential tremor and epilepsy [5–8]. With this, it remains an important clinical challenge to develop an efficient control technique with the ability to manipulate the neural synchrony. Recently, a number of methods have been proposed for the suppression of synchrony of the arrays of coupled oscillators in which oscillations are self-sustained, i.e. exist regardless of the applied noise [9,10]. The purpose of this paper is to demonstrate the possibility to manipulate the properties of the collective behavior of a network of units in which there is no dynamics without external perturbation, and any oscillations are induced merely by external sources of random noise.

2. Stochastic excitable network with delayed feedback

The control technique being probed is the Pyragas delayed feedback [11,12]. The controlling signal is constructed from the macroscopic mean field of the network, and the same signal is applied to all units. The model equations read [13]:

$$\begin{aligned}
\epsilon \dot{x}_i &= x_i - \frac{x_i^3}{3} - y_i + \gamma(M_X - x_i), \qquad (1)\\
\dot{y}_i &= x_i + a + \sqrt{2T}\xi_i(t) + K(M_Y(t-\tau) - M_Y(t)),\\
M_X &= \frac{1}{N}\sum_{i=1}^{N} x_i, \quad M_Y = \frac{1}{N}\sum_{i=1}^{N} y_i.
\end{aligned}$$

Here, N is the total number of units in the network; parameters $a = 1.05$ and $\epsilon = 0.01$ are fixed to ensure excitability of each unit; $\xi_i(t)$ is Gaussian white noise with zero mean and unity variance, and the noise sources in different elements are uncorrelated, i.e. $\xi_i(t)\xi_j(t+s) = \delta_{ij}\delta(s)$, where δ_{ij} is Kronecker delta and $\delta(s)$ is Dirac delta-function. T is the noise intensity which is the same in all units. Coupling between the units is realized through the mean field, when each unit experiences the same averaged input $M_X(t)$ from the rest of the network with the coupling strength γ. To allow for direct comparison of the effects of coupling and of the feedback with those in the earlier works, the network is coupled through $M_X(t)$ in the first of Eq. (1) like in [2], while the delayed feedback $K(M_Y(t-\tau) - M_Y(t))$ is applied in the second of Eq. (1) like in [14,15]. In the above, τ is the feedback delay and K is the feedback strength. There are two possible kinds of chaotic behavior

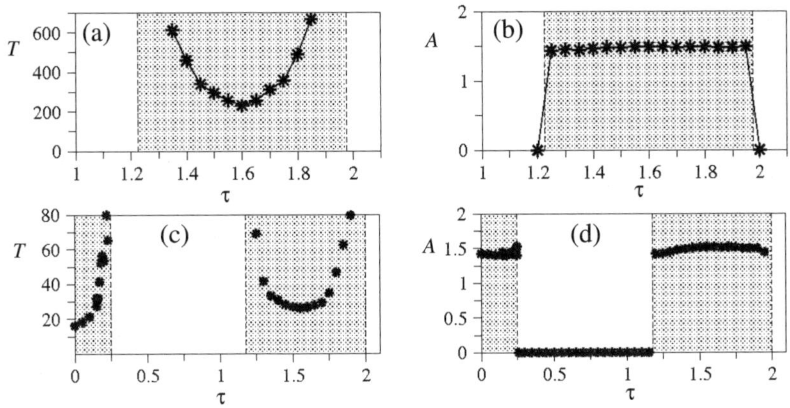

Fig. 1. Mean interspike intervals $\langle T \rangle$ and mean spiking amplitudes A as functions of delay τ as the feedback is applied with $\gamma = 0.1$: (a,b) $T = 0.00027$ (c,d) $T = 0.00028$.

of the mean field: those related to non-synchronous spiking in the network and those related to partially synchronized units. At $\gamma = 0.1$ these regimes can be observed e.g. at $T = 0.00027$ and at $T = 0.00028$, respectively.

We switch the delayed feedback on, fix its strength at $K = 0.1$ and examine the response of the network as a function of τ in the original states at $T = 0.00027$ and at $T = 0.00028$ with $\gamma = 0.1$. Equation (1) are numerically integrated with $N = 10000$ and the dynamics of the ensemble averages M_X and M_Y are studied. The collective response is characterized by the average interspike interval T and the amplitude A of the mean field M_X and is illustrated in Fig. 1.

One can see that where initially the network was desynchronized ($T = 0.00027$ (a,b)), there is a range of τ values at which the feedback is capable of inducing synchronization. The latter is detected by the finite value of T, as opposed to the infinitely large value when the mean field does not spike, and by the large amplitude A. The domains of synchronous spiking are shown as patterned areas in Fig. 1. At the same time, where the network was initially partially synchronized, ($T = 0.00028$ (c,d)), there are ranges of τ at which synchronization is suppressed (infinitely large $\langle T \rangle$). One can see that inside the domains of synchronization the time scale of the mean field oscillations depend on the value of τ and thus can be controlled by the feedback.

3. Cumulant equations with delay

Equation (1) without feedback ($K = 0$) were studied in detail in [2], and their behavior was qualitatively explained using the cumulant expansion of the probability density distribution (PDD) of the units, which was based on the assumption of the molecular chaos. Gaussian approximation was used to truncate the system of cumulant equations to five coupled cumulant equations whose dynamical behavior was qualitatively similar to that of the original stochastic network.

In this work we provide a qualitative explanation of the response of the stochastic network to delayed feedback control by deriving and analysing the cumulant equations with delay using the same two approximations. The cumulant equations read:

$$
\begin{aligned}
\epsilon\frac{dm_X}{dt} &= m_X - \frac{{m_X}^3}{3} - m_Y - m_X D_X, \\
\frac{dm_Y}{dt} &= m_X + a + K(m_Y(t-\tau) - m_Y), \\
\epsilon\frac{dD_X}{dt} &= 2\Big[D_X(1 - \gamma - {m_X}^2 - D_X) - D_{XY}\Big], \\
\frac{dD_Y}{dt} &= 2(D_{XY} + T), \\
\epsilon\frac{dD_{XY}}{dt} &= \epsilon D_X + D_{XY}(1 - {m_X}^2 - D_X - \gamma) - D_Y.
\end{aligned}
\tag{2}
$$

Here m_X and m_Y are the mean values of the distributions of the variables x and y, respectively, D_X and D_Y are their variances, and D_{XY} is their cross-variance which is the second moment of their joint distribution. As indicated in [2], the Gaussian approximation only provides a qualitative description of the effects in the network, while there is no quantitative agreement. With this, to compare the effects induced by the feedback in cumulant and in stochastic equations, we had to apply the feedback to topologically equivalent regimes. To do so, in Eq. (2) with $K = 0$ we chose such parameters T and γ with which they had the regimes similar to those of stochastic Eq. (1). Namely, at $\gamma = 0.1$ and $T = 0.001585$ Eq. (2) demonstrated subthreshold chaos similar to that in Figs. 1(a) and 1(b), while at $T = 0.001586$ they exhibited chaotic spiking.

It is known that the skeleton of a chaotic attractor is formed by the infinite number of unstable periodic orbits. Pyragas delayed feedback can stabilize such orbits if τ is equal to the orbit period and K is chosen appropriately. In addition, in [16] the effects of delayed feedback on a typical chaotic system were revealed for a large range of values of τ and K. It was

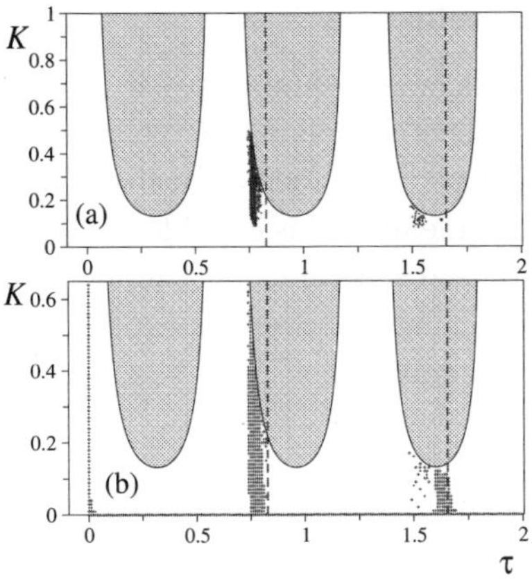

Fig. 2. Maps of regimes of cumulant equations with delayed feedback Eq. (2) on the plane τ and K as the feedback is applied when the original states of the cumulant equations were: (a) subthreshold chaotic oscillations at $T = 0.001585$; (b) chaotic spiking at $T = 0.001586$. Light grey areas inside parabolas indicate the regions of stability of the only fixed point; patterned green areas indicate spiking of cumulants; white areas indicate the absence of spiking in the cumulants.

found that in the plane (τ, K) domains can be found within which there are no stable oscillations at all, i.e. the fixed point is stabilized. Also, a range of bifurcations occur which arise as a result of feedback application.

In Fig. 2 the bifurcation diagrams of Eq. (2) in the plane (τ, K) are shown, which were obtained with the help of continuation technique using the free software DDEBIFTOOL [17] and also numerical simulation of oscillating solutions. The common feature of the diagrams is the large parabola-like curves of Andronov-Hopf bifurcation of the fixed point, above which the fixed point is stable. In the white areas below these curves the oscillations of the cumulants are subthreshold, either periodic or chaotic. Patterned green areas denote the regimes of spiking which can be periodic or chaotic. These regimes were found by numerical integration of the cumulant equations rather than by means of continuation technique. When comparing Figs. 1 and 2, one can notice that in both figures along the lines of fixed K there are domains in which all spiking of the mean field is suppressed, and those where spiking occurs. However, the agreement between the maps of regimes is only qualitative because Gaussian approximation does not give an accurate description of the stochastic network, and also because the initial regimes without the feedback were similar only in their nature but not in the dynamical detail.

4. Summary and conclusions

The purpose of the work was to find out if it was possible to manipulate the properties of the collective spiking of the stochastic network by using some macroscopic feedback. It is demonstrated that the Pyragas delayed feedback applied via the mean field is capable of destroying synchronization in a partially synchronized network. The action of the delayed feedback control on the large stochastic network is explained on a qualitative level by considering cumulant equations.

Acknowledgments

We gratefully acknowledge the fruitful discussions with Dr. Alexander Balanov. This work was supported by Engineering and Physical Sciences Research Council (UK).

References

1. B. Lindner, J. Garcia-Ojalvo, A. Neiman and L. Schimansky-Geier, *Physics Reports* **392**, 321 (2004).
2. M. Zaks, X. Sailer, L. Schimansky-Geier and A. Neiman, *CHAOS* **15**, p. 026117 (2005).
3. J. Samonds, J. Allison, H. Brown and A. Bonds, *Proc. Natl. Acad. Sci. U.S.A.* **101**, p. 6722 (2004).
4. A. Benucci, P. J. Verschure and P. König, *Phys. Rev. E* **70**, p. 051909 (2004).
5. F. Dreifuss and et al, *Epilepsia* **22**, p. 489 (1981).
6. P. Tass, *Phys. Rev. E* **66**, p. 036226 (2002).
7. P. Tass, M. G. Rosenblum, J. Weule, J. Kurths, A. Pikovsky, J. Volkmann, A. Schnitzler and H.-J. Freund, *Phys. Rev. Lett.* **81**, p. 3291 (1998).
8. P. Grosse, M. J. Cassidy and P. Brown, *Clin. Neurophysiol.* **113**, p. 1523 (2002).
9. M. Rosenblum and A. Pikovsky, *Phys. Rev. E* **70**, p. 041904 (2004).
10. O. Popovych, C. Hauptmann and P. Tass, *Phys. Rev. Lett* **94**, p. 164102 (2005).
11. K. Pyragas, *Phys. Lett. A* **170**, p. 421 (1992).
12. K. Pyragas, *Phys. Lett. A* **206**, p. 323 (1995).
13. S. Patidar, A. Pototsky and N. Janson, *New J. Physics* **11**, p. 073001 (2009).
14. N. Janson, A. Balanov and E. Schöll, *Phys. Rev. Lett.* **93**, p. 010601 (2004).
15. A. Balanov, N. Janson and E. Schöll, *Physica D* **199**, p. 1 (2004).
16. A. Balanov, N. Janson and E. Schöll, *Physical Review E* **71**, p. 016222 (2005).
17. K. Engelborghs, T. Luzyanina and G. Samaey, *Technical Report TW-330, Department of Computer Science, K.U. Leuven, Leuven, Belgium* (2001).

THE MODELING OF FUZZY SYSTEMS BASED ON LEE-OSCILLATORY CHAOTIC FUZZY MODEL (LoCFM)

MAX H. Y. WONG and JAMES N. K. LIU

Department of Computing, The Hong Kong Polytechnic University, Hong Kong

DENNIS T. F. SHUM and RAYMOND S. T. LEE

IATOPIA Research Centre, Hong Kong

This paper introduces a new fuzzy membership function — LEE-oscillatory Chaotic Fuzzy Model (LoCFM). The development of this model is based on fuzzy logic and the incorporation of chaos theory — LEE Oscillator. Prototype systems are being developed for handling imprecise problems, typically involving linguistic expression and fuzzy semantic meaning. In addition, the paper also examines the mechanism of the LEE Oscillator through analyzing its structure and neural dynamics. It demonstrates the potential application of the model in future development.

Keywords: Coupled oscillators, LEE Oscillator, retrograde signalling, chaotic fuzzy model, neural dynamics.

1. Introduction

Chaotic oscillator shows non-linear dynamic structures and hierarchy of a system. This paper introduces LEE Oscillator model and its enhancement LEE Oscillator (Retrograde Signalling) model. The structure and analysis of neural dynamics of both models are presented in Secs. 2 and 3. In Sec. 4, we propose a new type of fuzzy membership function — Lee-Oscillatory Chaotic Fuzzy Model (LoCFM) by implementing chaotic oscillator in fuzzy logic. It is to supplement type-1 fuzzy set on modelling uncertainty, we anticipate that LoCFM is suitable for classifying complexity problems in real world.

2. The review of LEE Oscillator

Research on neuroscience and brain science in recent years has observed various chaotic phenomena in brain functions [1] and behaviour of neurons are interactive triggering oscillation between excitatory and inhibitory neurons [2]. Based on these research findings, we have developed the LEE Oscillator and been able to simulated neural behaviour [3,4]. LEE Oscillator has shown outstanding results in pattern recognitions and explained progressive memory recall scheme by Rubin-vase experiment [3,4].

A LEE Oscillator consists of four neural dynamics of four constitutive neural elements: *u, v, w and z*, the neural dynamics of each of these constituent neurons are given by

$$u' = tanh(a_1 \cdot \mathrm{u} - \mathrm{a}_2 \cdot \mathrm{v} + \mathrm{I}) \quad (1)$$

$$v' = tanh(b_1 \cdot \mathrm{u} - \mathrm{b}_2 \cdot \mathrm{v}) \quad (2)$$

$$w = tanh(I) \quad (3)$$

$$z = (u' - v') \cdot e^{-kI^2} + w \quad (4)$$

where *u, v, w and z* are the state variables of the excitatory, inhibitory, input, and output neurons, respectively; *tanh()* is a hyperbolic tangent function; a_1, a_2, b_1 *and* b_2 are the weight parameters for these constitutive neurons; I is the external input stimulus; and k is the decay constant.

The largest Lyapunov Exponent (λ) of the LEE Oscillator is calculated by Wolf's algorithm, we had tested using two fixed sets of parameters with varying I in [-1, +1], those parameters are i) $a_1=5$, $a_2=5$, $b_1=1$, $b_2=1$ *and* $k=500$; and ii) $a_1=10$, $a_2=10$, $b_1=20$, $b_2=20$ *and* $k=500$. The Lyapunov Exponents for the former one is $0.000001 \leq \lambda \leq 3.3661$ and the latter one is $0.1817 \leq \lambda \leq 2.3200$.

3. The LEE Oscillator (Retrograde Signalling) model

The enhancement of the LEE Oscillator is derived from retrograde transport mechanism in axons, axonal transport (also named as axoplasmic flow) was discovered by Paul Weiss in 1948. Receptors, signalling proteins and enzymes for synthesis of neurotransmitter must be moved to distant axon terminals or dendrites from cell body is called anterograde transport. On the other hand, retrograde transport is a backward transmission mechanism which transmits neurotrophin from axon terminals to the cell body [5,6].

Neuroscientists have made an important discovery in recent years, retrograde signals influence neuronal survival, differentiation, homeostasis and plasticity [7]. The latest research on neuroscience pointed out that neurological diseases such as Alzheimer's disease and Down's syndrome are significantly related to malfunction of retrograde transport mechanism [8]. Equations (5) to (8) represent the neural dynamics of this model [9,10] and the meaning of all variables is as same as the previous model.

$$u' = tanh(a_1 \cdot \mathrm{u} - \mathrm{a}_2 \cdot \mathrm{v} + \mathrm{a}_3 \cdot \mathrm{z} + \mathrm{a}_4 \cdot \mathrm{I}) \quad (5)$$

$$v' = tanh(b_3 \cdot \mathrm{z} - \mathrm{b}_1 \cdot \mathrm{u} - \mathrm{b}_2 \cdot \mathrm{v} + \mathrm{b}_4 \cdot \mathrm{I}) \quad (6)$$

$$w = tanh(I) \quad (7)$$

$$z' = (v' - u') \cdot e^{-kI^2} + w \quad (8)$$

The Lyapunov Exponent of different types of oscillator are in positive values (Table 1). Figure 1 presents the neural dynamics of the oscillator in different parameters, ellipse orbits are shown in Figs. 1(a) and 1(c), spiral shell shape displayed in Fig. 1(b) and finally Fig. 1(d) demostrates similar behaviour as a Lissajous curve. We guess that the model is a hyperchaotic system, so different shapes were generated by inserting different parameters. Another research direction is why parameters changed in LEE Oscillator still keep its dynamics in two armed spiral pattern (not shown), but retrograde signalling model changed its dynamics according to parameters.

Table 1. Parameters use in different LEE Oscillator (RS) Models and their Lyapunov Exponents.

Parameter	Type A (Fig. 1(a))	Type B (Fig. 1(b))	Type C (Fig. 1(c))	Type D (Fig. 1(d))
a_1	0.6	1	0.55	1
a_2	0.6	1	0.55	1
a_3	-0.5	1	-0.5	1
a_4	0.5	1	0.5	1
b_1	-0.6	-1	-0.55	-1
b_2	-0.6	-1	-0.55	-1
b_3	-0.5	-1	0.5	-1
b_4	0.5	-1	-0.5	-1
k	50	50	50	300
Lyapunov Exponent (λ)	$0.2861 \leq \lambda \leq 1.0087$	$0.1643 \leq \lambda \leq 1.0716$	$0.2832 \leq \lambda \leq 1.0091$	$0.1643 \leq \lambda \leq 1.8580$

4. The prototype of chaotic fuzzy membership function — LEE-Oscillatory Chaotic Fuzzy Model (LoCFM)

As we all know human mind does not work in digital process, but computer does. Therefore, fuzzy logic is needed to link human's fuzzy feeling to computer through implementing probability and degree of membership associated with fuzzy set. It is doubt the possibility of modelling uncertainties, although Prof. ZADEH proposed the concept of type-2 fuzzy logic

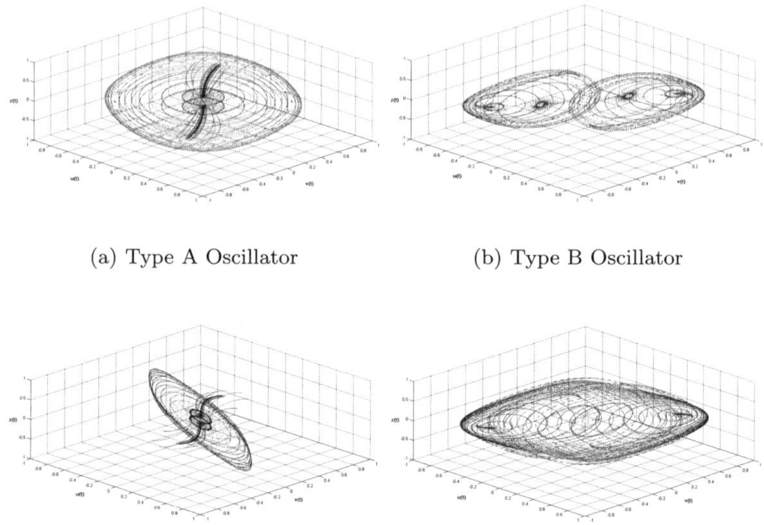

(a) Type A Oscillator

(b) Type B Oscillator

(c) Type C Oscillator

(d) Type D Oscillator

Fig. 1. 3D projection of oscillators in u, v *vs.* z

to strengthen the concept of uncertainty [11]. According to a number of research in brain science, scientists have found that our brain is in fact organized by chaos [1]. We conjecture relationship between chaotic patterns and fuzzy concepts in brain as follows. Brain operates in the rules of chaos theory by generating different neural dynamics signals, then fuzzy feelings are the results of the combination of different signals. These feelings let us realize the temperature of water, emotion or even cognition. Therefore, chaos theory may be one of the more suitable tools in bridging human and computer. For the above reasons and the successfulness of LEE Oscillator model, we proposed a theoretic work for a new series of fuzzy logic — LEE-Oscillatory Chaotic Fuzzy Model (LoCFM). We introduce two prototypes to show the integration between chaotic oscillator and fuzzy membership function to become the proposed models.

4.1. *Architecture*

Figure 2(a) and Eqs. (9) to (12) show the prototype of Chaotic Fuzzy Set — Number One (Prototype CF1):

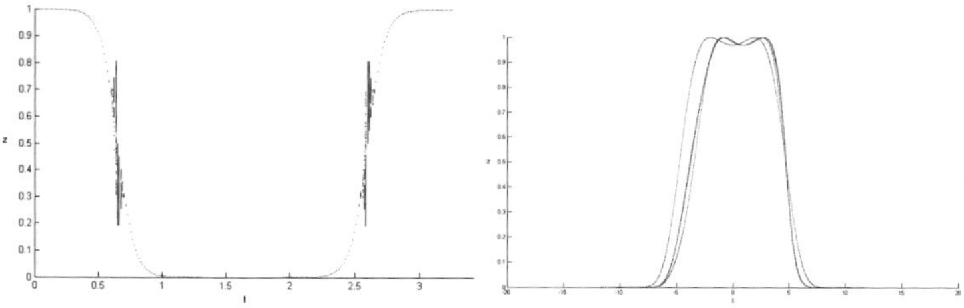

(a) Prototype CF1 (b) Prototype CF2

Fig. 2. LEE-oscillatory chaotic fuzzy model.

$$u' = tanh(a_1 \cdot \mathrm{u} - \mathrm{a}_2 \cdot \mathrm{v} + \mathrm{I}) \quad (9)$$

$$v' = tanh(b_1 \cdot \mathrm{u} - \mathrm{b}_2 \cdot \mathrm{v}) \quad (10)$$

$$w = tanh(I) \quad (11)$$

$$z = m \cdot ((u' - v') \cdot e^{-kI^2} + w) + n \quad (12)$$

Prototype CF1 had a minor modification based on Eq. (4), m *and* n are parameters of slope-intercept form, Fig. 2(a) shows the membership function of Prototype CF1. This prototype uses as similar as a shoulder membership function in type-1 fuzzy logic, but it is with chaotic regions.

Figure 2(b) and Eqs. (13) to (15) show the prototype of Chaotic Fuzzy Set — Number Two (Prototype CF2):

$$u' = a_1 \cdot \mathrm{u} - \mathrm{a}_2 \cdot \mathrm{v} + \mathrm{a}_3 \cdot \mathrm{z} + \mathrm{a}_4 \cdot \mathrm{I} \quad (13)$$

$$v' = b_3 \cdot \mathrm{z} - \mathrm{b}_1 \cdot \mathrm{u} - \mathrm{b}_2 \cdot \mathrm{v} + \mathrm{b}_4 \cdot \mathrm{I} \quad (14)$$

$$z' = e^{-\frac{[(u' \cdot v' \cdot k) - c]^2}{2\sigma^2}} \quad (15)$$

The Prototype CF2 is based on LEE Oscillator (Retrograde Signalling) model with two major modifications. First, the hyperbolic tangent function (*tanh*) is removed in order to compute input values which are smaller than -1 or larger than $+1$. Second, a Gaussian function is implemented in output neuron (z) for generating bell curves as membership function, c *and* σ are parameters of Gaussian function to control the position of the centre of the peak and the width of the *bell curve*. Figure 2(b) illustrates the Prototype CF2, which is used to extend the normal Gaussian fuzzy set to a chaotic Gaussian fuzzy set.

5. Summary

We have reported and analyzed LEE Oscillator model and its enhanced retrograde signalling model. Then, we have introduced a new concept of chaotic fuzzy set — LoCFM which provides an advanced paradigm in future computational intelligence techniques. The above models are required for further investigations and developments. First, the LEE Oscillator models are desired an in-depth study on chaotic dynamics. Second, exploring different chaotic fuzzy sets, such as implementing chaotic features into triangular and trapezoid shaped fuzzy sets. Third, towards a comprehensive theory of LoCFM, there are two major problems which need to be solved. i. How chaotic fuzzy models process by different operators including union, intersection and complement operations. ii. Inference engine and defuzzification process are required to be modified in order to suit with the LoCFM. Forth, a lot of experiments are desired to compare the performance and accuracy of LoCFM with fuzzy features.

References

1. H. Korn and W. J. Faure, *Comptes Rendus Biologies* **326**, 787 (2003).
2. K. Aihara and G. Matsumoto, *Forced oscillations and route to chaos in the Hodgkin-Huxley axons and squid giant axons*, in *Chaos in biological systems*, eds. h. Degn, A. V. Holden and L. F. Olsen (Plenum, New York, 1987), New York, pp. 121–131.
3. R. S. T. Lee, *IEEE Trans. Neural Networks* **15**, 1228 (2004).
4. R. S. T. Lee, *NN* **19**, 644 (2006).
5. K. E. Neet and R. B. Campenot, *CMLS Cellular and Molecular Life Science* **58**, 1021 (2001).
6. L. A. Freberg, *Discovering Biological Psychology* (Wadsworth Publishing, 2005).
7. L. S. Zweifel, R. Kuruvilla and D. D. Ginty, *Nature Reviews on Neuroscience* **6**, 615 (2005).
8. J. D. Copper, A. Salehi, J. D. Delcroix, C. L. Howe, P. V. Belichenko, J. Chua-Couzens, J. F. Kilbridge, E. J. Carlson, C. J. Epstein and W. C. Mobley, *Proc Natl Acad Sci USA* **98** (2001).
9. M. H. Y. Wong, R. S. T. Lee and J. N. K. Liu, Wind shear forecasting by chaotic oscillatory-based neural networks (conn) with lee oscillator (retrograde signalling) model, in *Proc. IEEE Int. Joint Conference on Neural Networks (IJCNN 2008)*, (Hong Kong, 2008).
10. K. M. Kwong, J. N. K. Liu, P. W. Chan and R. S. T. Lee, Using lidar doppler velocity data and chaotic oscillatory-based neural network for the forecast of meso-scale wind field, in *Proc. IEEE Int. Joint Conference on Neural Networks (IJCNN 2008)*, (Hong Kong, 2008).
11. L. A. Zadeh, *Information Sciences* **178**, 2751 (2008).

STOCHASTIC RESONANCE IN COUPLED BISTABLE SYSTEMS

A. KENFACK

Physikalische und Theoretische Chemie, Freie University in Berlin
Takustr. 3, 14195 Berlin, Germany
E-mail: kenfack@chemie.fu-berlin.de

K. P. SINGH

Department of Physics
Indian Institute of Science Education and Research
Mohali, Chandigarh 160019, India

We study two coupled driven and noisy bistable systems, focusing mainly on stochastic resonance (SR). In the absence of coupling, we found two critical damping parameters: one for the onset of SR, and another for which SR is optimum. We show that SR is governed by chaos in the weak coupling regime. Turning on the coupling, we found however that the strong coupling induces a coherence. Finally, we show that the system does not synchronize no matter what coupling.

Keywords: Stochastic resonance, chaos, coupled oscillators, synchronization.

1. Introduction

Among a large variety of phenomena which has been attracting researchers in coupled nonlinear systems over several decades, synchronization [1], chaos and bifurcations structures [2] are the most prominent. SR however has been mostly explored in 1D system [3] and in chemical reactions, bistable ring lasers, semiconductors devices, and mechanoreceptor cells in the tail of the grayfish. This now well-established effect requires three main ingredients: (i) a weak coherent signal, (ii) a noise source, and (iii) an energetic activation barrier. In the absence of noise, the signal should be weak enough such that the effect of signal-induced switching must not be observed. Likewise, the noise-induced switching should not be appreciable in the absence of the signal. It is the interplay of both the signal and the noise that results in a sharp enhancement of the power spectrum within a narrow range about the forcing frequency. This observation was explained by relating the forcing frequency with the switch rate (Kramer's rate) of the unperturbed system [4]. To distinguish this to the dynamical resonance, one speaks of SR. Due to its simplicity and robustness, SR has been implemented by mother nature on almost every scale, thus enabling interdisciplinary interest from physicists, geologists, engineers, biologists and medical doctors, who nowadays use it as an instrument for their specific purposes [5]. The first experimental observation of SR was performed while investigating the noise dependence of the spectral line of an ac-driven Schmitt-Trigger [6]. Although SR has been largely explored in various dynamical systems [3,5], little has been done for coupled stochastic systems [7–10]. The case of coupled underdampled stochastic bistable systems has hitherto not yet been considered.

In this paper, we demonstrate the constructive role of noise assisted by a weak signal in a coupled bistable system in which chaos plays a role. SR has been studied in such a system [8], but essentially overdamped.

2. The model system

The system that we are interested in is governed by the following dimensionless stochastic differential equations,

$$\ddot{x} = -\gamma\dot{x} + \frac{dV_1(x)}{dx} + k(y-x) + \sqrt{2D}\xi_1(t) + F(t) \tag{1}$$

$$\ddot{y} = -\gamma\dot{y} + \frac{dV_2(y)}{dy} - k(y-x) + \sqrt{2D}\xi_2(t) + F(t) \tag{2}$$

where k is the coupling strength, γ the damping parameter, D the noise intensity of two independent Gaussian white noise $\xi_i(t)$ and $\xi_j(t)$

$$\langle \xi_i(t)\xi_j(t')\rangle = 2\,D\,\delta_{ij}(t-t'), (i,j=1,2), \tag{3}$$

which are uncorrelated with zero-mean. The driving signal, $F(t) = A_0\cos(\Omega\,t+\Phi)$, is characterized by the amplitude A_0, the frequency Ω and the phase Φ. The potentials of the two subsystems $V_1(x) = a_1\,x^2/2 - b_1\,x^4/4$ and $V_2(y) = a_2\,y^2/2 - b_2\,y^4/4$ are sketched in Fig. 1(a), with $a_1 = b_1 = 1$ and $a_2 = 1, b_2 = 1.5$. This choice leads to two activation barrier energies $\Delta V_1 = 0.17$ and $\Delta V_2 = 0.25$. For the purposes of SR, we fix $A_0 < \Delta V_1, \Delta V_2$, so to avoid switchings that are due solely to the driving force. The relaxation frequencies of the two subsystems are equal to $\omega_1 = \omega_2 = \sqrt{2a_1}$. To allow for adiabatic driving, we set the modulation frequency smaller than the relaxation one, say $\Omega = \omega_1/20$. Considering y, with $D = 0$ and $\gamma = 0.25$, shown are vivid scenarios of no-switching with $A_0 = 0.1 < \Delta V_2$ Fig. 1(b), and switching with $A_0 = 0.15 > \Delta V_2$ Fig. 1(c). We note in passing that efforts have been put to understand the mechanism of SR in a 1D

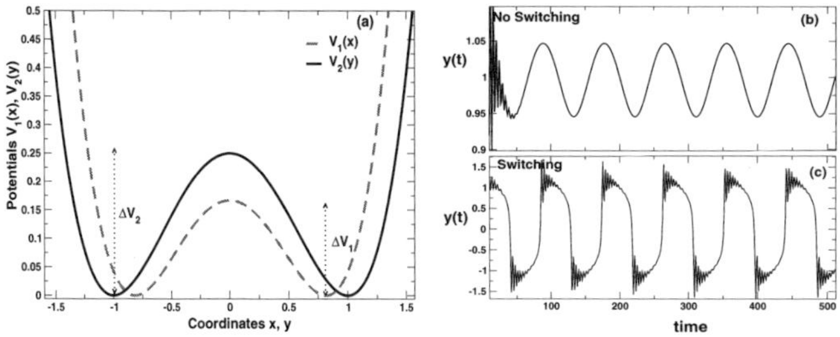

Fig. 1. (a) Potentials $V_1(x)$ (dashed) and $V_2(y)$ (solid), with $\Delta V_1 = 0.17$ and $\Delta V_2 = 0.25$, respectively. (b) Protypical scenarios of no-switching, $A_0 = 0.10$ (a) and switching, $A_0 = 0.50$ (b) of y for $k = 0.0, D = 0.0, \gamma = 0.25$.

system, Eq. (1) or Eq. (2) with $k = 0$, exhibiting a new type of SR [11–13]. Throughout, we set Φ to zero.

3. Stochastic resonance for uncoupled subsystems

Here, both subsystems are independent ($k = 0$), only the noise and the driving force are present. In order to study SR, we consider two subthreshold signal $A_0 = 0.10$ (a) and, $A_0 = 0.15$ (b) that do not alllow switching in the absence of noise. Note also that in the absence of the driving, the stochastic switching time scale which is characterized by the Kramer's rates, $\Gamma_{K_{x,y}} \propto \exp(-\Delta V_{1,2}/D)$, is too long due to the weakness of the white Gaussian noise, i.e. the noise only can not also induce switching. The time scale of switching being $1/\Gamma_{K_{x,y}}$, the time series for $D \in (0, 0.5)$ (not shown) do not exhibit any switching. When both the noise and the driving force are applied, the signal to noise ratio (SNR) is indeed a good candidate commonly used for evaluating the constructive role of noise. Figure 2 shows SNR in the weak damping regime $\gamma = 0.1$ (a) and in the strong one $\gamma = 0.75$ (b). In each panel, curves in black are for $\mathrm{A}_0 = 0.1$ while the ones in brown are for $\mathrm{A}_0 = 0.15$, where SNR_x is in solid and SNR_y in dashed. It turns out that the cooperative effect of noise and driving force does not show up for a weaker dissipation regime (a) where chaos is present. Exploring the dissipation as function of γ two critical values have been revealed, namely γ_{res} for which SR appears, and γ_{opt} for which SR is optimum. Figure 2 (c) depicts SNR_x for various values of γ. Here we found $\gamma_{res} = 0.08$ and $\gamma_{opt} = 0.5$. Finally the Lyapunov exponent, a good indicator of chaos in dynamical systems, has been plotted as function of γ in Fig. 2(d). This clearly confirms that chaos is present in the weak damping regime and may

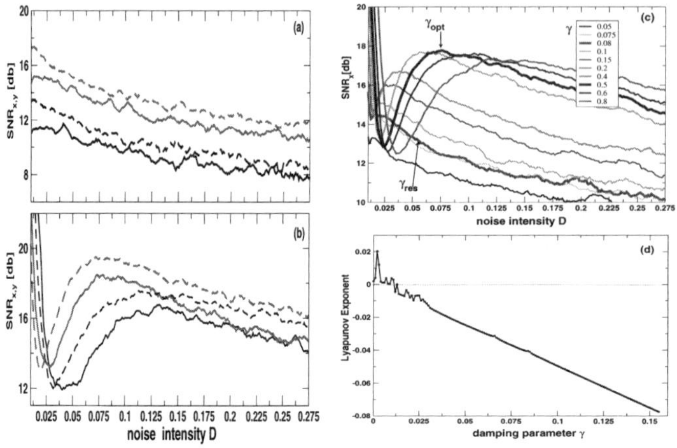

Fig. 2. $\mathrm{SNR}_{x,y}$ of x and y for $\gamma = 0.1$ (a) and $\gamma = 0.75$ (b). In each panel, black-solid (SNR_x) and black-dashed (SNR_y) for $\mathrm{A}_0 = 0.1$, and brown-solid (SNR_x) and brown-dashed (SNR_y) for $\mathrm{A}_0 = 0.15$. Unlike (a), resonances are clearly seen in (b). (c) SNR_x as function of D for $A_0 = 0.15$ and $k = 0.0$; $\gamma_{res} = 0.08$ and $\gamma_{opt} = 0.5$. (d) Lyapunov exponent of x as function of γ for $k = 0$, $D = 0$, and $A_0 = 0.15$; chaos for small γ.

prohibit the occurence of SR. A similar conclusion was drawn in Ref. 14 but in a noisy underdamped double-well potential. The mechanism preventing the appearance of SR is a topic of its own and will be published elsewhere.

4. Influence of the coupling parameter

4.1. *Stochastic resonance (SR)*

SR is essentially based on the exploration of the power spectra of subsystems $\bar{x}(\omega)$ and $\bar{y}(\omega)$. Because of the coupling, another quantity of interest is the coherence function defined as $\Gamma^2 = |S_{xy}(\omega)|/[S_{xx}(\omega)S_{yy}(\omega)]$ where $S_{xy}(\omega)$ is the cross spectrum of processes $x(t)$, $y(t)$ and $S_{xx}(\omega)$, $S_{yy}(\omega)$ are the power spectra of $x(t)$, $y(t)$, respectively. This quantity reaches unity in case both processes become coherent.

Figures 3(a)–3(b) shows SNR of the system as function of k for parameters of Figs. 2(a)–2(b). It turns out that the coupling has no influence on SNR. Remarkably, as k increases, SNR of both subsystems become identical at the stronger limit. Similarly the coherence Γ^2 (not shown) exhibits the same trend.

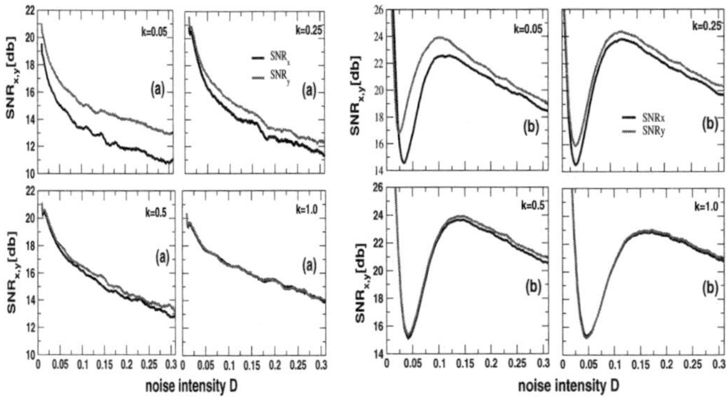

Fig. 3. $\mathrm{SNR}_{x,y}$ at a weak dissipation $\gamma = 0.1$ (a) and at a strong one $\gamma = 0.75$ (b) for values of k as indicated on panels.

4.2. *Synchronization*

To proceed we define the quantity $L^2(t) = (x(t) - y(t))^2 + (\dot{x}(t) - \dot{y}(t))^2$ which is a good measure of the synchronization [1]. Playing with k, no synchronization state has been achieved, even at the very strong limit that shows strong coherence. Figure 4 shows an example of the time series of $L(t)$, for $\mathrm{A}_0 = 0.15$, $\gamma = 1.0$, and $k = 1.0$. Similar outputs are found for any value of k, demonstrating that synchronization is not reached as $L(t)$ does not vanish. What makes this difficult to achieve is presumably not only because the two subsystems are topologically not identical, but also because they are non-deterministic. The opposite happens in deterministic coupled systems in which a strong coupling enforces the synchronization [15].

5. Conclusion

We have investigated the dynamics of two coupled driven and noisy bistable systems. SR which is central has been already considered in a similar system but overdamped [8]. Dealing first with the uncoupled system, we found two critical damping parameters; one indicating the threshold for the appearance of SR and another its optimum. Moreover we showed that the weak damping regime prohibits SR and that the non manifestation of SR is due to the presence of chaos. Then when the coupling is on, we found that SR is in general not affected. However, the strong coupling regime induces SNR of subsystems to match, thereby showing a very high coherence. Exploring the systems along the same lines, the synchronization has not been reached. The influence of the phase Φ will certainly help to get more insights.

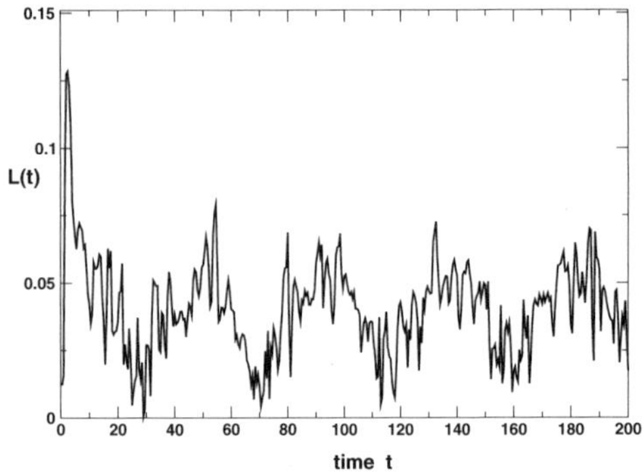

Fig. 4. Measure of synchronization $L(t)$ as function of time, for $k = 1$, $A_0 = 0.15$, and $\gamma = 0.75$. Signature of non synchronisation prevailing in the entire system.

Acknowledgments

We thank Dr. Benjamin Lindner for very useful discussions.

References

1. A. Pikovski, M. Rosenblum, and J. Kurths, in *Synchronization: A universal concept in nonlinear science* (Cambridge Univ. Press, Cambridge, 2001).
2. J. Kozlowski, U. Parlitz, W. Lauterborn, *Phys. Rev. E* **51**, 1861 (1995).
3. L. Gammaitoni et al., *Rev. Mod. Phys.* **70**, 223 (1998).
4. R. Benzi et al., *SIAM J. Appl. Math.* **43**, 565 (1983).
5. T. Wellens, V. Shatokhin, and A. Buchleitner, *Rep. Prog. Phys.* **67**, 45 (2004).
6. S. Fauve and F. Heslot, *Phys. Lett. A* **97**, 25 (1983).
7. A. R. Bulsara and G. Schmera, *Phys. Rev. E* **47**, 3734 (1993).
8. A. Neiman, L. Schimansky-Geier, *Phys. Lett.* **A**379, (1995).
9. V. M. Gandhimathi et al., *Phys. Lett. A* **360**, 279 (2006).
10. V. S. Anishchenko, V. Astakhov, A. Neiman, T. Vadivasova, L. Schimansky-Geier, in *Nonlinear Dynamics of Chaotic and Stochastic Systems* (Springer, Berlin, 2007).
11. N. G. Stocks et al., *J. Phys. A* ***26***, L385 (1993).
12. M. I. Dykman et al., *J. Sat. Phys.* **70**, 479 (1993).
13. Y.- M. Kang, J.-X. Xu, and Y. Xie, *Phys. Rev. E* **68**, 036123-1 (2003)
14. L. Schimansky and H. Herzel, *J. Stat. Phys.* **70**, 141 (1993)
15. U. E. Vincent, A. Kenfack, A. N. Njah, and O. Okinlade, *Phys. Rev. E* **72**, 056213 (2005);U. E. Vincent and A. Kenfack, *Phys. Scr.* **77**, 045005 (2008)

PART C

Multistability in natural and laboratory-scale nonlinear systems

CONTROL OF NOISY MULTISTABLE SYSTEMS BY PERIODIC PERTURBATION

B. K. GOSWAMI

Laser and Plasma Technology Division
Bhabha Atomic Research Centre
Mumbai 400085, India
E-mail: binoy@barc.gov.in

A. GELTRUDE, S. EUZZOR, K. AL NAIMEE, R. MEUCCI and F. T. ARECCHI

Istituto Nazionale di Ottica Applicata
50125 Florence, Italy

Recent theoretical investigations with Toda oscillator and the Lorenzs paradigm of chaotic systems demonstrate that the periodic perturbation may be successful in such complex stochastic multistable scenario. These features have found robust foundation from further experimental validations with an analog circuit of Lorenz equations, individually driven by white Gaussian noise and pink noise.

Keywords: Control of multistability, Lorenz equations, noise, analog circuit.

1. Introduction

Noisy disturbances, commonly observed among natural and laboratory-scale systems, can govern the multistable dynamics in many complex ways [1]. The role of noise depends on its strength and nature, basin sizes, the nature of basin boundaries [2], and the proximity of the operating point to any local or global bifurcation point in the parameter space. In general, the smaller the basin of attraction is, the more susceptible the system is to move out of the attractor. In contrast, the larger basins will be able to hold the system inside relatively for stronger noise [3]. When the noise is so strong that no basin can hold the system for long, it leads to intermittent hopping among coexisting attractors. The other important noise-induced phenomena include stochastic resonance [4], coherence resonance [5] and modification of the onset of various local bifurcations, crises.

In general, multistability in the presence of noisy disturbances is undesirable if the device has to remain at any given attractor, say for any designed applications in thermal hydraulics, communication. Equally important scenario is bistability in cardiac arrhythmia that could be life-endangering. For some further detail discussions in these contexts and references, we request the readership to Refs. [6,7]. The well-known method of controlling multistability are (i) the application of noise and feedback [8–10], and (ii) periodic perturbation, in the form of modulation of any system parameter [11] or introduction of a driving force [12]. The feedback based control mechanism has certain practical limitations toward wider and reproducible applications. In contrast, the periodic perturbation control mechanism is easily realizable, purely deterministic, and therefore, reliably reproducible. The periodic perturbation makes the undesirable attractor chaotic and simultaneously transforms the invariant manifolds of the neighboring boundary saddle, leading to homoclinic tangency and boundary crisis of the chaotic attractor. The control mechanism has been originally demonstrated with practically noise-free systems like CO2 and doped fibre lasers and analog circuits. However, no control mechanism can claim its robustness unless tested in the presence of noise. Indeed, recent theoretical [13] as well as experimental [14] demonstrations have endorsed very strongly the applicability of this simple yet novel mechanism in controlling noise-induced multistable dynamics. In this paper, we summarize the recent experiments on the multistable Lorenz circuit individually in the presence of white Gaussian noise and pink noise.

2. Control of white Gaussian noise driven multistable Lorenz circuit

The circuit diagram of the analog circuit of Lorenz equations is shown in Fig. 1 where $R_1 = 75k\Omega$, $R_2 = 7.5k\Omega$, $R_3 = 2.89k\Omega$ $R_4 = 28.125k\Omega$, $R_5 = 30k\Omega$ and $C = 6.8$nF. We have used LT1114 operational amplifier ICs for analog integrations (denoted by I1, I2 and I3) and inversions (denoted by '-1'), and MLT04 ICs for analog multiplications (denoted by cross symbols inside circles). The noise generator is shown by V_{noise} and the periodic control modulation by $V_{\text{control}} = A\sin(2\pi\nu t)$. In the absence of noise and control signals, the circuit equations can be transformed to Lorenz equations with $\sigma = 10$, $\rho = 25.95$, $\beta = 8/3$. For $2.87k\Omega < R_3 < 2.91k\Omega$, the circuit exhibits simultaneous coexistence of the chaotic attractor and two stable steady states, denoted by S_+ (for $X > 0$) and S_- (for $X < 0$). The operating point $R_3 = 2.89k\Omega$ is inside the multistable regime. The basin of each steady state is determined by the surrounding unstable periodic orbit (UPO). In the experiments with the circuit alone (i.e., not driven by any external noise generator), we have observed destruction of chaotic attractor by the periodic modulation of system parameters.

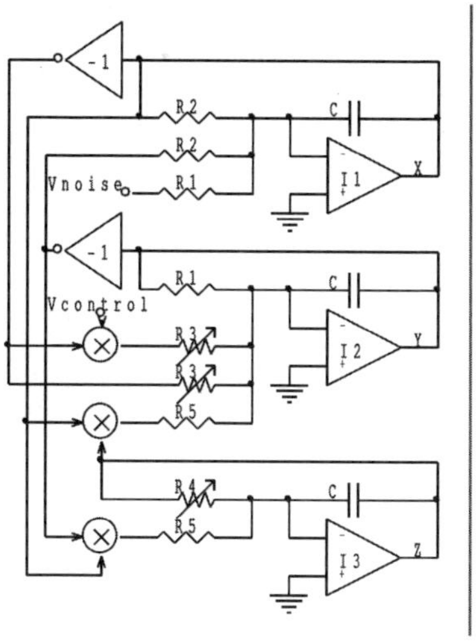

Fig. 1. Analog circuit of stochastic Lorenz equations under periodic parameter modulation.

The star symbols in Fig. 2(a) show the crisis threshold control amplitude ($A_{\min}$) at various control frequencies. The threshold amplitude is minimal when the control frequency is close to 400 Hz and 1.6 kHz. With these information about the noise-free case, we now introduce white noise to the circuit while the control is switched off. Figures 2(b) and 2(c) illustrate some statistical features of the white noise generator. One million time-series data of noise (sampling period 20 microseconds) have been used for such analysis. Figure 2(b) shows that the noise magnitude lies in the range [-6.5 V:6.5 V] that represents strong noise as the absolute magnitudes of X, Y and Z lie in general within 3 V. The entire range of noise magnitude has been divided into 25 intervals and the probability distribution over these intervals has been computed and illustrated by filled circles. The solid line in this plot denotes the Gaussian fit with standard deviation 2.55 V and zero mean. Figure 2(c) demonstrates the corresponding uniform broad-band Fourier spectrum. We analyze the dynamics in the phase-space of such strong noise-driven circuit. A suitably large region in the phase-space ($-1.5 < X < 1.5$, $0 < Z < 3$) is selected around the chaotic attractor and the stable steady states. This phase-space region is uniformly divided into ($N_1 = 25 \times 25$) cells. We record the time series of X and Z voltages for suitably large time-interval, say $N_2 = 10^6$ time steps (each of 20 microseconds duration), and compute the total number

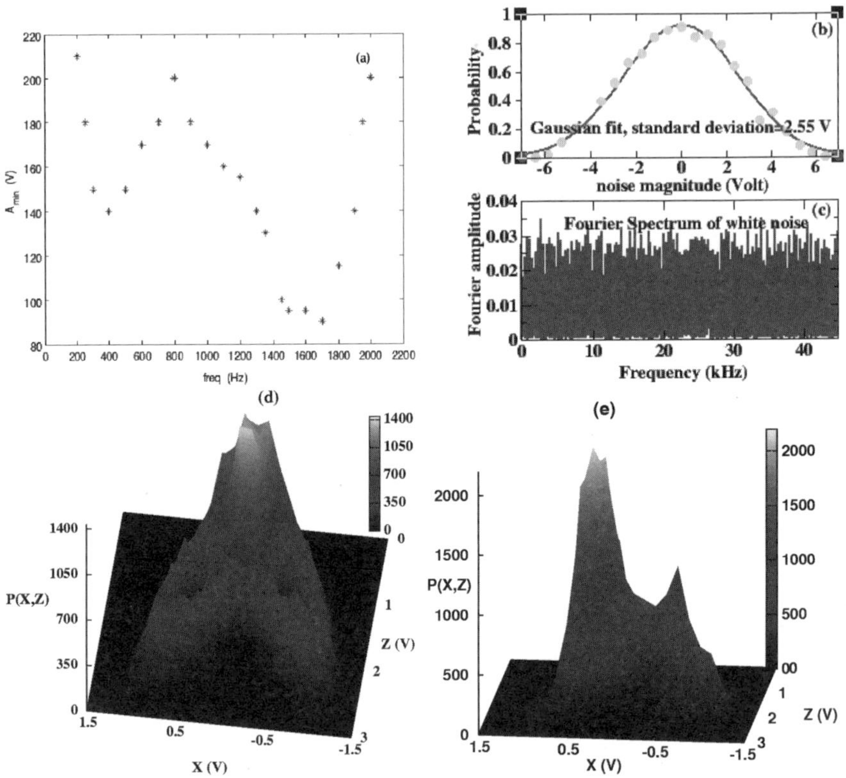

Fig. 2. (a) Threshold control amplitude ($A_{\rm min}$) versus control frequency (ν), shown by star symbols. (b) Probability distribution of the white noise: experimental data are shown by filled circle symbols and the corresponding Gaussian fit by solid line. (c) Fourier spectrum of white noise. (d) and (e) represent the probability distributions for the uncontrolled noisy and controlled noisy circuits respectively. (d) Without control, probability is highest around the saddle ($X = Y = Z = 0$) inside the chaotic attractor and negligible in the basins of steady-state attractors. (e) However, in the presence of control modulation, the circuit dynamics is essentially confined around one steady state while the occupation probability in the chaotic attractor exhibits a sharp decrease.

of visits in each cell. The occupation probability density $D(X, Z)$ per unit cell around the point (X, Z) is defined by $D(X, Z) = P(X, Z)/(N_1N_2)$ where $P(X, Z)$ denotes the number of time-steps the system remains inside the cell around (X, Z) point. While comparing the probability distribution of the uncontrolled noisy dynamics [Fig. 2(d)] with that of the controlled scenario [Fig. 2(e)], we keep the values of N_1 and N_2 unchanged. The probability distribution [for convenience, represented by $P(X, Z)$] in Fig. 2(d) vividly reveals the stochastically driven motion in the basin of the chaotic attractor. In particular, the system stays maximum probable period around the saddle ($X = Y = Z = 0$). This feature can be explained as follows: The speed of convergence to the saddle via stable manifold or divergence away from the saddle via unstable manifold decreases substantially in the vicinity of the saddle resulting in the sharp increase of the occupation probability. However, away from the saddle, the system moves relatively fast inside the basin of the chaotic attractor, including occasional spiraling around the UPOs, before approaching back again to the saddle via the stable manifold. The prominent two holes in the probability distribution profile indicate that the system rarely stays inside the basins of steady states. If the system ever enters any such basin, noise is strong enough to eject the system very fast and put it back inside the chaotic attractor.

Such a scenario can also be changed remarkably when the control modulation is switched on with appropriately set parameter values. Figure 2(e) demonstrates such a case. The control frequency is set at $\nu = 1.70$ kHz where the crisis threshold amplitude is minimum. Also, the control amplitude is set at ($A_c = 200$ mV), higher than the crisis threshold. The phase-space probability distribution demonstrates a completely contrasting scenario with respect to the uncontrolled case. The probability density around the steady state S_+ is now much larger than that around the chaotic attractor in general and the saddle in particular. The underlying phenomenon behind such significant transformation is the control perturbation induced boundary crisis of the chaotic attractor. In particular, the sharp reduction of occupation period around the saddle ($X = Y = Z = 0$) is a consequence of a homoclinic tangency. This is because the system is driven away by noise whenever the trajectory is close to any tangency points and therefore the system does not get adequate opportunity to approach the saddle.

3. Control of pink noise driven multistable Lorenz circuit

Here we analyze the effect of control modulation in the presence of pink noise. Figure 3(a) shows a typical time series that clearly suggests rare occurrence of small-amplitude noise. Figure 3(b) describes the corresponding (grossly) symmetric probability distribution with the minimum at the centre and the most probable magnitude (absolute value) of noise in the interval between 0.6V to 1V. This is in contrast with Gaussian white noise. Besides, the Fourier spectrum in Fig. 3(c) shows a maximum in the low frequency range and an over all decrease along the high frequency range. Thus the spectrum is also qualitatively different from that of white noise. We compute the phase-space occupation probability distribution in the same

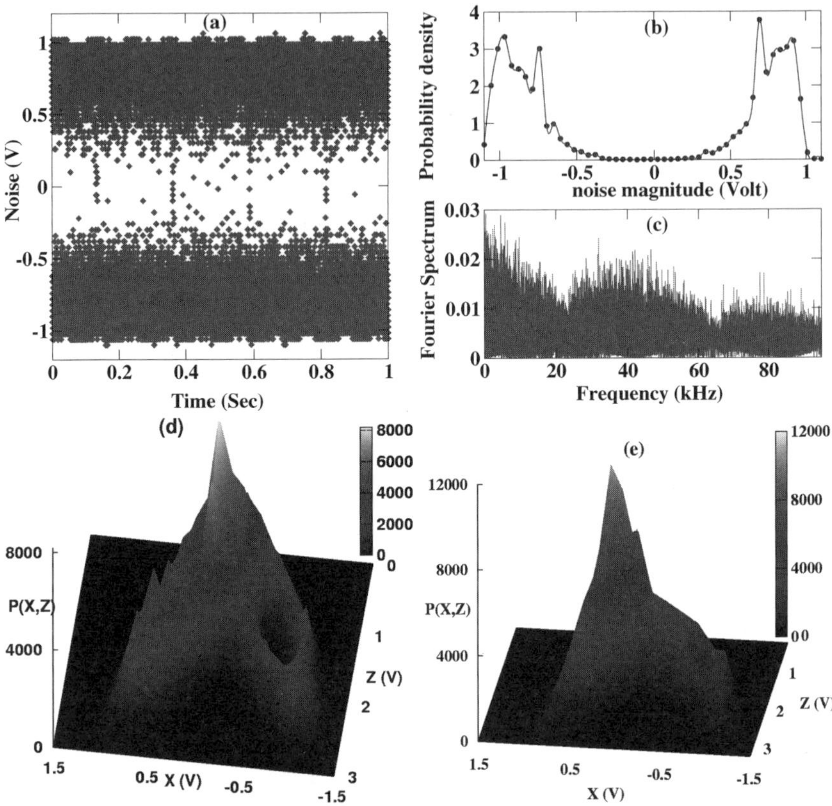

Fig. 3. (a) Time series, (b) Probability distribution, and (c) Fourier spectrum of pink noise. (d) and (e) represent the probability density of phase-space occupation for uncontrolled noisy and controlled noisy circuits respectively. (d) Without control, probability is highest around the saddle ($X = Y = Z = 0$) and negligible in the basins of the steady-state attractors. (e) In the presence of control modulation, one steady state becomes most preferred attractor while the occupation density sharply reduces in the chaotic attractor.

manner as followed in the case of white noise. The circuit dynamics is analyzed for $N_2 = 500000$ time-steps. Figure 3(d) illustrates the probability distribution of the pink-noise induced circuit dynamics. It reveals that the circuit remains in the chaotic attractor. In particular, the probability is maximum around the saddle ($X = Y = Z = 0$). Also, the occupation density is again minimum around the steady states. These features are similar with the white noise case. As we introduce the control modulation suitably, the situation changes drastically. Fig. 3(e) shows the probability distribution in the presence of control modulation ($A_c = 130$ mV; $\nu = 1.7$

kHz). Noticeably the scenario is again completely contrasting to the uncontrolled case. P(X,Z) is maximum around 'S_+' steady state and the occupation probability inside the chaotic attractor, in particular, around the saddle, has predominantly gone down. Thus, we have demonstrated control of stochastic multistable scenario quite successfully.

In conclusion, in the experimental demonstrations, we report the recent experiments carried out with an analog circuit of Lorenz equations, individually driven by white Gaussian noise and pink noise. In both cases, the experiments establish the robustness of this novel control mechanism in resolving stochastic multistability. One reason of such magnificent success of the control mechanism could be the fact that noise does not deter crisis. On the contrary, noise may advance the onset of such crisis [15,16].

References

1. M. Franaszek and L. Fronzoni, *Phys. Rev. E* **49**, 3888 (1994).
2. S. Kraut and U. Feudel, *Phys. Rev. E* **66**, 015207(R) (2002).
3. S. Kraut, U. Feudel and C. Grebogi, *Phys. Rev. E* **59**, 5253 (1999).
4. C. Masoller, *Phys. Rev. Lett.* **88**, 034102 (2002).
5. L. Gammaitoni, P. Hanggi, P. Jung and F. Marchesoni, *Rev. Mod. Phys.* **70**, 223 (1998).
6. B. K. Goswami and A. N. Pisarchik, *Int. J. Bifurcation Chaos Appl. Sci. Eng.* **18**, 1645 (2008).
7. U. Feudel, *Int. J. Bifurcation Chaos Appl. Sci.* **18**, 1607 (2008).
8. L. Poon and C. Grebogi, *Phys. Rev. Lett.* **75**, 4023 (1995).
9. U. Feudel and C. Grebogi, *Chaos* **7**, 597 (1997).
10. Y.-C. Lai, *Phys. Lett. A* **221**, 375 (1996); Y.-C. Lai and C. Grebogi, *Phys. Rev. E* **54**, 4667 (1996); E. E. N. Macau and C.Grebogi, *ibid.* **59**, 4062 (1999).
11. A. N. Pisarchik and B. K. Goswami, *Phys. Rev. Lett.* **84**, 1423 (2000).
12. B. K. Goswami, *Phys. Rev. E* **78**, 066208 (2008).
13. B. K. Goswami, *Phys. Rev. E* **76**, 016219 (2007).
14. B. K. Goswami, S. Euzzor, K. Al Namee, A. Geltrude and F. T. Arecchi, *Phys. Rev. E* **80**, 016211 (2009).
15. J. C. Sommerer, E. Ott and C. Grebogi, *Phys. Rev. A* **43**, 1754 (1991).
16. J. C. Sommerer, W. L. Ditto, C. Grebogi, E. Ott and M. L. Spano, *Phys. Rev. Lett.* **66**, 1947 (1991).

MULTI-STABILITY AND TRANSIENT CHAOTIC DYNAMICS IN SEMICONDUCTOR LASERS WITH TIME-DELAYED OPTICAL FEEDBACK

J. ZAMORA-MUNT*, C. MASOLLEr† and J. GARCÍA-OJALVO‡

Departament de Física i Enginyeria Nuclear, Universitat Politècnica de Catalunya
Edif. GAIA, Rambla de Sant Nebridi s/n, Terrassa 08222, Barcelona, Spain
*E-mail: *jordi.zamora.munt@upc.edu*
† cristina.masoller@upc.edu
‡ jordi.g.ojalvo@upc.edu
http://donll.upc.edu/

We investigate numerically the chaotic transient dynamics of a semiconductor laser with time-delayed optical feedback. This transient behavior is seen in the low-frequency fluctuations (LFF) regime occurring when the laser is pumped close to threshold and is subjected to moderate feedback strengths. We characterize the transient dynamics in terms of its duration, the time intervals between consecutive dropouts, and the total number of dropouts in each transient event. We statistically analyze these quantities as a function of the noise strength and other system parameters related to different nonlinear mechanisms of light-matter interaction inside the laser active medium.

Keywords: Semiconductor lasers, laser diodes, chaotic transients, low-frequency fluctuations, optical feedback, time-delayed systems, multi-stability.

1. Introduction

Semiconductor lasers with an external cavity show delay-induced multi-stability and a rich variety of dynamical regimes. One of these regimes is characterized by low frequency fluctuations (LFFs), which modulate the fast (picosecond) intensity pulses in the form of a slowly varying envelope that exhibits sudden dropouts followed by a slow recovery process. This LFF behavior has been frequently considered a sustained regime, but some studies [1–3] have shown that for a large range of realistic laser parameters the regime is transient: the LFFs eventually disappear as the system trajectory reaches a stable fixed point and remains there. We present a statistical analysis of this transient dynamics for different parameters, in particular we analyze the statistics of the number of dropouts per transient (i.e., the number of intensity dropouts that occur before the system trajectory finds a stable fixed point) and the statistics of the time interval between consecutive dropouts. Common characteristics are found for parameters related to the laser active medium. Our results are in qualitatively good agreement with previous work on the sustained LFF regime, which suggests that the chaotic attractors have similar statistical properties in the two regimes.

2. Model

A phenomenological model frequently used to simulate a semiconductor laser with optical feedback is the Lang-Kobayashi (LK) model [4]. The differential rate-equations for a single-mode laser are given in terms of the complex electric field E and the carrier number N as follows

$$\begin{aligned}\frac{dE}{dt} &= k(1+i\alpha)[G(E,N)-1]E + \kappa_{fb}E(t-\tau)e^{-i\omega_0\tau} + \sqrt{D}\xi(t),\\ \frac{dN}{dt} &= \gamma_N[J - N - G(E,N)|E|^2], \end{aligned} \tag{1}$$

where k is the decay rate of the field inside the cavity, α is the linewidth enhancement factor, κ_{fb} is the strength of the optical feedback, τ is the round-trip time of the light in the external cavity, ω_0 is the frequency of the solitary laser, D is the spontaneous emission rate and ξ is a Gaussian white noise with $\langle \xi(t)\xi(t')\rangle = \delta(t-t')$. The carrier decay rate is represented by γ_N, and J stands for the injection current normalized to the solitary laser threshold. We can take into account saturation effects inside the active media using a nonlinear gain given by $G(E,N) = N/(1+\varepsilon|E|^2)$ where ε is the gain saturation coefficient. This term is related to different nonlinear light-matter effects such as carrier heating, carrier diffusion and spatial hole burning.

3. Transient time statistics of low frequency fluctuations

In our simulations we used typical parameters frequently employed in the LK model. Specifically, we considered $\alpha = 3$, $k = 300$ ns^{-1}, $\kappa_{fb} = 30$ ns^{-1}, $\tau = 6.667$ ns, $\omega_0\tau = 0$ rad, $\gamma_N = 1$ ns^{-1}, $J = 1.02$, $\varepsilon = 0$ and $D = 10^{-4}$ ns^{-1}, unless specified otherwise. The initial conditions are chosen in the solitary laser steady state with a random contribution

$$E(t) = E_s e^{i\phi_0} + \eta\xi(t), \; for \; -\tau \le t \le 0 \tag{2}$$

$$N(0) = N_s + \rho\zeta \tag{3}$$

where $\eta = \rho = 10^{-3}$. At time $t = 0$, the feedback is switched on and after a time $T_{\rm LFF}$ the system reaches the stationary state and remains there permanently, provided the noise is not too large. In our simulations, we decide the end of the transient time when the standard deviation of the intensity fluctuations decreases below a certain threshold. These fluctuations are measured inside a time window of 1.8 μs, much larger that the characteristic fluctuation time of the fast picosecond dynamics. A typical time trace of this transient phenomenon is shown in Fig. 1(a). The final

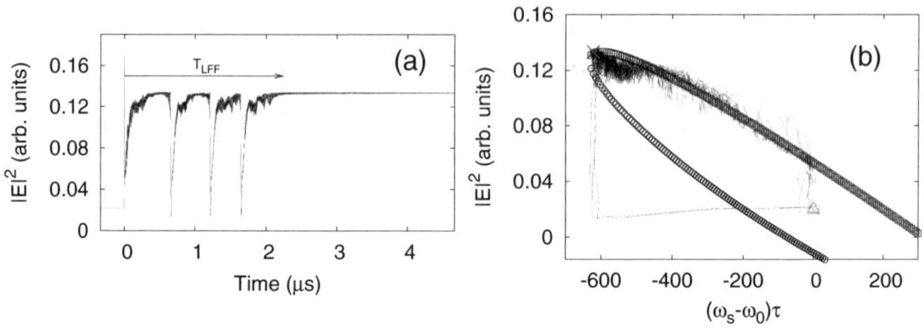

Fig. 1. (a) A typical time trace for the filtered intensity, displaying three LFF dropouts. A filter with a cut-off frequency of 120MHz is used to obtain this time trace. (b) Filtered global trajectory in the phase space of intensity and phase difference of the electric field for the trajectory shown in (a); squares are the nodes and circles are the anti-nodes. The triangle and the cross mark the initial conditions and the final state respectively.

state of the laser is given by one of the coexisting fixed point solutions of Eq. (1), known as External Cavity Modes (ECMs), which are given by:

$$\omega_s\tau = \omega_0\tau - \kappa_{fb}\tau\sqrt{1+\alpha^2}\sin(\omega_s\tau + \tan^{-1}\alpha) \tag{4}$$

$$N_s = \frac{1-(\kappa_{fb}/k)\cos(\omega_s\tau)}{(1+\varepsilon)} + \frac{J\varepsilon}{1+\varepsilon} \tag{5}$$

$$|E_s|^2 = \frac{J-N_s}{N_s-(J-N_s)\varepsilon} \tag{6}$$

These ECMs are shown in Fig. 1(b) in the phase space of intensity and phase difference. A typical global trajectory in this phase space begins at $(\omega_s-\omega_0)\tau$ =0 (the triangle in Fig. 1(b)). Then the trajectory is attracted to the stable ECMs for low values of the phase difference and high values of the intensity, but in this region the system can reach the vicinity of one of the unstable ECMs (anti-nodes) and it can be ejected towards the phase-space region of low intensities, after which the process starts again. The LFF dropout dynamics ends when the system's trajectory finds one of the stable ECMs (the cross in Fig. 1(b)).

The LFF transient lasts for a time that depends strongly on the specific random initial conditions, as expected in a chaotic transient. We now discuss the statistical properties of the transient time, $\langle T_{\rm LFF}\rangle$, as a function of noise

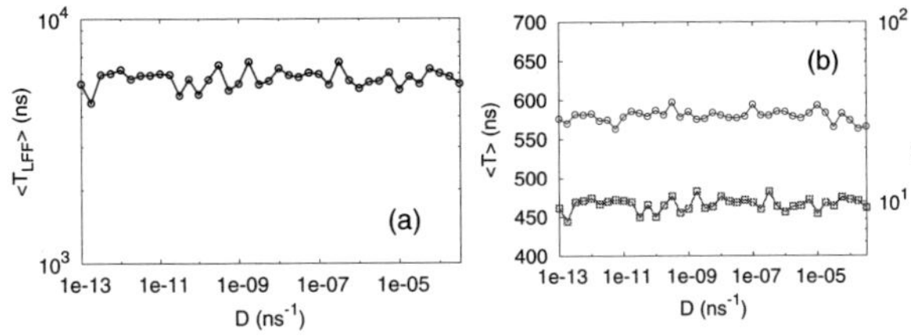

Fig. 2. (a) Average transient time as a function of the noise intensity. (b) Average time interval between consecutive dropouts (circles, left axis) and average number of dropouts per transient (squares, right axis) as a function of the noise intensity; note the logarithmic scale in the right axis.

and different laser and feedback parameters. As can be seen in Fig. 2(a) the average transient time does not depend on the noise intensity. This behavior was verified for a large range of parameters, with the same result. The transient can be understood as a sequence of N_{LFF} dropouts that are spaced at an average time $\langle T \rangle$. These values are shown in the Fig. 2(b), where we plot both the average time between dropouts per transient ($\langle T \rangle$, left axis) and the average number of dropouts per transient ($\langle N_{\mathrm{LFF}} \rangle$, right axis). For large enough number of dropouts, the product of these two quantities corresponds to the average transient time, $\langle T_{\mathrm{LFF}} \rangle$. It can be seen in Fig. 2(a) that neither $\langle N_{\mathrm{LFF}} \rangle$ nor $\langle T \rangle$ depend on the noise intensity.

We have studied the properties of the transient time for different laser parameters. We found a very similar behavior for increasing gain saturation coefficient, linewidth enhancement factor and normalized injection current. In the upper row of Fig. 2 (plots a-c) we show that all these parameters increase the transient time. The bottom row of that figure (plots d-f) represents the contributions of the average dropout number, $\langle N_{\mathrm{LFF}} \rangle$, and the average inter-dropout time ,$\langle T \rangle$, to the transient time. The figure shows that $\langle N_{\mathrm{LFF}} \rangle$ increases monotonically with the three laser parameters considered, while $\langle T \rangle$ decreases. In particular, the decrease in the time interval between consecutive dropouts, $\langle T \rangle$, for increasing injection current J [Fig. 2(f)] agrees qualitatively well with the well-known behavior observed experimentally in the sustained LFF regime [5,6]. Note that even though the two contributors to $\langle T_{\mathrm{LFF}} \rangle$, namely $\langle N_{\mathrm{LFF}} \rangle$ and $\langle T \rangle$, behave complementarily, their trends do not cancel out since the increase of $\langle N_{\mathrm{LFF}} \rangle$ is supra-exponential for the three parameters, while $\langle T \rangle$ decreases only linearly [cf the different scaling of the left and right y-axes in Figs. 2(d)–2(f)].

Different results are found when investigating the influence of the feedback parameters. Figure 3(a) shows that the transient time increases nearly

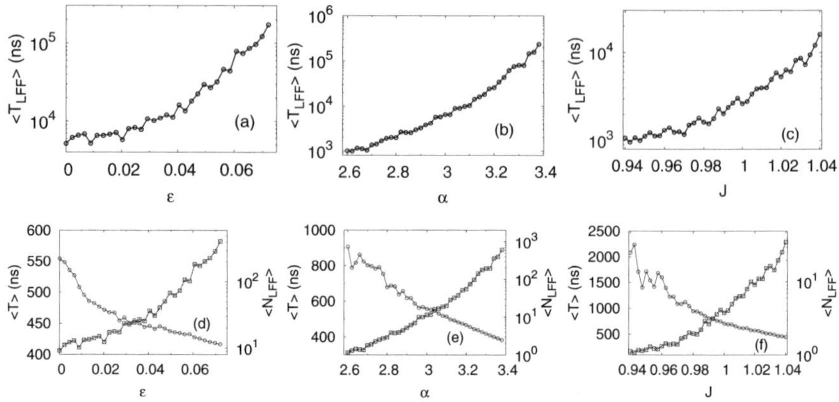

Fig. 3. Averaged transient time as a function of the saturation coefficient (a), linewidth enhancement factor (b), and normalized injection current (c). The bottom row shows the dependence of the average time interval between consecutive dropouts (circles, left axis) and average number of dropouts per transient (squares, right axis) on the same laser parameters; note the logarithmic scale in the right axis.

exponentially with the delay time, τ. This behavior is a consequence of the exponential increase of $\langle N_{\mathrm{LFF}} \rangle$ [Fig. 3(c)] and is re-inforced by $\langle T \rangle$, which in this case also increases monotonically with the feedback delay time. This latter effect is in qualitatively well agreement with Refs. [7,8]. On the other hand, for increasing feedback strengths, unlike the other parameters studied above, the transient time decreases monotonically [Fig. 3(b)]. Analysing $\langle N_{\mathrm{LFF}} \rangle$ and $\langle T \rangle$ in Fig. 3(d) we see that these quantities depend on κ_{fb} oppositely to the laser parameters (Fig. 2). However, $\langle N_{\mathrm{LFF}} \rangle$ still dominates the behavior of the transient time, since for increasing feedback strength it decreases nearly exponentially, while $\langle T \rangle$ increases nearly linearly. The latter behavior is in qualitative good agreement with Refs. [5,8].

In conclusion, our numerical results indicate that the transient LFFs are basically a deterministic phenomenon that does not depend on the noise intensity, and that for a large range of laser parameters can be sustained by various nonlinear effects in the laser active medium. The chaotic attractor of the transient LFFs has statistical properties similar to those of the stationary LFF regime.

Acknowledgments

Stimulating discussions with J. Tiana-Alsina and M. C. Torrent are gratefully acknowledged. This research was supported in part by U.S. Air Force

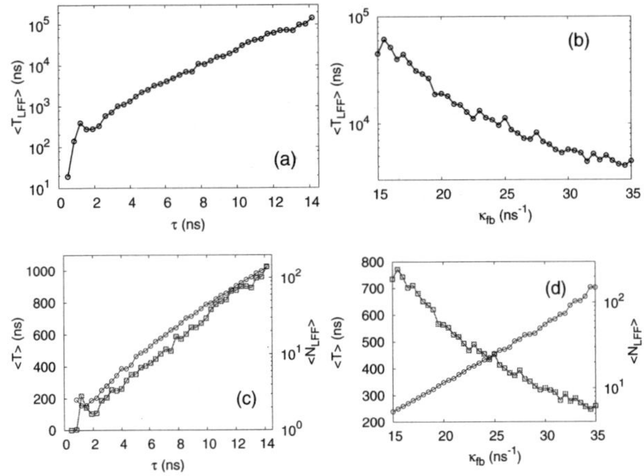

Fig. 4. Average transient time as a function of the external round-trip time (a) and feedback strength (b). The bottom row shows the dependence of the average time interval between consecutive dropouts (circles, left axis) and average number of dropouts per transient (squares, right axis); note the logarithmic scale in the right axis.

Office of Scientific Research under grant FA9550-07-1-0238, the Spanish Ministerio de Educacion y Ciencia through project FIS2009-13360, and the Agencia de Gestio d'Ajuts Universitaris i de Recerca (AGAUR), Generalitat de Catalunya, through project 2009 SGR 1168.

References

1. T. Heil, I. Fischer, W. Elsäßer, J. Mulet and C. R. Mirasso, *Opt. Lett.* **24**, 1275 (1999).
2. R. Davidchack, Y.-C. Lai, A. Gavrielides and V. Kovanis, *Physica D* **145**, p. 130 (2000).
3. A. Torcini, S. Barland, G. Giacomelli and F. Marin, *Phys. Rev. A* **74**, p. 063801 (2006).
4. R. Lang and K. Kobayashi, *IEEE J. Quantum Electron.* **16**, p. 347 (1980).
5. Y. Hong and K. Shore, *Opt. Lett.* **30**, p. 3332 (2005).
6. D. Sukow, J. Gardner and D. Gauthier, *Phys. Rev. A* **56**, p. R3370 (1997).
7. J. M. Buldú, J. García-Ojalvo and M. C. Torrent, *Phys. Rev. E* **69**, p. 046207 (2004).
8. Y. Hong and K. A. Shore, *IEEE J. Quantum Electron.* **41**, 1054 (2005).

MULTISTABILITY IN A SEMICONDUCTOR LASER SUBJECT TO OPTICAL FEEDBACK FROM A FABRY-PEROT FILTER

H. ERZGRÄBER

School of Engineering, Mathematics and Physical Sciences
University of Exeter, Exeter EX4 4AU, United Kingdom
E-mail: h.erzgraber@exeter.ac.uk

B. KRAUSKOPF

Department of Engineering Mathematics
University of Bristol, Bristol BS8 1TR, United Kingdom

We study the structure of the multistable continuous wave emission region of a semiconductor laser subject to coherent optical feedback from a Fabry-Perot filter. Key parameters organizing the degree of multistability are uncovered, and they include the feedback phase and the frequency detuning.

Keywords: Delay differential equation, optical feedback, bifurcation analysis.

1. Introduction

Semiconductor lasers find many applications, for example, in optical telecommunication. Because of their high material gain they are very sensible to external perturbations, which may lead to instabilities and possibly even chaotic laser emission. We show here that this sensitivity can also lead to a complex structure of multistable continuous wave (cw) and oscillatory emission. More specifically, we consider a semiconductor laser that is subject to coherent optical feedback from a Fabry-Perot filter. Filtered optical feedback (FOF) is a frequently used set-up to control the dynamics of a laser via the spectral properties of the feedback light, which are determined by the filter detuning and the filter width. The system can be modelled by rate equations for the complex-valued laser field $E(t)$, the complex-valued filter field $F(t)$, and the real-valued laser inversion $N(t)$. In normalized form [1,2] the model is given by,

$$\begin{aligned}\dot{E}(t) &= (1+i\alpha)E(t)N(t)+\kappa F(t)\,,\\ T\dot{N}(t) &= P-N(t)-(1+2N(t))|E(t)|^2\,,\\ \dot{F}(t) &= \Lambda E(t-\tau)e^{-iC_p}+(i\Delta-\Lambda)F(t)\,.\end{aligned}\tag{1}$$

This model takes into account the time delay τ of the feedback light that arises from the propagation through the feedback loop. Hence, the model Eq. (1) takes the form of a system of delay differential equations; here time is rescaled with respect to the photon decay time, which is typically in the order of picoseconds. The laser parameters are the linewidth enhancement factor α, the ratio between electron and photon decay time T, and the pump parameter P. The feedback is characterized by the feedback strength κ, the feedback phase C_p, the linewidth of the filter Λ, and the detuning $\Delta = \Omega_F - \Omega_0$ of the filter from the laser; here Ω_F is the filter frequency and Ω_0 is the laser frequency. In accordance with experimental practice [3], we change the detuning in Eq. (1) by changing Ω_0 while keeping Ω_F fixed. Importantly, the feedback phase C_p takes into account the phase that the laser field accumulates as it propagates through the feedback loop. The parameter were set to the realistic values that can be found in Table 1.

Equation (1) model the dynamics of a semiconductor laser subject to coherent optical feedback from a Fabry-Perot filter. Of particular interest in this system is the additional control over the feedback field in terms of the width Λ and the detuning Δ of the filter. Its dynamics has been studied experimentally and mathematically for example in Refs. [1–5]. In this system the laser receives maximum feedback intensity when its optical frequency equals the center frequency of the filter. An alternative operation of the filter, known as non-invasive feedback, has been studied, for example, in Refs. [6,7]; in this case the feedback intensity has its minimum when the laser operates at the center frequency of the filter.

A key observation is that Eq. (1) have certain symmetries, and this has influence on the structure of its solutions. In particular, there is the S^1-symmetry

$$(E,\, F,\, N) \to (e^{i\varphi}E,\, e^{i\varphi}F,\, N)\,, \tag{2}$$

which corresponds to a rotation of the optical fields of the laser E and the filter F by an arbitrary angle $\varphi \in [0, 2\pi)$ in the complex plane. Moreover, there is the parameter symmetry

$$(E,\, F,\, N;\, C_p) \to (E,\, F,\, N;\, C_p + 2\pi)\,, \tag{3}$$

that involves the feedback phase C_p.

Table 1. Summary of the parameters and their normalized values.

Symbol	Meaning	Value
α	Linewidth enhancement parameter	5
T	Ratio between electron and photon decay time	100
P	Pump parameter	3.5
τ	Delay time	300
κ	Feedback strength	0.01
Λ	Linewidth of the Fabry-Perot filter (HWHM)	0.07
Ω_F	Center frequency of the Fabry-Perot filter	-0.07
Ω_0	Optical frequency of the laser	free
C_p	Feedback phase	free

2. External filtered modes

A continuous wave solution of system (1) characterizes a state where the FOF laser exhibits a constant intensity emission. It has the form

$$(E(t);\ N(t);\ F(t)) = (|E_s|e^{i\omega_s t};\ N_t;\ |F_s|e^{(i\omega_s t + i\phi)}), \tag{4}$$

where the laser field and the filter field oscillate with the same frequency ω_s and constant amplitudes $|E_s|$, $|F_s|$. There is a phase difference ϕ, and the laser inversion N_s is a constant. We call these solutions external filtered modes (EFMs) of the FOF laser. Mathematically, they correspond to group orbits of the S^1-symmetry Eq. (2) of the FOF laser system. In what follows, we analyse the stability of the EFMs as a function of the laser frequency Ω_0, which changes Δ in Eq. (1), and as a function of the feedback phase C_p. A numerical bifurcation analysis with DDE-BIFTOOL [8] will reveal saddle-node (S) and Hopf (H) bifurcation curves that form the stability boundary of the EFMs as well as codimension-two points that indicate changes of the stability boundary.

3. Stability analysis

Figure 1 shows the stability of the external filtered modes (EFMs) of the FOF laser system as a function of the laser frequency Ω_0. Stable EFMs are plotted thick and unstable EFMs are plotted thin; they change stability at saddle-node (+) and Hopf (∗) bifurcations. In a saddle-node bifurcation either a pair of EFMs collide and disappear, or new pair of EFMs is created. In a Hopf bifurcation an EFM looses stability and a periodic orbit is created; physically the laser intensity starts to oscillate. The EFMs (4) can be plotted in different representations. In particular, Fig. 1(a) shows the laser intensity $|E_S|^2$, which undergoes only relatively small variations when Ω_0 is changed. Figure 1(b) shows the filter intensity $|F_s|^2$; it is large around $\Omega_0 = 0$, implying high feedback intensity. The filter intensity $|F_s|^2$ is close to zero for sufficient large and sufficient small laser frequency Ω_0. Figure 1(c) shows the actual frequency ω_s of the FOF laser system, which also varies as a function of Ω_0. Finally, Fig. 1(d) shows the phase Φ between the laser field E and the filter field F. For large and small values of Ω_0 the phase changes very rapidly as Ω_0 is varied. Only the part of the EFM

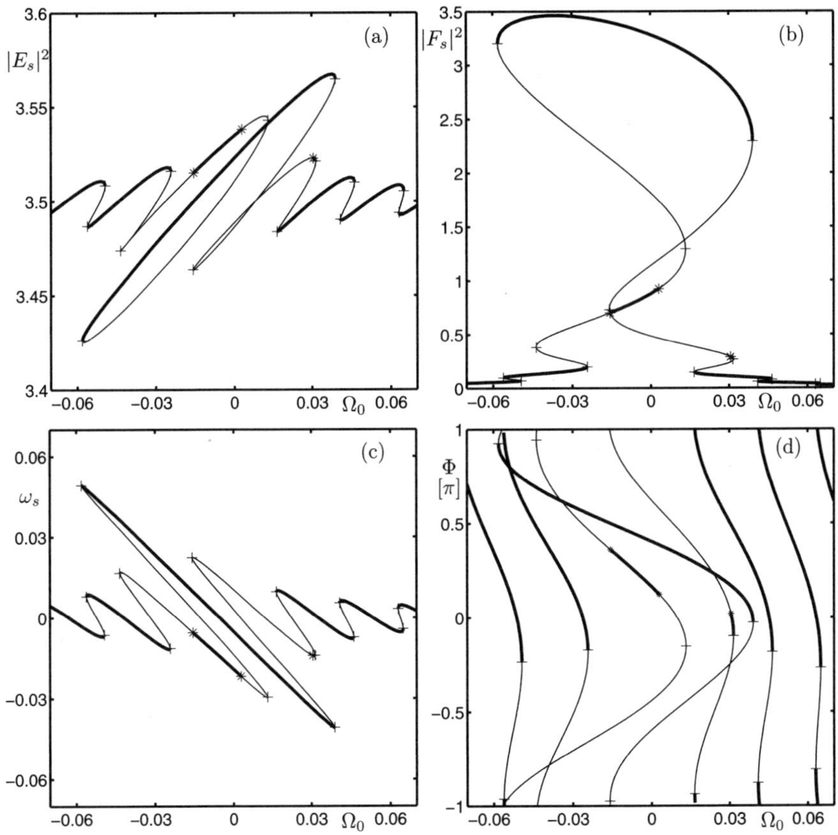

Fig. 1. One-parameter bifurcation analysis of the EFMs as a function of the laser frequency Ω_0. EFMs are stable along thick parts of curves and unstable along thin parts; also shown are saddle-node (+) and Hopf (∗) bifurcations. The individual panels show different representations of the same branch of EFMs: panel (a) shows the laser intensity $|E_s|^2$, panel (b) the feedback intensity $|F_s|^2$, panel (c) the FOF laser frequency ω_s, and panel (d) the phase shift Φ between the laser field and the filter field.

branch around $\Omega_0 = 0$, which corresponds to a high filter intensity $|F_s|^2$, exhibits less rapid variations. Indeed from Fig. 1 we can already identify several parameter intervals with bistable EFMs.

We now extend our stability analysis to two parameters by including the feedback phase C_p. Figure 2 shows a two-parameter bifurcation analysis in the (Ω_0, C_p) parameter plane. The stability regions of the EFMs are bounded by saddle-node (S) and Hopf (H) bifurcation curves; changes in the boundary are marked by codimension-two saddle-node Hopf (SH), Bogdanov-Takens (BT), and double Hopf (HH) points. Figure 2(a) shows the covering space of the 2π-periodic parameter C_p, where the gray shaded regions indicate stable EFMs. Because of the symmetry Eq. (3), each curve and region can be shifted by multiples of 2π. The result of this is shown in Fig. 2(b) where we now can restrict C_p to a fundamental 2π interval; for convenience we dropped the labeling. Since the 2π copies from Fig. 2(a) overlap each other we find multistable EFM regions. In particular, the white region indicates one stable EFM, the light gray region two stable EFMs, and the dark gray regions three stable EFMs.

As Fig. 2 shows, we find a considerable degree of multistability between different EFMs. In fact, changing other feedback parameters, such

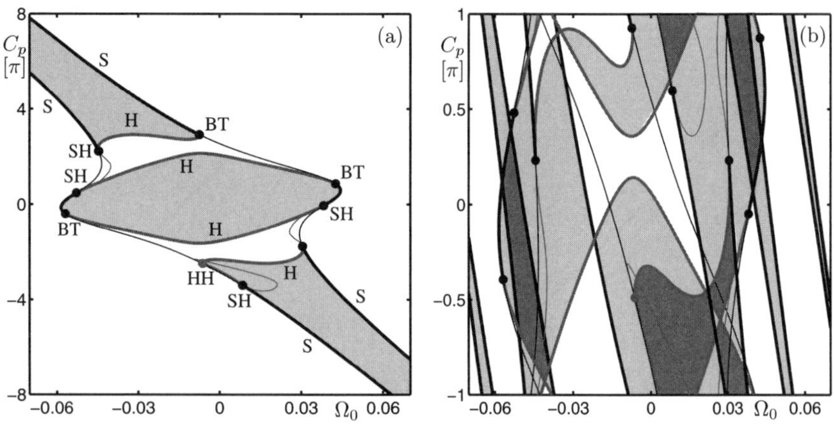

Fig. 2. Two-parameter bifurcation analysis of the EFMs in the (Ω_0, C_p)-plane. Shown are saddle-node (S) and Hopf (H) bifurcation curves; codimension-two points are saddle-node Hopf (SH), Bogdanov-Takens (BT) and double Hopf (HH). Panel (a) shows the covering space of the parameter C_p over several intervals of 2π; in this panel gray shaded regions indicate stable EFMs. Panel (b) shows a fundamental 2π-interval of C_p; in this panel white regions indicate one stable EFM, light gray regions two stable EFMs, and dark gray regions three stable EFMs.

as the feedback strength κ and the filter linewidth Λ, can have substantial influence on the multistable EFM regions [3,9]. Finally, we mention that bifurcating periodic solutions occur stable in large regions that give rise to an additional degree of multistability [10].

4. Conclusions

We studied the multistable continuous wave emission of a semiconductor laser subject to coherent optical feedback from a Fabry-Perot filter. Saddle-node bifurcations create an increasing number of external filtered modes as the laser frequency get close to the center frequency of the filter. The stability regions of those external filtered modes overlap which leads to multistability of between up to three external filtered modes. Here we concentrated on the stability of the external filtered modes. We found several codimension-two saddle-node Hopf, Bogdanov-Takens and double Hopf points that indicate a complex bifurcation structure of periodic orbits. For example, Refs. [3,10] contain a more comprehensive bifurcation analysis in dependence on different system parameters, which also includes the study of coexisting periodic orbits.

References

1. M. Yousefi and D. Lenstra, *IEEE Journal of Quantum Electronics* **35**, 970 (1999).
2. K. Green and B. Krauskopf, *Optics Communications* **258**, 243 (2006).
3. H. Erzgräber, B. Krauskopf and D. Lenstra, *SIAM Journal on Applied Dynamical Systems* **6**, 1 (2007).
4. A. P. A. Fischer, M. Yousefi, D. Lenstra, M. W. Carter and G. Vemuri, *Phys. Rev. Lett.* **92**, p. 023901 (2004).
5. G. Hek and V. Rottschafer, *IMA J Appl Math* **72**, 420 (2007).
6. V. Z. Tronciu, H.-J. Wünsche, M. Wolfrum and M. Radziunas, *Physical Review E* **73**, p. 046205 (2006).
7. T. Dahms, P. Hövel and E. Schöll, *Physical Review E* **78**, p. 056213 (2008).
8. K. Engelborghs, T. Luzyanina and G. Samaey, *DDE-BIFTOOL v. 2.00 user manual: a Matlab package for bifurcation analysis of delay differential equations.*, Technical Report TW-330, Department of Computer Science (K. U. Leuven, Leuven, 2001).
9. H. Erzgräber and B. Krauskopf, *Optits Letters* **32**, 2441 (2007).
10. H. Erzgräber, D. Lenstra, B. Krauskopf, A. P. A. Fischer and G. Vemuri, *Physical Review E* **76**, p. 026212 (2007).

EXPERIMENTAL PHASE CONTROL OF A FORCED CHUA'S CIRCUIT

GIOVANNI CHESSARI[1], STEFANO EUZZOR[2], LUIGI FORTUNA[1], MATTIA FRASCA[1], RICCARDO MEUCCI[2] and FORTUNATO T. ARECCHI[2]

[1] *Università degli Studi di Catania, DIEES, Faculty of Engineering, Catania, Italy*
[2] *Istituto Nazionale di Ottica Applicata, Firenze, Italy*

In this paper the suitability of the phase control technique to control the forced Chua's circuit is demonstrated. According to this technique a further sinusoidal term is introduced in the forced Chua's circuit at the same frequency of the original driving signal, but with a different phase. The experimental results discussed in the paper show that the phase difference between the two sinusoidal terms can act as a control parameter for the circuit.

Keywords: Chaos, chaos control, Chua's circuit, phase control.

1. Introduction

The techniques to control chaos can be classified in feedback and open loop methods [1]. Feedback methods usually allow the system to be stabilized in any of the unstable periodic orbits lying in the chaotic attractor, but require fast and accurate response to work properly. On the other hand, open loop techniques usually exploit the effect of some (small) perturbations added to the system to modify the final state of the controlled dynamics.

Non-feedback methods have been mainly used to suppress chaos in periodically driven dynamical systems:

$$\dot{\mathbf{x}} = \mathbf{f}(\mathbf{x}, \lambda) + \mathbf{F}\cos(\omega t) \tag{1}$$

where $\mathbf{x}$, $\mathbf{f}$ and $\mathbf{F}$ are vectors of the m-dimensional phase space, and λ is a parameter of the system. The main idea of these non-feedback methods is to apply a harmonic perturbation either to some of the parameters of the systems

$$\dot{\mathbf{x}} = \mathbf{f}(\mathbf{x}, \lambda(1 + \varepsilon\cos(r\omega t + \varphi))) + \mathbf{F}\cos(\omega t) \tag{2}$$

or as an additional forcing

$$\dot{\mathbf{x}} = \mathbf{f}(\mathbf{x}, \lambda) + \mathbf{F}\cos(\omega t) + \mathbf{u}\cos(r\omega t + \varphi) \tag{3}$$

where $\mathbf{u}$ is a conveniently chosen unitary vector.

The effectiveness of this type of methods has been tested experimentally in different works [2,3]. In the first, where these non-feedback method was explored, the numerical and experimental explorations were essentially focused on the role played by the perturbation amplitude and the resonance condition r, but the role of the phase difference φ was hardly explored. However, in the Ref. [3], it was observed that the phase difference φ between the periodic forcing and the perturbation had certain influence on the dynamical behavior of the system. Furthermore, in the Ref. [4], the authors have shown that φ plays a crucial role on the global dynamics of the system. The type of control based on varying the phase difference φ in search of a desired dynamical behavior is known as the *phase control* technique [5]. In many systems a correct choice of the phase allows to suppress chaos with a very small amplitude harmonic perturbation. Phase control has been applied to the Duffing system [4,6,7], to control intermittency in a CO_2 laser [8] and to avoid escapes of a Helmholtz oscillator [9].

Both non-feedback and feedback techniques have been applied to the Chua's circuit [10]. In this paper we investigate the suitability of the phase control technique to suppress chaos in a driven Chua's circuit.

2. The Chua's circuit

In this paper, phase control is applied to the Chua's circuit, which can be described by the following dimensionless equations [10]:

$$\begin{aligned} \dot{x} &= \alpha[y - h(x)] \\ \dot{y} &= x - y + z \\ \dot{z} &= -\beta y - \gamma z \end{aligned} \tag{4}$$

with $h(x) = m_1 x + 0.5(m_0 - m_1)(|x+1| - |x-1|)$. From Eq. (4) the classical double scroll attractor shown by the Chua's circuit is obtained for the following parameters: $\alpha = 9$, $\beta = 14.286$, $\gamma = 0$, $m_0 = -1/7$, $m_1 = 2/7$.

Many different implementations have been proposed in literature to realize the Chua's circuit [10]. We focused on the so-called implementation based on State Controlled Cellular Nonlinear Network (SC-CNN) [11]. Without discussing the details of the SC-CNN implementation of Chua's circuit, we wish to briefly recall here the main ideas underlying this circuit. The SC-CNN implementation of Chua's circuit is essentially a compact and robust implementation of Chua's circuit based on operational amplifiers. For each state variables of Eq. (4) an algebraic adder operational amplifier and an RC filter are designed. The nonlinearity is implemented by exploiting the natural saturation of a further operational amplifier so that four operational amplifiers are needed to realize the whole circuit. In such a way an implementation where all the state variables can be easily accessible is obtained.

Although the Chua's circuit has been originally designed as an autonomous circuit, several studies have dealt with the case in which a forcing term is added to this circuit. For instance, in [12] the sinusoidal forcing is introduced by adding a new branch in the Chua's circuit (originally consisting of only two capacitors, an inductor, a passive resistor and a nonlinear resistor). The experimental investigation of this circuit carried on in [12] revealed a great variety of bifurcation sequences. In particular, period-adding bifurcations, quasi-periodicity, hysteresis and intermittent behaviors have been observed.

Another interesting result is the possibility of controlling many of these phenomena by adding a further sinusoidal generator in series with the previous one. In [13], for example, a second sinusoidal forcing with a different frequency is used. Starting from a chaotic behavior in the absence of the second forcing, Murali and Lakshmanan demonstrate that periodic orbits (for instance of period-3) can be stabilized adding a sinusoidal term with a small amplitude. In the following, we show that similar behavior can be obtained by using two sinusoidal terms at the same frequency but different phase frequencies.

3. Phase control of the Chua's circuit

In the forced Chua's circuit considered in this paper the sinusoidal term is added to the second equation of Eq. (4). A further sinusoidal term is then added to this equation. This "control" signal has the same frequency of the driving signal, but has a different phase. The Chua's circuit driven by a periodic forcing can be modelled by the following dimensionless equations:

$$\begin{aligned} \dot{x} &= \alpha[y - h(x)] \\ \dot{y} &= x - y + z + A_m \sin(2\pi f_m t) - A_c \sin(2\pi f_c t + \phi) \\ \dot{z} &= -\beta y \end{aligned} \tag{5}$$

with $h(x) = m_1 x + 0.5(m_0 - m_1)(|x+1| - |x-1|)$ and where $A_m sin(2\pi f_m t)$ represents the main driving signal and $A_c sin(2\pi f_c t + \phi)$ with $f_c = f_m$ the further harmonic perturbation used to control chaos. Our analysis has been carried out experimentally. As introduced above, the Chua's circuit has been implemented using the state variable approach described in [10], which has the advantage of an easy and flexible implementation of the forcing terms. The experimental results are discussed in the next section.

4. Experimental results

In our experiments, the amplitude of this signal is kept constant to a small value $A_c = 20mV$ (compared to $A_m = 240mV$) and the phase difference φ is varied. The other parameters of the system have been fixed to: $f_m = 4088Hz$, $\alpha = 7.85$, $\beta = 12.195$, $m_0 = -1/7$, and $m_1 = 2/7$. For such choice of parameters, in the absence of the main driving signal (i.e., $A_m = 0$) the system shows a chaotic behavior with a single scroll attractor. When the driving signal is applied (i.e., $A_m = 240mV$), the chaotic attractor of the system is a double scroll attractor. The addition of the small periodic control signal with zero phase does not significantly affect the circuit dynamics. However, tuning the phase difference parameter φ influences the dynamical behavior of the system, and there exist suitable values for which a stable limit cycle behavior is obtained.

Figure 1 shows the double scroll chaotic attractor exhibited by the driven Chua's circuit (5) when the phase difference is $\varphi = 0°$. Figure 2 shows the effect of a control signal with a phase difference equal to $\varphi = 85°$: a period behavior is obtained (in particular, a stable period-3 limit cycle). Finally, Fig. 3 shows the bifurcation diagram experimentally obtained. The bifurcation diagram shows the local minima of the state variable y with respect to φ. It can be observed that, in general, the phase parameter modulates the maximum amplitude of the state variable y, and that there exist several windows of periodic behavior, thus confirming the suitability of the phase control method to suppress chaos in a driven Chua's circuit.

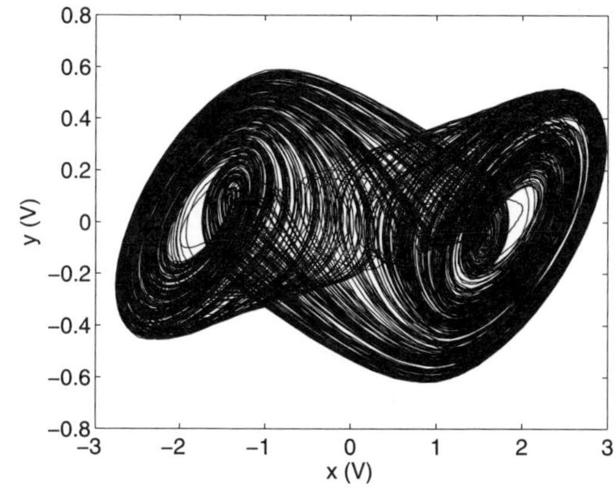

Fig. 1. Projection of the chaotic attractor obtained for $\phi = 0°$ on the phase plane $x - -y$.

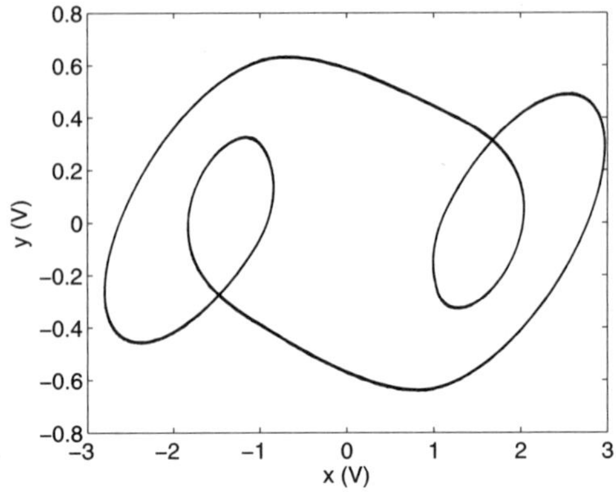

Fig. 2. Projection of the chaotic attractor obtained for $\phi = 85°$ on the phase plane $x - -y$.

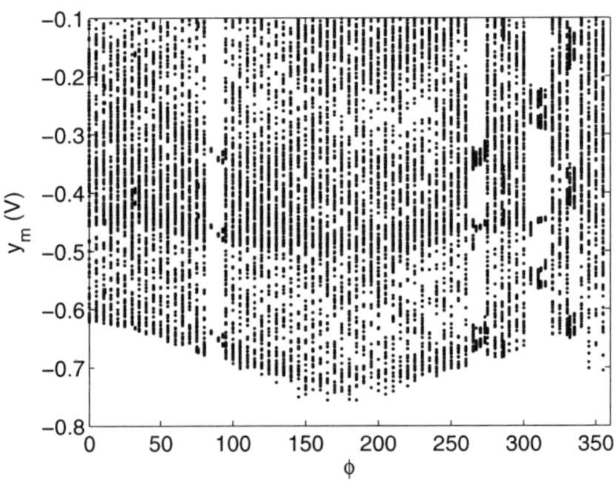

Fig. 3. Experimental bifurcation diagram with respect to the parameter ϕ.

5. Conclusions

We have shown the robustness and the general nature of the phase control technique in the sense that the experimental implementation in the Chua's circuit, i.e., one of the most investigated circuits in nonlinear dynamics, confirms the results.

The role of the phase in selecting the final dynamical state is very important from a control point of vies, since there is a large variety of situations in which the modulation of the accessible parameters might be limited, and φ is an additional degree of freedom that may be very useful.

In summary, we have shown that the phase control scheme is very versatile and useful in a wide variety of dynamical situations: to suppress chaos, to control global dynamics as in the case of intermittency in chaotic systems close to a crisis or to avoid escapes in open dynamical systems.

References

1. S. Boccaletti, C. Grebogi, Y.-C. Lai, H. Mancini, D. Maza, Phys. Rep. 329 (2000), 103.
2. R. Lima and M. Pettini, Phys. Rev. A. 41 (1990), 726.
3. R. Meucci, W. Gadomski, M. Ciofini, and F.T. Arecchi, Phys. Rev. E, 49 (1994), R2528.
4. Z. Qu, G. Hu, G. Yang, G. Qin, Phys. Rev. Lett. 74 (1995), 1736.
5. F.T. Arecchi, W. Gadomski and R. Meucci, Recent Advances in Laser Dynamics: Control Synchronization, 41-78, 2008 Ed. A. N. Pisarchik, Research Signpost, Kerala, India.
6. J. Yang, Z. Qu, G. Hu, Phys. Rev. E 53 (1996), 4402.
7. S. Zambrano, E. Allaria, S. Brugioni, I. Levya, R. Meucci, M.A.F. Sanjuan, F.T. Arecchi, Chaos 16 (2006), 013111.
8. S. Zambrano, I.P. Marino, S. Euzzor, R. Meucci, F.T. Arecchi, M.A.F. Sanjuan, Phys. Rev. E 74 (2006), 016202.
9. J.M. Seoane, S. Zambrano, S. Euzzor, R. Meucci, F.T. Arecchi, M.A.F. Sanjuan, Phys. Rev. E 78 (2008), 016205.
10. L. Fortuna, M. Frasca, M.G. Xibilia, *Chua's Circuit Implementations: Yesterday, Today and Tomorrow*, World Scintific Publishing Company, 2009.
11. P. Arena, S. Baglio, L. Fortuna, G. Manganaro, IEEE Trans. Circuits and Systems I, 42 (1995), 123.
12. K. Murali, M. Lakshmanan, IEEE Trans. Circuits and Systems I, 39 (1992), 264.
13. K. Murali, M. Lakshmanan, IEEE Trans. Circuits and Systems I, 40 (1993), 836.

PART D

Linear and matritial algebra, open problems related to control theory

DISTURBANCE DECOUPLING FOR SINGULAR SYSTEMS BY PROPORTIONAL AND DERIVATIVE FEEDBACK AND PROPORTIONAL AND DERIVATIVE OUTPUT INJECTION

M. I. GARCIA-PLANAS

Departament de Matemàtica Aplicada I
Universitat Politècnica de Catalunya
08038 Barcelona, Spain
E-mail: maria.isabel.garcia@upc.edu

We study the disturbance decoupling problem for linear time invariant singular systems. We give necessary and sufficient conditions for the existence of a solution to the disturbance decoupling problem with or without stability via a proportional and derivative feedback and proportional and derivative output injection that also makes the resulting closed-loop system regular and/or of index at most one. All results are based on canonical reduced forms that can be computed using a complete system of invariants that can be implemented in a numerically stable way.

Keywords: Singular systems, equivalence relation, disturbance decoupling.

1. Introduction

We consider linear and time-invariant continuous singular systems of the form

$$\left\{ \begin{array}{ll} E\dot{x}(t) & = Ax(t) + Bu(t) + Gg(t), \;\; x(t_0) = x_0, \;\; t \geq 0 \\ y(t) & = Cx(t), \end{array} \right. \tag{1}$$

where $E, A \in M_n(C)$, $B \in M_{n\times m}(C)$, $C \in M_{p\times n}(C)$, $G \in M_{n\times q}(C)$ and $\dot{x} = dx/dt$. The term $g(t)$, $t \geq 0$, represents a disturbance, which may represent modeling or measuring errors, noise, or higher order terms in linearization. Singular systems arise naturally in circuits design, mechanical multibody systems and a large variety of the applications (see [5] and [6], for example), and they have been studied under different points of view. The problem of constructing feedbacks and/or output injections that suppress this disturbance in the sense that $g(t)$ does not affect the input-output behavior of the system is analyzed. In the case of standard state space systems the disturbance decoupling problem has been largely studied (see [1],[7],[8] for example), This problem for singular systems has also been studied (see [2], [4] for example). In this paper we study the disturbance decoupling problem for singular systems that can be stated as follows: Find necessary and sufficient conditions under which we can choose state and derivative feedback as well state and derivative output injection such that, the matrix pencil $(E + BF_E^B + F_E^C C, A + BF_A^B + F_A^C C)$ is regular of index at most one and $C(s(E + BF_E^B + F_E^C C) - (A + BF_A^B + F_A^C C))^{-1}G = 0$.

We assume, without loss of generality, that $\operatorname{rank} B = m$, $\operatorname{rank} G = q$, and $\operatorname{rank} C = p$. If this is not the case, then this can be easily achieved, by removing the nullspaces and appropriate renaming of variables.

In the sequel we will use the following notations:

I_n denotes the n-order identity matrix, N denotes a nilpotent matrix in its reduced form $N = \operatorname{diag}(N_1, \dots, N_t)$, $N_i = \begin{pmatrix} 0 & I_{n_i-1} \\ 0 & 0 \end{pmatrix} \in M_{n_i}(C)$, J denotes the Jordan matrix $J = \operatorname{diag}(J_1, \dots, J_t)$, $J_i = \operatorname{diag}(J_{i_1}, \dots, J_{i_s})$, $J_{i_j} = \lambda_i I_{i_j} + N$, L denotes the diagonal matrix $L = \operatorname{diag}(L_1, \dots, L_q)$, where $L_j = \begin{pmatrix} I_{n_j} & 0 \end{pmatrix} \in M_{n_j \times (n_j+1)}(C)$, and R denotes the diagonal matrix $R = \operatorname{diag}(R_1, \dots, R_p)$, where $R_j = \begin{pmatrix} 0 & I_{n_j} \end{pmatrix} \in M_{n_j \times (n_j+1)}(C)$.

We represent systems of the form (1) as quadruples of matrices (E, A, B, C) in the case of disturbance do not appear or it is not considered, and a quintuples of matrices (E, A, B, C, G) otherwise.

2. Reduced form

We recall that, given a singular system (not necessarily square) using standard transformations in state, input and output spaces $x(t) = Px_1(t)$, $u(t) = Ru_1(t)$, $y_1(t) = Sy(t)$, premultiplication by an invertible matrix $QE\dot{x}(t) = QAx(t) + Qu(t)$ making feedbacks $u(t) = u_1(t) - Vx(t)$, $u(t) = u_1(t) - U\dot{x}(t)$ as well as output injections $u(t) = u_1(t) - Wy(t)$, $u(t) = u_1(t) - Z\dot{y}(t)$, it is possible to reduce to $E_r\dot{x}_1(t) = A_r x_1(t) + B_r u_1(t) + G_1$, $y_1 = C_r x(t)$ where

$$E_r = \begin{pmatrix} I_1 & & & & & & \\ & I_2 & & & & & \\ & & I_3 & & & & \\ & & & I_4 & & & \\ & & & & N_1 & & \\ & & & & & L_1 & \\ & & & & & & L_2^t \\ & & & & & & 0 \end{pmatrix},$$

$$A_r = \begin{pmatrix} N_2 & & & & & & \\ & N_3 & & & & & \\ & & N_4 & & & & \\ & & & J & & & \\ & & & & I_5 & & \\ & & & & & R_1 & \\ & & & & & & R_2^t \\ & & & & & & 0 \end{pmatrix}$$

$$B_r = \begin{pmatrix} B_1 & 0 & 0 \\ 0 & B_2 & 0 \\ 0 & 0 & 0 \\ 0 & 0 & 0 \\ 0 & 0 & 0 \\ 0 & 0 & 0 \\ 0 & 0 & 0 \\ 0 & 0 & B_3 \end{pmatrix}, \quad C_r = \begin{pmatrix} C_1 & 0 & 0 & 0 & 0 & 0 & 0 \\ 0 & 0 & C_2 & 0 & 0 & 0 & 0 \end{pmatrix}$$

or

$$E_r = \begin{pmatrix} I_1 & & & & & & & \\ & I_2 & & & & & & \\ & & I_3 & & & & & \\ & & & I_4 & & & & \\ & & & & N_1 & & & \\ & & & & & L_1 & & \\ & & & & & & L_2^t & 0 \end{pmatrix},$$

$$A_r = \begin{pmatrix} N_2 & & & & & & & \\ & N_3 & & & & & & \\ & & N_4 & & & & & \\ & & & J & & & & \\ & & & & I_5 & & & \\ & & & & & R_1 & & \\ & & & & & & R_2^t & 0 \end{pmatrix}$$

$$B_r = \begin{pmatrix} B_1 & 0 \\ 0 & B_2 \\ 0 & 0 \\ 0 & 0 \\ 0 & 0 \\ 0 & 0 \\ 0 & 0 \end{pmatrix}, \quad C_r = \begin{pmatrix} C_1 & 0 & 0 & 0 & 0 & 0 & 0 & 0 \\ 0 & 0 & C_2 & 0 & 0 & 0 & 0 & 0 \\ 0 & 0 & 0 & 0 & 0 & 0 & 0 & C_3 \end{pmatrix}$$

and

i) (I_1, N_2, B_1, C_1) is an n_1-completely controllable-observable system.
ii) (I_2, N_3, B_2) is an n_2-completely controllable non observable system.
iii) (I_3, N_4, C_2) is an n_3-completely observable non controllable system.
iv) (I_4, J) is an n_4-system having only finite zeroes.
v) (N_1, I_5) is an n_5-system having only non transferable infinite zeroes.
vi) (L_1, R_1), completely singular systems of n_6 rows.
vii) (L_2^t, R_2^t) completely singular systems of n_7 rows respectively.
viii) $B_3 = \mathrm{diag}(I_{n_{8_1}}, 0_{n_{8_2}})$ or $C_3 = \mathrm{diag}(I_{n_{9_1}}, 0_{n_{9_2}})$.

The regular part of the system is maximal among all possible reductions of the system decomposing it in a regular part and a singular part and not all parts i),..., viii) necessarily appears in the decomposition of a system.

The proof is based in the following proposition:

Proposition 2.1. *Two quadruples of matrices (E_i, A_i, B_i, C_i) are equivalent under equivalence relation considered if and only if the matrix pencils $\lambda \begin{pmatrix} E_i & B_i & 0 \\ C_i & 0 & 0 \\ 0 & 0 & 0 \end{pmatrix} + \begin{pmatrix} A_i & 0 & B_i \\ 0 & 0 & 0 \\ C_i & 0 & 0 \end{pmatrix}$ are strictly equivalent.*

Based on reduced form, the system (1), is reduced to the following independent subsystems:

$$\begin{cases} \dot{x}_1 & = N_2 x_1 + B_1 u_1 + G_1 g_1 \\ y_1 & = C_1 x_1 \end{cases} \tag{2}$$

$$\{\ \dot{x}_2 = N_3 x_2 + B_2 u_2 + G_2 g_2 \tag{3}$$

$$\begin{cases} \dot{x}_3 & = N_4 x_3 + G_3 g_3 \\ y_3 & = C_2 x_3 \end{cases} \tag{4}$$

$$\{\ \dot{x}_4 = J x_4 + G_4 g_4 \tag{5}$$

$$\{\ N_1 \dot{x}_5 = x_5 + G_5 g_5 \tag{6}$$

$$\{\ L_1 \dot{x}_6 = R_1 x_6 + G_6 g_6 \tag{7}$$

$$\{\ L_2^t \dot{x}_7 = R_2^t x_7 + G_7 g_7 \tag{8}$$

$$\{B_3 u_3 = 0 \quad \text{or} \quad \{C_3 x_8 = 0. \tag{9}$$

Systems from (2) to (6) are regular and (7), (8) and (9) are completely singular and there are not feedbacks, derivative feedbacks, output injections and derivative output injections regularizing partially or totally the systems (7), (8) and (9).

3. The disturbance decoupling problem

In this section we will use the reduced form for the system in order to analyze the disturbance decoupling problem.

Proposition 3.1. *Consider a system of the form (1). The system can be regularized by means an state and derivative feedback as well as an state and a derivative output injection with index at most one, if and only if the reduced form does not contain parts vi), vii), and viii), and if it contains v) then the nilpotent matrix* N_1 *is the zero matrix.*

Proof. It suffices to observe that a system is regularisable if and only if the reduced form is regularisable and the index of the system is the index of matrix N_1. □

Let $H(\lambda) = \lambda \begin{pmatrix} E & B & 0 \\ C & 0 & 0 \\ 0 & 0 & 0 \end{pmatrix} + \begin{pmatrix} A & 0 & B \\ 0 & 0 & 0 \\ C & 0 & 0 \end{pmatrix}$ be a pencil associate to the system (E, A, B, C).

Theorem 3.1. *Consider a system of the form (1). The system can be regularized by means an state and derivative feedback as well as an state and derivative output injection with index at most one if and only if i)* $r_1 - r_0 \geq n$, *ii)* $s_k \leq 2(r_B - t)$ *and iii)* $l_k \leq 2(r_C - t)$, *where* $r_0 = rank \begin{pmatrix} E & B \\ C & 0 \end{pmatrix}$, s_k *is the number of column minimal indices of the pencil* $H(\lambda)$, $r_B = rank B$, l_k *is the number of row minimal indices of the pencil* $H(\lambda)$, $r_C = rank C$, $t = r_n - r_{n-1} - n$ *and* $r_\ell = rm\ rank\, M_\ell, \forall \ell \geq 1$

$$M_\ell = \begin{pmatrix} E & B & & & & & & \\ C & 0 & & & & & & \\ A & 0 & E & B & & & & \\ & & C & 0 & & & & \\ & & A & 0 & & & & \\ & & & & \ddots & & & \\ & & & & & E & B & & \\ & & & & & C & 0 & & \\ & & & & & A & 0 & E & B \\ & & & & & & & C & 0 \end{pmatrix}$$

$\in M_{(\ell+1)(n+p)\times(\ell+1)(n+m)(C)}$.

Proof. It suffices to observe that the controllable and observable subsystem joint with subsystem (N_1, I_5), correspond to the infinite zeros of the pencil associate. Controllable non observable subsystem corresponds to the column singular part of the pencil and observable non controllable subsystem corresponds to the row singular part of the pencil. □

Using quadruples in its reduced form, extending the equivalence to the quintuples of matrices (i.e. $QG = \overline{G}$) and taking into account [2], lemma 2.4, we have the following proposition.

Proposition 3.2. *Assume* $\overline{G} = \begin{pmatrix} G_1 & \dots & G_5 \end{pmatrix}^t$ *according to the subsystems (2), ..., (6). Let* $s \in C$ *such that* $\det(sI_{n_1} - N_2) \neq 0$, $\det(sI_{n_2} - N_3) \neq 0$, $\det(sI_{n_3} - N_4) \neq 0$, $\det(sI_n - J) \neq 0$ *and* $\det(sN_1 I_{n_5}) \neq 0$, *(it exists because of regularity of the subsystems (2),..., (6)). Then*

i) $C_1(sI_{n_1} - N_2)^{-1}G_1 = 0$ *if and only if rank* $\begin{pmatrix} sI_{n_1}-N_2 & G_1 \\ C_1 & 0 \end{pmatrix} = n_1$,
ii) $(sI_{n_2} - N_3)^{-1}G_2 = 0$ *if and only if* $G_2 = 0$
iii) $C_1(sI_{n_3} - N_4)^{-1}G_3 = 0$ *if and only if rank* $\begin{pmatrix} sI_{n_3}-N_4 & G_3 \\ C_2 & 0 \end{pmatrix} = n_3$
iv) $(sI_{n_4} - J)^{-1}G_4 = 0$ *if and only if* $G_4 = 0$
v) $(sN_1 - I_{n_5})^{-1}G_5 = 0$ *if and only if* $G_5 = 0$

As a consequence we have.

Corollary 3.1. *Let* (E, A, B, C, G) *be a quintuple of matrices in its reduced form, and we assume* $\overline{G} = \begin{pmatrix} G_1 & \dots & G_5 \end{pmatrix}^t$ *according to the decomposition of the system. If* $G_2 = 0$, $G_4 = 0$, $G_5 = 0$, *rank* $\begin{pmatrix} sI_{n_1}-N_2 & G_1 \\ C_1 & 0 \end{pmatrix} = n_1$ *and rank* $\begin{pmatrix} sI_{n_3}-N_2 & G_3 \\ C_1 & 0 \end{pmatrix} = n_3$, *then the given system is trivially disturbance decoupled.*

The disturbance decoupling problem is called with stability if one imposes the additional constraint that the close-loop $(E+BF_E^B+F_E^C C)\dot{x}(t) = (A+BF_A^B+F_A^C C)x(t)+Bu(t)+Gg(t)$, $y(t) = Cx(t)$ system is stable. Remember that a singular system is stable if and only if the spectrum of the system lies in C^{-1}.

Proposition 3.3. *Given a singular system* (E, A, B, C). *There exist a proportional and derivative feedback as well as a proportional and derivative output injection such that the close-loop system* $(E + BF_E^B + F_E^C C, A + BF_A^B + F_A^C C, B, C)$ *is stable (and we call stable under proportional and derivative feedback, and proportional and derivative output injection) if and only if rank* $\begin{pmatrix} sE-A & B \\ C & 0 \end{pmatrix} = n$, $\forall s \in C^+$.

Proof. The spectrum of a system coincides with the spectrum of the associate pencil, and the spectrum is invariant under equivalence relation. □

As a consequence we have.

Corollary 3.2. *Let* (E, A, B, C, G) *be a quintuple of matrices in its reduced form, and we assume* $\overline{G} = \begin{pmatrix} G_1 & \dots & G_5 \end{pmatrix}^t$ *according to the decomposition of the system. If* $G_2 = 0$, $G_4 = 0$, $G_5 = 0$, *rank* $\begin{pmatrix} sI_{n_1}-N_2 & G_1 \\ C_1 & 0 \end{pmatrix} = n_1$, *rank* $\begin{pmatrix} sI_{n_3}-N_4 & G_3 \\ C_1 & 0 \end{pmatrix} = n_3$ *and* $\sigma(J) \subset C^{-1}$. *Then the given system is trivially disturbance decoupled with stability.*

References

1. M. Günther and U. Feldmann, The DAE-index in electric circuit simulation, *Math. Comput. Simulation*, **39** (1995), pp. 573-582.
2. Hou M., *Descriptor Systems: Observer and Fault Diagnosis*, Fortschr-Ber. VDI Reihe 8, Nr. 482 (VDI Verlag, Düsseldorf, FRG, 1995).
3. A. Ailon, A solution to the disturbance decoupling problem in singular systems via analogy with state-space systems, *Aut. J. IFAC*, **29** (1993), pp. 1541-1545.
4. A. S. Morse and W. M. Wonham, Decoupling and pole assignment by dynamic compensation, *SIAM J. Control*, **8** (1970), pp. 317-337.
5. M. Rakowski, *Transfer function approach to disturbance decoupling problem*, Linear Algebra for Control Theory, B. W. P. Van Dooren, ed., IMA Vol. Math. Appl. 62 (Springer-Verlag, Berlin, 1994), pp. 159-176.
6. D. Chu and V. Mehrmann, Disturbance Decoupling for Descriptor Systems by state feedback, *Siam J. Control Optim.* **38**, 6 (2000), pp. 1830-1858.
7. L. R. Fletcher and A. Asaraai, On disturbance decoupling in descriptor systems, *SIAM J. Control Optim.*, **27** (1989), pp. 1319-1332.

ENUMERATION OF FEEDBACK EQUIVALENCE CLASSES OF LINEAR SYSTEMS OVER A COMMUTATIVE RING VS. PARTITIONS OF ELEMENTS OF A MONOID

M. V. CARRIEGOS* and M. M. LÓPEZ-CABECEIRA

Departamento de Matemáticas, Universidad de León
León, Spain
*E-mail: *miguel.carriegos@unileon.es*

The problem of finding all feedback equivalence classes of Brunovsky and locally Brunovsky linear systems defined on a commutative ring is related with combinatorial problem of visiting all partitions of elements in a given monoid.

Keywords: Feedback classification, projective module, enumeration, K-theory.

1. Introduction

The theory of linear control systems over a commutative ring R goes back to the models of [1] for delay systems. The main example in our study will be the ring of continuous real functions $R = \mathcal{C}(K)$ defined on a compact topological space K (which was introduced in the control theory framework in [2] as model for studying parametriced families of systems). Rings of continuous functions also apply to the geometric study of differential deformations of linear systems (see [3]).

This paper deals with the feedback classification of linear systems over a commutative ring. To be concise, we are interested in the enumeration of all feedback classes of reachable linear systems. For general reading on the subject we refer to [4], [5], [6], [7], [8] and references therein.

Geometric properties of commutative ring R are crucial: In fact we prove that the task to enumerate all feedback classes of reachable linear systems over R^n is equivalent to the task to enumerate all direct-sum decompositions of R^n as an element of monoid $\mathbf{P}(R)$ of finitely generated projective R-modules. Besides, in the case of $R = \mathcal{C}(K)$ being the ring of real continuous functions defined on a compact space K, the monoid $(\mathbf{P}(R), \oplus)$ is equivalent to monoid of finite dimensional real vector bundles over K (see [9]).

The description of monoid $\mathbf{P}(R)$ is in general very difficult. To avoid this, we also provide a new *stable* feedback equivalence relation and conjecture that this new relation may be studied by working with the Grothendieck's group completion $K_0(R)$ of monoid $\mathbf{P}(R)$.

The paper is organized in the following manner. In Sec. 2 we review some well known facts about linear systems over a commutative ring. Main invariants under feedback are also revised. Section 3 is devoted to study Brunovsky case: This is the case of constant linear systems over a field, and it is solved by studying classical partitions of an integer. Section 4 deals with the more general case of locally Brunovsky linear systems: This is the case of projective invariants or, if $R = \mathcal{C}(K)$, the case of smooth invariants. We solve the problem by studying direct-sum decompositions of elements of monoid $\mathbf{P}(R)$. The cases of real circle and real sphere are specially focused. Section 5 is devoted to describe a dynamic study of feedback equivalence related to the Grothendieck's group completion $K_0(R)$ of monoid $\mathbf{P}(R)$.

2. Preliminaries

Let R be a commutative ring with identity element $1 \neq 0$.
i) A linear system over R is given by a linear rule (or right hand side) on the form $x^+ = Ax + Bu$ where $x \in X$ are states, $u \in U$ are inputs, and x^+ is the time-derivative or time-shift in the sequential case. Sets of states X and of inputs U are R-modules while maps A and B are R-linear maps.

$$\Sigma : \boxed{\begin{array}{ll} U & \\ \quad\searrow^{B} & \\ & X \to^{A} X \end{array}} \tag{1}$$

ii) Above system Σ and an analogous system Σ' are said to be Feedback Equivalent if one can bring one of them into the another by a finite composition of the following Basic Feedback Actions: Isomorphisms $Q : U \to U'$ in the input module, isomorphisms $P : X \to X'$ in the state module and feedback actions $F : X \to U$ which transform (A, B) to system $(P(A + BF)P^{-1}, PBQ)$.
iii) Partial reachability linear map given by

$$\varphi_i^{\Sigma} = \begin{pmatrix} B & AB & \cdots & A^{i-1}B \end{pmatrix} : U^{\oplus i} \longrightarrow X \tag{2}$$

is a feedback invariant, up to equivalence, associated to Σ (see [6] and [7]). Therefore, we obtain our main set of feedback invariants:
iv) Quotient modules

$$N_{i+1}^{\Sigma}/N_i^{\Sigma} = \mathrm{Im}(B, AB, ..., A^iB)/\mathrm{Im}(B, AB, ..., A^{i-1}B) \tag{3}$$

are feedback invariants, up to isomorphism, associated to system Σ.

3. n-dimensional Brunovsky systems and partitions of integer n

i) A linear system $\Sigma = (A, B)$ is a Brunovsky linear system (BLS), if it is equivalent to a Brunovsky canonical form (BCF).

In the case of $R = \mathbb{K}$ being a field, a BLS is just a reachable linear system. Then one has the following result:
ii) The task to enumerate all feedback classes of reachable linear systems over $\mathbb{K}^n$ is equivalent to the classical task to enumerate all partitions of integer n. So, classical enumeration algorithms [10] may be directly translated from partitions of a given integer n to reachable systems over $\mathbb{K}^n$.

The problem can also be attacked in the case of linear systems such that all its invariants are free defined over R such that finitely generated projective R-modules are free:
iii) Assume that R is projectively trivial (i.e. all projective R-modules are free), then linear system $\Sigma = (A, B)$ over R^n is equivalent to a BCF if and only if all invariant R-modules $N_{i+1}^{\Sigma}/N_i^{\Sigma}$ are free.

So, the problem of enumerating all feedback classes of reachable linear systems with free invariants over a projectively trivial ring is actually equivalent to the problem of enumerating all BCFs and thus, equivalent to the problem of enumerating all partitions of the integer n.
iv) The key is that, in the case of reachable linear systems over a field or, in the more general framework of projective-free rings, if all the R-modules $N_{i+1}^{\Sigma}/N_i^{\Sigma}$ are free, then they are really a complete set of invariants verifying that decompositions

$$X = N_1^{\Sigma} \oplus (N_2^{\Sigma}/N_1^{\Sigma}) \oplus \cdots \oplus (N_s^{\Sigma}/N_{s-1}^{\Sigma}) \qquad (4)$$

are in one-to-one correspondence with the set of partitions of integer n in decreasing sequences or, equivalently, by all the Ferrers diagrams of integer n. For example, if we set $n = 4$ we have the following Ferrers diagrams visited following the reverse lexicographical order:

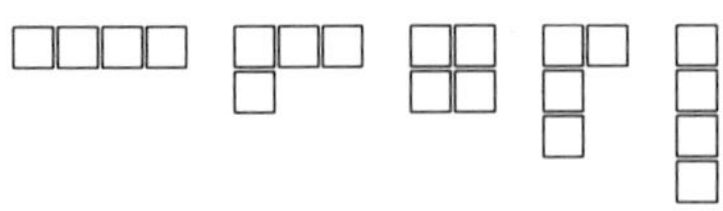

equivalently, the partitions of $4 \in \mathbb{N}$: (4),$(3,1)$,$(2,2)$,$(2,1,1)$ and $(1,1,1,1)$.

The description of all types of BLSs over R does not depend on R but on the dimension n of free state module X. In fact, there are exactly $p(n)$ BLSs over a free module $X \cong R^n$, with $p(n)$ the number of partitions of n.

4. Locally Brunovsky systems and partitions of an element of a monoid

Let Σ be a reachable linear system over R^n.
i) If $N_{i+1}^{\Sigma}/N_{i+1}^{\Sigma}$ are projective, then Σ is locally Brunovsky (see [7]).
ii) The feedback classification problem for locally BLSs over R^n is equivalent to the problem of characterization of all possible decompositions of finitely generated R-modules U and X on the form $U = Q \oplus P_1$ and $X = P_1 \oplus P_2 \oplus \cdots \oplus P_s$, with the only restriction to solve the system of equations is that P_{i+1} must be a direct summand of P_i for all i.

On the other hand, if we are not worried about he generators of input space (i.e. we allow ancillary blank inputs), the following result applies:

Theorem 4.1. *Enumerating all feedback equivalence classes of reachable linear systems over X allowing ancillary inputs is equivalent to enumerating all decompositions $X = P_1 \oplus \cdots \oplus P_s$, with P_{i+1} direct summand of P_i.*

iv) Let's denote by $p_R(n)$ the number of non-isomorphic decompositions of R^n, while $\tilde{p}_R(n)$ denotes the number of non-isomorphic decompositions $R^n \cong P_1 \oplus \cdots \oplus P_s$ with P_{i+1} direct summand of P_i. Note that, if R is projectively trivial, then $\tilde{p}_R(n) = p_R(n) = p(n)$ is the number of partitions of integer n, but in general $\tilde{p}_R(n) \leq p_R(n)$.

Anyway, one needs to know exactly the monoid $(\mathbf{P}(R), \oplus)$ of isomorphism classes of finitely generated R-modules in order to give the complete classification of locally BLSs. The full description of $(\mathbf{P}(R), \oplus)$ is a great task. Obviously, if finitely generated projectives are free, then $(\mathbf{P}(R), \oplus)$ is isomorphic to $(\mathbb{N} \cup \{0\}, +)$, but in general it is not true.

If $R = \mathcal{C}(K)$ is the ring of continuous functions defined on a compact topological space K, then $M = (\mathbf{P}(R), \oplus)$ depends on the topology of K. For instance, if $K = \mathbb{S}^1$, then M is the commutative monoid generated by the symbols R (trivial vector bundle) and P (Möbius Strip) modulo the relation $P \oplus P = R \oplus R = R^2$ (see [12]). In other words, if $K = \mathbb{S}^1$, then $(\mathbf{P}(\mathcal{C}(\mathbb{S}^1)), \oplus)$ is the commutative monoid given, in terms of generators and relations, by $\langle R, P : P^2 = R^2 \rangle$. Thus, a feedback class of an n-dimensional locally BLS over $\mathbb{S}^1$ is determined by a partition of $R^n \in \langle R, P : P^2 = R^2 \rangle$. Allowed partitions for $n = 3$ (see Theorem 4.1) are $R^3 \cong R^2 \oplus R \cong (R \oplus P) \oplus P \cong R \oplus R \oplus R$ and hence $\tilde{p}_{\mathbb{S}^1}(n) = 4$ (though $p_{\mathbb{S}^1}(n) = 5$).

Example 4.1. We compute the number of classes in terms of R and n for some cases: projectively trivial ring, the ring of continuous functions defined over $\mathbb{S}^1_{\mathbb{R}}$ and the ring of real continuous functions defined over $\mathbb{S}^2_{\mathbb{R}}$.

n	1	2	3	4	5	6	$\cdots$
$p_{\mathbb{N}}(n)$	1	2	3	5	7	11	$\cdots$
$\tilde{p}_{\mathbb{S}^1}(n)$	1	3	4	9	11	24	$\cdots$
$\tilde{p}_{\mathbb{S}^2}(n)$	1	2	3	∞	∞	∞	$\cdots$

(5)

Computations are as follows: For the case of partitions of integer $p_{\mathbb{N}}(n)$ we have the usual Euler theory (see [10]). If $R = \mathbb{S}^1_{\mathbb{R}} = \mathbb{R}[\sin\theta, \cos\theta]$, then $\mathbf{P}(R) = \langle a, b\rangle / \{ab = ba, a^2 = b^2\}$ and the calculation of $p_{\mathbb{S}^1}(n)$ can be performed by using "colored" Ferrers Diagrams [11]. In particular, it is not hard to prove that $p_{\mathbb{S}^1}(n) < \infty$, but it may be interesting to evaluate the asymptotic behavior $\mathcal{O}(p_{\mathbb{S}^1}(n))$ in terms of the dimension n of state-space. Finally, if $R = \mathbb{S}^2_{\mathbb{R}} = \mathbb{R}[x, y, z]/(x^2+y^2+z^2-1)$, then $\mathbf{P}(R)$ is the monoid [12]

$$(\mathbf{P}(R), \oplus) = (\{(0,0), (1,0), (2,\alpha), (n,\beta) : \alpha \in \mathbb{Z}, \beta \in \mathbb{Z}_2, n \geq 3\}, \maltese)$$

where $\maltese$ operates as follows: If $a + c \leq 2$ then $(a, b)\maltese(c, d) = (a + c, b + d)$, while if $a + c \geq 3$ one has $(a, b)\maltese(c, d) = (a + c, b + d \operatorname{mod} 2)$. Thus, $p_{\mathbb{S}^2}(n)$ can be directly computed for $n \leq 3$. To check that $p_{\mathbb{S}^2}(n) = \infty$ for $n \geq 4$ only note that for $n \geq 2$ and all j we have $(2n, 0) = (2, 2j)\maltese \cdots \maltese(2, 2j)$ and $(2n + 1, 0) = (3, 0)\maltese(2, 2j)\maltese \cdots \maltese(2, 2j)$.

v) Note that, in the general case a recursive procedure calculating $P_M(n)$ and in particular $\tilde{P}_M(n)$ is needed.

5. Future work: Dynamic study and $K_0(R)$

The usual dynamic study of a system (A, B) (dynamic stabilization, see [13], for example) allows to introduce ancillary variables in the system

$$(\widehat{A}, \widehat{B}) = (\mathbf{0}, \mathbf{1}) \oplus (A, B) = \left(\left(\begin{smallmatrix} 0 & 0 \\ 0 & A \end{smallmatrix}\right), \left(\begin{smallmatrix} 1 & 0 \\ 0 & B \end{smallmatrix}\right)\right) \tag{6}$$

Definition 5.1. (A generalization of dynamic study) We say that system (A, B) is stably equivalent to system (A', B') if there exists a BLS (I, J) such that $(I, J) \oplus (A, B)$ and $(I, J) \oplus (A', B')$ are equivalent.

Theorem 5.1. *Feedback invariants of augmented system splits:*

$$\frac{N_{i+1}^{(I,J)\oplus(A,B)}}{N_i^{(I,J)\oplus(A,B)}} \cong \frac{N_{i+1}^{(I,J)}}{N_i^{(I,J)}} \oplus \frac{N_{i+1}^{(A,B)}}{N_i^{(A,B)}} \tag{7}$$

As consequence, we conjecture that two locally BLSs are stably equivalent if and only if their feedback invariants lie in the same class in $K_0(R)$.

Therefore, feedback equivalence would be related to the study of some kind of partitions in $\mathbf{P}(R)$, while stable feedback equivalence deals with the study of some kind of partitions in $K_0(R)$.

Groups $K_0(R)$ are described for the case of spheres of any dimension both in the real, complex and quaternion cases (see [12], [14]). Hence, in order to describe all the stable feedback classes for the case of systems over the ring of continuous functions over real or complex spheres, we only need to compute partitions in the following K_0-groups depending on the dimension n of the sphere:

$n(\text{mod } 8)$	1	2	3	4	5	6	7	8
$K_0(\mathcal{C}(\mathbb{S}^n_{\mathbb{R}}))$	$\mathbb{Z} \oplus \mathbb{Z}_2$	$\mathbb{Z} \oplus \mathbb{Z}_2$	$\mathbb{Z}$	$\mathbb{Z} \oplus \mathbb{Z}$	$\mathbb{Z}$	$\mathbb{Z}$	$\mathbb{Z}$	$\mathbb{Z} \oplus \mathbb{Z}$

$n(\text{mod } 2)$	1	2
$K_0(\mathcal{C}(\mathbb{S}^n_{\mathbb{C}}))$	$\mathbb{Z}$	$\mathbb{Z} \oplus \mathbb{Z}$

References

1. A. S. Morse, Ring models for delay-differential systems, *Automatica*, **12** (1976), pp. 529–531.
2. R. Bumby, E. D. Sontag, H. J. Sussmann and W. V. Vasconcelos, Remarks on the pole-shifting problem over rings, *J. of P. and Appl. Alg.*, **20** (1981), pp. 113–127.
3. J. Ferrer, Mª I. García-Planas and F. Puerta, Brunowsky local form of a holomorphic family of pairs of matrices, *LAA*, **253** (1997), pp. 175–198.
4. J. W. Brewer, J. W. Bunce and F. S. VanVleck, *Linear systems over commutative rings* (Marcel Dekker, New York, 1986).
5. P. A. Brunovsky, A classification of linear controllable systems, *Kybernetika*, **3** (1970), pp. 173–187.
6. J. A. Hermida-Alonso, P. Pérez and T. Sánchez-Giralda, Brunovsky's canonical form for linear dynamical systems over commutative rings, *LAA*, **233** (1996), pp. 131–147.
7. M. V. Carriegos, On the local-global decomposition of linear control systems, *Communications in Nonlinear Sci. and Num. Sim.*, **9** (2003), pp. 149–156.
8. M. V. Carriegos and T. Sánchez-Giralda, Canonical forms for linear dynamical systems over commutative rings: The local case, in *Ring Theory and Algebraic Geometry* (Marcel Deker, New York, 2001), pp. 113–131.
9. R. G. Swan, Vector Bundles and Projective Modules, *Transactions of The Amererican Mathematical Society*, **105**, 2 (1962), pp. 264–277.
10. D. E. Knuth, *The art of computer programming. Prefascicle 3B: Sections 7.2.1.4-5 "Genetrating all partitions"* (Addison-Wesley, 2004).
11. M. V. Carriegos, A study of control systems from the geometry of ring of scalars, *IPACS Electronic Library* (2007), `http://lib.physcon.ru`.
12. J. Rosenberg, *Algebraic K-theory and its applications*, (Springer, Berlin, 1994).
13. J. A. Hermida-Alonso, M. M. López-Cabeceira and M. T. Trobajo, When are dynamic and static feedback equivalent?, *LAA*, **405** (2005), pp. 74–82.
14. C. A. Weibel, *An Introduction to algebraic K-Theory* (e-book in progress), `http://www.math.rutgers.edu/weibel`.

CONTROLLABILITY OF TIME-INVARIANT SINGULAR LINEAR SYSTEMS

M. I. GARCIA-PLANAS*, S. TARRAGONA and A. DÍAZ

Departament de Matemàtica Aplicada I
Universitat Politècnica de Catalunya
08038 Barcelona, Spain
*E-mail: *maria.isabel.garcia@upc.edu*

We consider triples of matrices (E, A, B), representing singular linear time invariant systems in the form $E\dot{x}(t) = Ax(t) + Bu(t)$, with $E, A \in M_n(C)$ and $B \in M_{n\times m}(C)$, under proportional and derivative feedback.

Structural invariants under equivalence relation characterizing singular linear systems are used to obtain conditions for controllability of the systems.

Keywords: Singular systems, proportional and derivative feedback, controllability.

1. Introduction

We consider linear and time-invariant continuous singular systems of the form

$$E\dot{x}(t) = Ax(t) + Bu(t), \quad x(t_0) = x_0, \tag{1}$$

where $E, A \in M_n(C)$, $B \in M_{n\times m}(C)$, and $\dot{x} = dx/dt$, that we will represent as a triples of matrices (E, A, B), and we will denote by M, the set of this kind of triples.

Singular systems also called descriptor systems, generalized systems or diferential/algebraic systems, are found in engineering systems such as electrical , chemical processing circuit or power systems among others, and they have attracted interest in recent years.

In this paper, we present a collection of structural invariants which we will call *controllability indices of the triple* in terms of ranks of certain matrices associated to the triple, that permit us to give the explicit form of reduced triple, without knowing the transformation matrices reducing the triple. As a corollary, a necessary and sufficient condition for controllability of the triple is deduced also in terms of the rank of a certain matrix.

We recall that the author L. Dai [1] studied the controllability character for singular systems but he does not consider feedback and derivative feedback in the equivalence relation, only considers basis change in the state space, input space and premultiplication the system by invertible matrices.

The problem to obtain structural invariants permitting to conclude conditions for controllability, was largely studied for standard linear systems under several equivalence relations that can be considered ([2], [3], [4] for example).

In the sequel we identify triples of matrices (E, A, B) with rectangular matrices $\begin{pmatrix} E & A & B \end{pmatrix}$ in order to use matrix expressions.

2. Feedback equivalence

Different useful and interesting equivalence relations between generalized systems have been defined. We deal with the equivalence relation $(E', A', B') = (QEP + QBF_E, QAP + QBF_A, QBR)$. with $P \in Gl(n; C)$, $Q \in Gl(n; C)$, $R \in Gl(m; C)$, $F_A, F_E \in M_{m\times n}(C)$, that is to say the equivalence relation accepting one or more, of the following standard transformations: basis change in the state space, input space, feedback, derivative feedback and premultiplication by an invertible matrix.

Definition 2.1. Two triples (E', A', B') and (E, A, B) in M are called equivalent if, and only if, there exist matrices $P \in Gl(n; C)$, $Q \in Gl(n; C)$, $R \in Gl(m; C)$, $F_E, F_A \in M_{m\times n}(C)$, such that

$$(E', A', B') = (QEP + QBF_E, QAP + QBF_A, QBR),$$

or in a matrix form

$$\begin{pmatrix} E' & A' & B' \end{pmatrix} = Q \begin{pmatrix} E & A & B \end{pmatrix} \begin{pmatrix} P & 0 & 0 \\ 0 & P & 0 \\ F_E & F_A & R \end{pmatrix}$$

It is easy to check that this relation is an equivalence relation.

Making use of the following notations.

- I_n denotes the n-order identity matrix,
- $N = \text{diag}(N_1, \dots, N_\ell)$, with $N_i = \begin{pmatrix} 0 & I_{n_i-1} \\ 0 & 0 \end{pmatrix} \in M_{n_i}(C)$,
- $J = \text{diag}(J_1, \dots, J_t)$, $J_i = \text{diag}(J_{i_1}, \dots, J_{i_s})$, with $J_{i_j} = \lambda_i I + N$,
- $L = \text{diag} = (L_1, \dots, L_q)$, $L_j = \begin{pmatrix} I_{n_j} & 0 \end{pmatrix} \in M_{n_j\times(n_j+1)}(C)$,
- $R = \text{diag}(R_1, \dots, R_p)$, $R_{n_j} = \begin{pmatrix} 0 & I_{n_j} \end{pmatrix} \in M_{n_j\times(n_j+1)}(C)$.

Theorem 2.1 (García-Planas, 2009). *Let (E, A, B) be a triple. Then, it is equivalent to*

$$\left(\begin{pmatrix} E_1 & \\ & S_E \end{pmatrix}, \begin{pmatrix} A_1 & \\ & S_A \end{pmatrix}, \begin{pmatrix} B_1 \\ 0 \end{pmatrix} \right), \tag{2}$$

where (E_1, A_1, B_1) *is a regularizable triple in its Kronecker reduced form (see [5]), concretely*

$$(E_1, A_1, B_1) = \left(\begin{pmatrix} I_1 & & \\ & I_2 & \\ & & N_2 \end{pmatrix}, \begin{pmatrix} N_1 & & \\ & J & \\ & & I_3 \end{pmatrix}, \begin{pmatrix} B_1 \\ 0 \\ 0 \end{pmatrix} \right)$$

The triple (I_1, N_1, B_1), *is a controllable standard system in its Kronecker reduced form,* $(I_2, J, 0)$ *corresponds to the finite zeros of the triple and* J *in its Jordan reduced form,* $(N_2, I_3, 0)$ *corresponds to the infinite zeros of the triple and* N_2 *in its Jordan reduced form. The triple* $(S_E, S_A, 0)$ *is the strictly singular part of the system in its Kronecker reduced form:*

$$\left(\begin{pmatrix} L_1 & & \\ & L_2^t & 0 \end{pmatrix}, \begin{pmatrix} R_1 & & \\ & R_2^t & 0 \end{pmatrix}, \begin{pmatrix} c0 \\ 0 \end{pmatrix} \right)$$

A complete system of invariants to obtain the canonical reduced form can be found in [5].

3. Controllability

We recall that a system is called controllable (see [1]) if, for any $t_1 > 0$, $x(0) \in R^n$ and $w \in R^n$, there exists a control input $u(t)$ such that $x(t_1) = w$.

Equivalently, we have the following theorem:

Theorem 3.1 (Dai, 1989). *A system* $(E, A, B) \in M$ *is controllable if and only if*

$$\begin{aligned} &\mathit{rank} \begin{pmatrix} E & B \end{pmatrix} = n, \\ &\mathit{rank} \begin{pmatrix} sE - A & B \end{pmatrix} = n, \ \textit{for all } s \in C. \end{aligned}$$

First of all, we prove that the controllability is preserved by the equivalence relation considered.

Proposition 3.1. *The controllability character is invariant under equivalence relation considered.*

Proof. Let (E, A, B), (E', A', B') two equivalent triples. So, there exist matrices $Q, P \in Gl(n; C)$, $R \in Gl(m; C)$ and $F_E, F_A \in M_{m \times n}(C)$ such that

$$\begin{pmatrix} E' & A' & B' \end{pmatrix} = Q \begin{pmatrix} E & A & B \end{pmatrix} \begin{pmatrix} P & 0 & 0 \\ 0 & P & 0 \\ F_E & F_A & R \end{pmatrix}.$$

Then

$$\operatorname{rank} \begin{pmatrix} E' & B' \end{pmatrix} = \operatorname{rank} Q \begin{pmatrix} E & B \end{pmatrix} \begin{pmatrix} P & \\ F_E & R \end{pmatrix} = \operatorname{rank} \begin{pmatrix} E & B \end{pmatrix},$$

and

$$\operatorname{rank}\begin{pmatrix} sE' - A' & B' \end{pmatrix} = \operatorname{rank} Q \begin{pmatrix} sE - A & B \end{pmatrix} \begin{pmatrix} & P & \\ sF_E - F_A & & R \end{pmatrix} =$$

$$= \operatorname{rank}\begin{pmatrix} sE - A & B \end{pmatrix}. \qquad \square$$

This proposition permits, if necessary, to take an equivalent triple in its canonical reduced form.

Proposition 3.2. *A necessary condition for controllability is that the system be standardizable. That is to say, the triple is equivalent to* (I, N, B_1).

Proof. It suffices to take a triple (E, A, B) in its reduced form and compute $\operatorname{rank}\begin{pmatrix} E & B \end{pmatrix}$, and $\operatorname{rank}\begin{pmatrix} sE - A & B \end{pmatrix}$. $\square$

We consider the following matrices

$$M_i(C) \in M_{(i+1)n\times(in+2im)}(C).$$

$$M_1 = \begin{pmatrix} E & B & 0 \\ A & 0 & B \end{pmatrix}, \qquad M_2 = \begin{pmatrix} E & B & 0 & 0 & 0 & 0 \\ A & 0 & B & E & B & 0 \\ 0 & 0 & 0 & A & 0 & B \end{pmatrix}, \qquad \dots$$

$$M_\ell = \begin{pmatrix} E & B & 0 & 0 & 0 & 0 & 0 & 0 & 0 & \\ A & 0 & B & E & B & 0 & 0 & 0 & 0 & \\ 0 & 0 & 0 & A & 0 & B & E & B & 0 & \\ 0 & 0 & 0 & 0 & 0 & 0 & A & 0 & B & \\ & & & & & & & & & \ddots \end{pmatrix}.$$

Definition 3.1. We consider the following numbers $r = (r_1, \dots, r_\ell, \dots)$, where $r_i = \operatorname{rank} M_i$, $\forall i = 1, 2, \dots.$

Proposition 3.3. *In the set M of singular systems, the r_i numbers are invariant under the equivalence relation considered.*

Proof. Let (E, A, B), (E', A', B') be two equivalent triples in M, then, there exist matrices $P, Q \in Gl(n; C)$, $R \in Gl(m; C)$, $F_E, F_A \in M_{m\times n}(C)$ such that

$$\begin{pmatrix} E' & A' & B' \end{pmatrix} = Q \begin{pmatrix} E & A & B \end{pmatrix} \begin{pmatrix} P & 0 & 0 \\ 0 & P & 0 \\ F_E & F_A & R \end{pmatrix}$$

So,

$$\begin{aligned} {r_1}' &= \operatorname{rank}\begin{pmatrix} E' & B' & 0 \\ A' & 0 & B' \end{pmatrix} = \operatorname{rank}\begin{pmatrix} Q & 0 \\ 0 & Q \end{pmatrix}\begin{pmatrix} E & B & 0 \\ A & 0 & B \end{pmatrix}\begin{pmatrix} P & 0 & 0 \\ F_E & R & 0 \\ F_A & 0 & R \end{pmatrix} = \\ &= \operatorname{rank}\begin{pmatrix} E & B & 0 \\ A & 0 & B \end{pmatrix} = r_1. \end{aligned}$$

Calling $\mathbf{Q} = \begin{pmatrix} Q & & \\ & \ddots & \\ & & Q \end{pmatrix}$ and $\mathbf{P} = \begin{pmatrix} \begin{smallmatrix} P & 0 & 0 \\ F_E & R & 0 \\ F_A & 0 & R \end{smallmatrix} & & \\ & \begin{smallmatrix} P & 0 & 0 \\ F_E & R & 0 \\ F_A & 0 & R \end{smallmatrix} & \\ & & \ddots \end{pmatrix}$

$$r_\ell{}' = \operatorname{rank} \begin{pmatrix} \begin{smallmatrix} E' & B' & 0 & 0 & 0 & 0 \\ A' & 0 & B' & E' & B' & 0 \\ 0 & 0 & 0 & A' & 0 & B' \end{smallmatrix} & \\ & \ddots \end{pmatrix} = \operatorname{rank} = \mathbf{Q} \left(\begin{smallmatrix} E & B & 0 \\ A & 0 & B \end{smallmatrix} \right) \mathbf{P} =$$

$$= \operatorname{rank} \begin{pmatrix} \begin{smallmatrix} E & B & 0 & 0 & 0 & 0 \\ A & 0 & B & E & B & 0 \\ 0 & 0 & 0 & A & 0 & B \end{smallmatrix} & \\ & \ddots \end{pmatrix} = r_\ell .$$

□

Theorem 3.2. *A triple (E, A, B) is controllable if and only if $r_{n-1} = n^2$.*

Proof.

Suppose now that the triple is controllable, taking into account proposition 3.2, the triple is equivalent to an standard one in the form (I, N, B_1). Then, computing r_{n-1} in this equivalent reduced form, we obtain

$$r_{n-1} = (n-1)n + \operatorname{rank} \begin{pmatrix} B_1 & NB_1 & \dots & N^{n-1}B_1 \end{pmatrix}.$$

We observe that $r_{n-1} = n^2$, if and only if, (N, B_1) is controllable.

Conversely, suppose $r_{n-1} = n^2$, we have $\operatorname{rank} \begin{pmatrix} E & B \end{pmatrix} = n$, because $\begin{pmatrix} E & B \end{pmatrix}$ correspond to the first n-row block matrix in the $n^2 \times ((n-1)n + 2(n-1)m)$ matrix $\begin{pmatrix} \begin{smallmatrix} E & B & 0 & 0 & 0 & 0 \\ A & 0 & B & E & B & 0 \\ 0 & 0 & 0 & A & 0 & B \end{smallmatrix} & \\ & \ddots \end{pmatrix}$ having full rank. So the triple is standardizable, and we can compute r_{n-1} using the reduced form (I_n, A_1, B_1). Finally, it suffices to observe that the standardizable triple (E, A, B) is controllable, if and only if, (A_1, B_1) is controllable:

$$\begin{aligned} &\operatorname{rank} \begin{pmatrix} E & B \end{pmatrix} = \operatorname{rank} \begin{pmatrix} I_n & B_1 \end{pmatrix} = n, \\ &\operatorname{rank} \begin{pmatrix} sE - A & B \end{pmatrix} = \operatorname{rank} \begin{pmatrix} sI_n - A_1 & B_1 \end{pmatrix} = n, \ \forall s \in C. \end{aligned}$$

□

Finally, we define a collection of numbers that permit us to deduce the controllability indices of a controllable triple.

We call $r_0 = \operatorname{rank} B$, and we define the ρ-numbers in the following manner.

Definition 3.2.

$$\rho_0 = r_0, \ \rho_1 = r_1 - r_0 - n, \ \rho_2 = r_2 - r_1 - n, \ \dots, \ \rho_s = r_{s-1} - r_s - n.$$

It is obvious the following proposition.

Proposition 3.4. *The ρ-numbers are invariant under equivalence relation considered.*

Proposition 3.5. *The controllability indices $[k_1, \dots, k_p]$, of a controllable triple, are the conjugate partition of $[\rho_0, \rho_1, \dots, \rho_s]$.*

Proof. It suffices to compute the ρ-numbers of a equivalent reduced form to the controllable triple. □

We observe that if the (E, A, B) is controllable then $k_1 + \ldots + k_p = n$ and $p = \rho_0 = \operatorname{rank} B$.

Theorem 3.3. *Let (E, A, B) be a controllable triple with controllability indices $[k_1, \dots, k_p]$. Then the triple can be reduced to (I_n, A_1, B_1) with*

$$A_1 = \begin{pmatrix} N_1 & & \\ & \ddots & \\ & & N_p \end{pmatrix}, B_1 = \begin{pmatrix} B_1^1 & & \\ & \ddots & \\ & & B_p^1 \end{pmatrix}$$

and $N_i = \begin{pmatrix} 0 & 1 & & 0 \\ & \ddots & \ddots & \\ & & & 1 \\ & & & 0 \end{pmatrix} \in M_{k_i}(C), B_i^1 = \begin{pmatrix} 0 \\ \vdots \\ 0 \\ 1 \end{pmatrix} \in M_{k_i \times 1}(C).$

Proof. It suffices to observe that the controllability indices of the triple (E, A, B) coincide with Kronecker indices of the pair (A_1, B_1). □

References

1. L. Dai, *Singular Control Systems*, (Springer Verlag, New York, 1989).
2. M. I. García-Planas, M.D. Magret, An alternative System of Structural Invariants for Quadruples of Matrices, *Linear Algebra and its Applications* **291**, (1-3) (1999), pp. 83-102.
3. A.S. Morse, Structural invariants of linear multivariable systems, *SIAM J. Contr.* **11**, (1973), pp. 446-465.
4. J. S. Thorp, The Singular Pencil of Linear Dynamical System, *System. Int. J. Control*, **18**, (3) (1973), pp. 557-596.
5. M. I. García-Planas, A Complete system of structural invariants for singular systems under proportional and derivative feedback. *Int. J. Contemp. Math. Scinces.* **4**, (21) (2009) pp. 1049-1057.

PART E

Localization of oscillations in dynamical systems and control of oscillatory delayed-coupled networks

CONTROL OF OSCILLATIONS IN MANUFACTURING NETWORKS

ALEXANDER Y. POGROMSKY

Eindhoven University of Technology, Eindhoven, The Netherlands
E-mail: a.pogromsky@tue.nl
http://www.tue.nl

BORIS ANDRIEVSKY

Institute for Problems of Mechanical Engineering of Russian Academy of Sciences
Saint Petersburg, Russia
E-mail: boris.andrievsky@gmail.com
http://www.ipme.ru

JACOBUS E. ROODA

Eindhoven University of Technology
Eindhoven, The Netherlands
E-mail: j.e.rooda@tue.nl
http://www.tue.nl

The paper is devoted to the supression of the manufacturing network oscillation, induced by the combined influence of control saturation, input signal fluctuation and the presence of the integral component in the control law. A feedback-feedforward observer-based control strategy is proposed, significantly reducing the magnitude of oscillation.

Keywords: Oscillations control, manufacturing systems.

1. Introduction

The production control of manufacturing systems, i.e. how to control the production rates of machines such that the system tracks a certain customer demand while keeping a low inventory level, has been a field of interest for several decades. Simple discrete-event manufacturing systems can be controlled by policies such as PUSH, CONWIP or Kanban (see e.g. [1]). However, as manufacturing systems become more complex, these policies become less effective. A more structured approach for the control of manufacturing systems was proposed in the 1980's, i.e. supervisory control theory in [2], which is also based on a discrete-event description of the manufacturing system. A disadvantage of this approach, however, is that if it comes to the control of large manufacturing systems (or networks of such systems), supervisory control is not very suitable due to the high level of detail they deal with, which causes the corresponding control problem to grow intractably large.

Therefore, there is need for a simple, straightforward control strategy for manufacturing systems, that does not rely on predictions of the future demand. In [3,4] such a strategy is derived by using feedback control of continuous systems. A simple PI controller is used to set the production rate in the ODE model of a manufacturing system such that the production meets a certain demand. The combination of control action saturation and the integrator in the PI controller leads to a phenomenon called *integrator windup*. When the actuator saturates, the effective control signal cannot exceed some value, which affects the system behavior and therefore again the control signal. As a result of this, the closed-loop performance of the system can deteriorate, and in some situations the system can even become unstable. By adding a so-called anti-windup controller to the system, this loss of performance can be counteracted by "turning off" the integrator in the controller when the machine saturates. In this paper and a different solution to this problem is provided, based on employing the observer in the framework of feedforward-feedback control strategy of [5,6]. The paper is organized as follows. The problem statement and the continuous-time representation of a line of manufacturing machines are given in Sec. 2. Decentralized control strategy for a manufacturing line is proposed in Sec. 3. The numerical example is given in Sec. 4, where the simulation results in continuous domain and discrete-event representation are given. Concluding remarks and the future work intentions are presented in Sec. 5.

2. Problem statement

Consider a line of N manufacturing machines $\mathbf{M}_1$, $\mathbf{M}_2$, $\ldots, \mathbf{M}_N$, which are separated by buffers $\mathbf{B}_{j-1,j}$, $j = 1, \ldots, N$ with infinite capacity, see Fig. 1. The first machine $\mathbf{M}_1$ is supplied by raw material, the Nth machine $\mathbf{M}_N$ produces finished product. Each machine $\mathbf{M}_j$ takes out a raw product from the corresponding input buffer $\mathbf{B}_{j-1,j}$ and puts a processed product to the output buffer $\mathbf{B}_{j,j+1}$. In what follows suppose that there is always sufficient raw material to feed the first machine, i.e. that the buffer $\mathbf{B}_{0,1}$ is never exhausted.

Following [4–7], at the stage of the control law design a *continuous approximation* of the discrete-event manufacturing machine is used. Assume that a manufacturing machine produces items continuously in time $t \in \mathbb{R}$ with a certain *production rate* $u_j(t) \in \mathbb{R}$, where $j = 1, \ldots N$ is a number

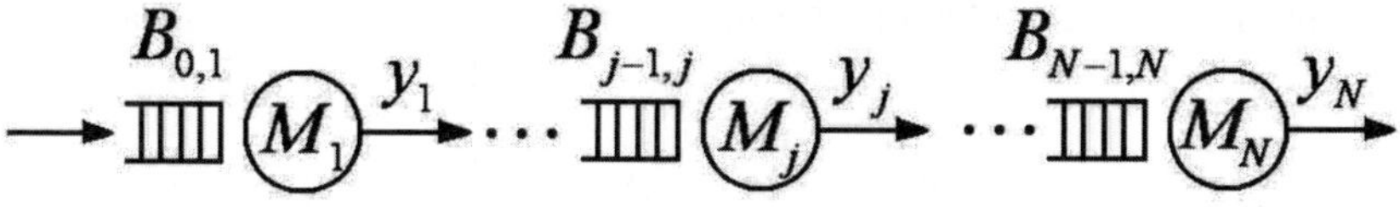

Fig. 1. Schematics of a line of N manufacturing machines. $\mathbf{M}_j$ – machines, $\mathbf{B}_{i,j}$ – buffers, $i, j = 1, \ldots, N$.

of the machine. The total amount of items produced by jth machine is described by a continuous variable $y_j(t) \in \mathbb{R}$. Interaction between the machines is described by the *buffer content* variables $w_j(t) = \max(y_{j-1} - y_j, 0)$, $j = 2, \ldots, N$. The case of $w_j(t) = 0$ means absence of the row material in the input buffer of jth machine and, therefore, the machine $\mathbf{M}_j$ work is suspended. The above reasons lead to the following continuous model of the manufacturing machine:

$$\dot{y}_j(t) = \begin{cases} u_j(t), & \text{if } w_j(t) > 0, \\ 0, & \text{otherwise,} \end{cases} \tag{1}$$

where $t \in \mathbb{R}$ stands for continuous time argument; $j = 1, \ldots, N$ is a machine number. The production rates u_j are bounded by $u_{\max}$ due to machine capacity limitation. In the sequel we assume, without loss of generality, that all the machine capacities in the line have the same upper bound $u_{\max}$. Since the production rates u_j can not also be negative, the following bounds are valid for $u_j(t)$:

$$0 \leq u_j(t) \leq u_{\max}, \; j = 1, \ldots, N, \; \forall t \geq 0. \tag{2}$$

Inequalities (1) lead to a *saturation effect* in the system. This effect restricts the production rate, and complicates design of the controller and the system performance analysis.

Summarizing, we obtain the following manufacturing line model

$$\begin{cases} \dot{y}_1(t) = u_1(t), \\ \dot{y}_2(t) = u_2(t) \cdot \operatorname{sgn}(w_2(t)), \\ \cdots\cdots \\ \dot{y}_N(t) = u_N(t) \cdot \operatorname{sgn}(w_N(t)), \end{cases} \tag{3}$$

where $\operatorname{sgn}(z) = \big(1, \text{ if } z > 0 \mid 0, \text{ otherwise}\big)$.

The control aim is tracking the non-decreasing *reference production* variable $y_d(t)$. Since the finished product of the manufacturing line is an output of the Nth machine (see Fig. 1), the system accuracy is expressed in terms of the reference error $e(t) \equiv e_N(t) = y_d(t) - y_N(t)$. Let us represent the demand $y_d(t)$ as a sum of a linear on time t function and a casual term as follows

$$y_d(t) = y_{d,0} + v_d\, t + \varphi(t), \tag{4}$$

where $y_{d,0}$ denotes the bias in the production demand, v_d is a constant, representing the *average desired production rate*, $\varphi(t)$ is a bounded function, describing fluctuation of the production demand from the linear trend $y_{d,0} + v_d t$. This fluctuation may be caused by market seasonal variations, for example. Suppose that $\varphi(t)$ has a "zero mean" in a some sense because its averaged value may be referred to $y_{d,0}$.

3. Control strategy

3.1. *Wind-up effect for the case of PI-control and input saturation*

Since the demand (2) has a part $v_d t$ that is linear, it can be argued by means of the final value theorem from linear control theory that for $\varphi(t) \equiv 0$ a controller with integral action should be used to track the error $e(t) = y_d(t) - y(t)$ to zero. The simplest controller with integral action is a PI controller for which the controller output at time t is given by:

$$u(t) = k_P e(t) + k_I \int_0^t e(\tau) d\tau, \tag{5}$$

with k_P and k_I the controller parameters. Using the Routh-Hurwitz stability criterion, it can be concluded that the closed loop system is stable iff k_P and $k_I 0$ are both positive. A specific choice of these parameters has to be made based on *performance* criteria, for instance certain demands for the sensitivity and complementary sensitivity functions.

To demonstrate the windup effect let us consider the following numerical example.

Let the single manufacturing machine be modeled by (1), the control signal is bounded by $u_{\max} = 1$, the demand y_d is given by (2) and has the following parameters: $y_{d,0} = 0$, $v_d = 0.75$. Fluctuation signal $\varphi(t)$ in (2) has a harmonic form, $\varphi(t) = \varphi_0 \sin(\omega t)$, where $\varphi_0 = 2.5$, $\omega = 0.2$ s^{-1}. Let us apply the pole placement technique to find the PI-controller (5) parameters for the case of non-saturated (linear) system. It may be easily checked that choice of the gains $k_I = 9$ S^{-1}, $k_P = 5$ of PI-controller (5) ensures the Butterworth distribution of the closed-loop system eigenvalues $s_{1,2}$ as $s_{1,2} = -2.5 \pm 1.66i$, $|s_{1,2}| = 3$ s^{-1}. In the absence of the control signal saturation, the close-loop system has a trancient time about one second and, in the steady-state mode, the error signal magnitude $|e(t)|_{\max} \approx 0.011$.

The system behavior is dramatically changed due to saturation in control, as it is evident from Fig. 2, where the simulation results for the considered saturated system are depicted. The tracking error $e(t)$ in this case

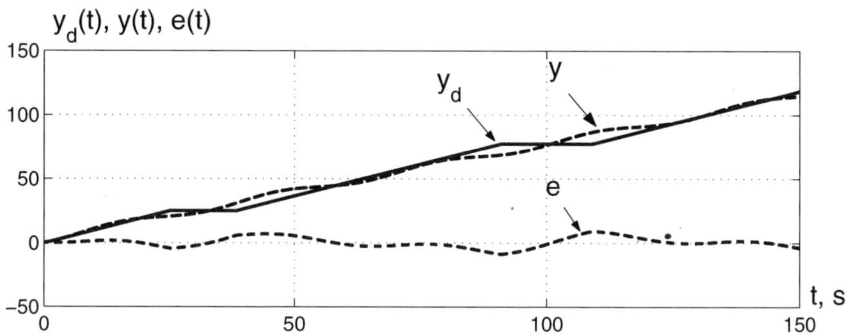

Fig. 2. Windup effect in the manufacturing process; $u_{\max} = 1$, $y_d(t) = 0.75t + 2.5\sin(0.2t)$, $k_I = 9\ \mathrm{s}^{-1}$, $k_P = 5$.

has a form of irregular oscillations of the magnitude about 10.

The windup-caused oscillations may be reduced by means of the anti-windup control, see e.g. [8] for details. The PI-controller with an anti-windup control strategy for a single manufacturing machine is proposed and thoughtfully studied in [4,7]. This controller ensures asymptotically vanishing tracking error $e(t)$ for constant $\varphi(t)$ and, also, independence of the asymptotic system behavior of the initial conditions if fluctuations and disturbances take effect on the system (the so called "*convergence property*".[a]

Supression of system oscillation may be also ensured by means of the observer-based control strategy proposed in [5,6]. Controller of [5,6] implements a proportional (P-) control law with an estimator of the average desired production rate. Absence of the integral component makes possible to avoid the anti-windup compensator of [4,7] in the controller.

[a] Recall that this property means that the system, being excited by a bounded input, have a unique bounded globally asymptotically stable steady-state solution, see [9,10] for details.

3.2. *Observer-based feedback controller for a single machine*

To start, let us recall the observer-based feedback control strategy of [5,6] for a single machine.

Assuming that $y(t)$, v_d may be measured and used to form the control signal $u(t)$, the following feedforward-feedback control law may be used

$$u(t) = \mathrm{sat}_{[0,u_{\max}]}\big(k_p e(t) + v_d\big), \tag{6}$$

where $e(t) = y_d(t) - y(t)$ denotes the tracking error, k_p is the controller parameter (a *proportional gain*), $\mathrm{sat}(\cdot)$ denotes the *saturation function* $\mathrm{sat}_{[a,b]}(z) = \min\big(b, \max(a,z)\big)$. It may be easily seen that, in the absence of control saturation, the control strategy (3) leads to asymptotically vanishing error $e(t)$ for linear on t demand $y_d(t)$. Assuming that only the error signal $e(t)$ can be measured and used to form the control action, in [5] was proposed to replace v_d by its estimate $\hat{r}(t)$, provided by the *observer*, which employs only available signals $e(t) = y_d(t) - y(t)$ and $u(t)$. Luenberger's design method [11] leads to the following *reduced-order observer*

$$\begin{cases} \dot{\sigma}(t) = -\lambda\sigma(t) - \lambda^2 e(t) + \lambda u(t) \\ \hat{r}(t) = \sigma(t) + \lambda e(t), \end{cases} \tag{7}$$

where $\lambda > 0$ is the observer parameter (observer gain), setting the transient time for the estimation procedure.

Finally, the control action $u(t)$ takes the form

$$u(t) = \mathrm{sat}_{[0,u_{\max}]}\big(k_P e(t) + \hat{r}(t)\big), \tag{8}$$

where $e(t) = y_d(t) - y(t)$, $\hat{r}(t)$ is governed by (7). Equations (7), (8) describe the first-order feedback controller. The control signal $u(t)$ is calculated based on the error $e(t)$ measurement only. The gains $k_P > 0$ and $\lambda > 0$ are the controller parameters.

Let us continue the above given example. Choice the control gain $k_P = 5$ and $\lambda = 25\ \mathrm{s}^{-1}$ in (7), (8) lead to the error signal shown in Fig. 3 (solid line), where for the sake of comparability the error signal of the PI-controlled system (1), (5) is also depicted (dash-dotted line).

The simulation results demonstrate that the oscillations magnitude for the observer-based controller (7), (8) is about five times less than that for the PI-controller (1), (5).

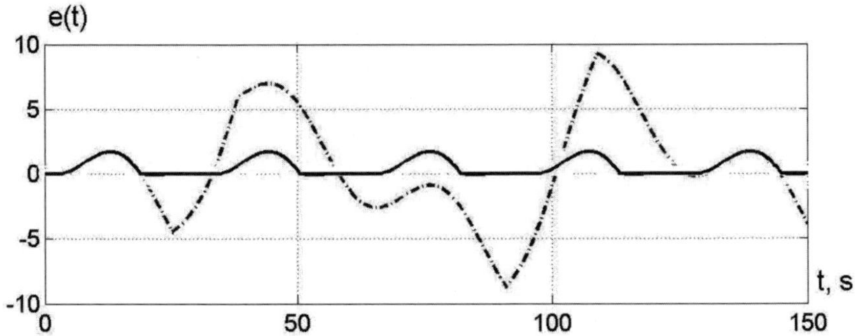

Fig. 3. Tracking error $e(t)$ for the system with the observer-based controller (7), (8) (solid line) and PI-controller (1), (5) (dash-dotted line); $u_{\max} = 1$, $y_d(t) = 0.75t + 2.5\sin(0.2t)$.

3.3. *Control strategy for a line of machines*

The control strategy (7), (8) was extended to control of a line of the manufacturing mashines in [12]. The direct usage of (7), (8) for each machine is rather unpractical, because in this case the buffer contents are not taken into account, which may lead to exhaustion of some buffers or, alternatively, to stacking in buffers an extra amount of material. Besides, from implementation reasons, it is desirable to organize interactions between the neighboring machines only and avoid transferring the reference signal to each machine. Due to these reasons, the following modification of the control strategy (7), (8), intended to control of a manufacturing line has been proposed in [12]. The desirable constant level of the buffer contents $w_d > 0$ was introduced and the "penalty" term $k_w\big(w_d - w_{j+1}(t)\big)$, where $k_w > 0$ is a certain gain (designed parameter) to jth control action $u_j(t)$ was added. The following demand signal for jth machine ensuring equality $y_{j-1}(t) = y_j(t) + w_d$ in the steady-state nominal regime was used. This leads to the following control strategy for the line of manufacturing machines.

Take the control law for Nth machine in the form (7), (8), namely let the control signal $u_N(t)$ be calculated as

$$\begin{cases} u_N = \mathrm{sat}_{[0,u_{\max}]}\big(k_p\varepsilon_N + \hat{r}_N\big), \\ \dot{\sigma}_N(t) = -\lambda\sigma_N(t) - \lambda^2 e(t) + \lambda u_N(t), \\ \hat{r}_N(t) = \sigma_N(t) + \lambda e(t), \end{cases} \tag{9}$$

where $\varepsilon_N(t) \equiv e(t) = y_d(t) - y_N(t)$ is the reference error. Take the control law for machine $\mathbf{M}_j$, $j=1,\ldots,N-1$ in the following form:

$$\begin{cases} u_j = \mathrm{sat}_{[0,u_{\max}]}\big(k_p\varepsilon_j+\hat{r}_j + k_w(w_d - w_{j+1})\big), \\ \varepsilon_j(t) = w_d + \varepsilon_{j+1}(t) - w_{j+1}(t) \\ \dot{\sigma}_j(t)=-\lambda\sigma_j(t)-\lambda^2\varepsilon_j(t)+\lambda u_j(t), \\ \hat{r}_{N-1}(t)=\sigma_{N-1}(t)+\lambda\varepsilon_{N-1}(t), \end{cases} \tag{10}$$

where $w_{j+1}(t) = y_j(t)-y_{j+1}(t)$; w_d is the buffer contents demand. Formulas (9), (10) recursively specify the distributed controller for a line of $N \geq 2$ manufacturing machines.

4. Numerical example

Consider the manufacturing line from $N = 4$ machines. Let us take the following numerical values of the system parameters [5,6]: $u_{\mathrm{p,max}} = 1.0$, $k_{\mathrm{p}} = 5$, $\lambda = 25$. Choose $k_w = 30$, $w_d = 1$. Consider again the system behavior for the case of $y_d(t)=0.75t+2.5\sin(0.2t)$. The simulation results are plotted in Fig. 4, where the tracking errors $e_j(t)$, $j = 1,\ldots,N$ for the line of $N = 4$ manufacturing machines with the PI-controller (1), (5) and the observer-based controller (7), (8) are depicted.

5. Conclusions

Supression of the manufacturing network oscillation, induced by the combined influence of control saturation, input signal fluctuation and presence of the integral component in the control law is considered. Efficiency of a feedback-feedforward observer-based control strategy in reducing oscillation magnitude for a line of the manufacturing machines is demonstrated.

Future work is aimed to studying the discrete-event implementation of the proposed control (the first attempt is presented in [12]) and in generalisation of the proposed control strategy to manufacturing networks of more general topology. The problems of coping with imprecisions, missing data and delays in the system will be also considered.

Acknowledgments

The work was done when the second author was with the Eindhoven University of Technology.

Partly supported by C4C project, by De Nederlandse Organisatie voor Wetenschappelijk Onderzoek (NWO), ref. # B 69-113 and by the Russian Foundation for Basic Research (RFBR), Proj. # 09-08-00803.

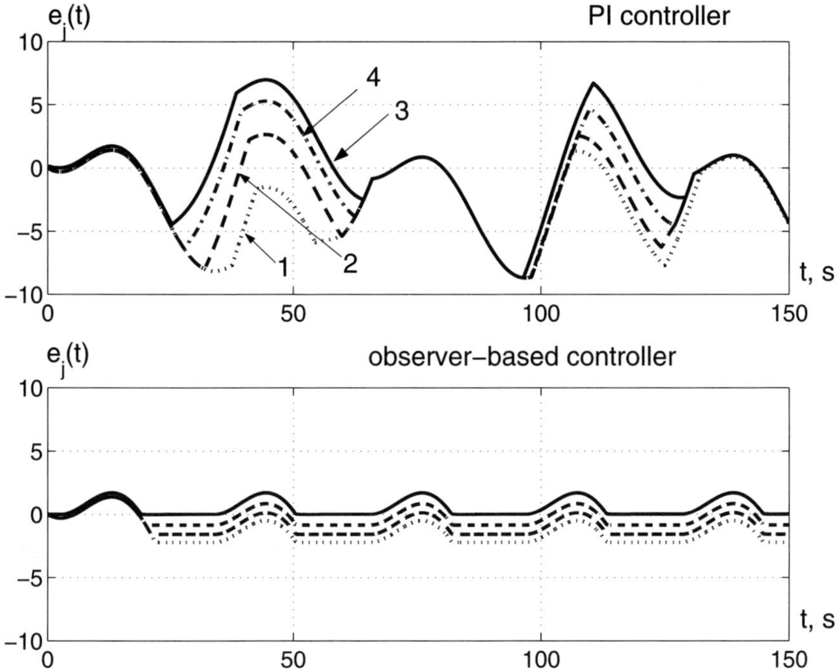

Fig. 4. Tracking errors $e_j(t)$ for the line of $N = 4$ manufacturing machines with the PI-controller (1), (5) and the observer-based controller (7), (8); $u_{\max} = 1$, $y_d(t) = 0.75t + 2.5\sin(0.2t)$, $w_d = 1$. e_1 – dotted line, e_2 – dashed line, e_3 – dash-dotted line, e_4 – solid line.

References

1. W. J. Hopp and M. L. Spearman, *Factory physics*, second edn. (McGraw-Hill, New York, 2000).
2. P. Ramadge and W. Wonham, *SIAM Journal on Control and Optimization* **25**, 206 (1987).
3. W. A. P. Van den Bremer, R. A. Van den Berg, A. Y. Pogromsky and J. E. Rooda, Anti-windup based approach to the control of manufacturing machines, in *Proc. 17th IFAC World Congress*, (Seoul, Korea, 2008).
4. R. A. Van den Berg, Performance analysis of switching systems, PhD thesis, Eindhoven University of Technology, (Eindhoven, 2008).
5. B. Andrievsky, A. Y. Pogromsky and J. E. Rooda, Observer-based production control of manufacturing machines, in *Proc. 13th IFAC Symposium on Information Control Problems in Manufacturing (INCOM09)*, (IFAC, Moscow, Russia, June 3 – 5, 2009).
6. A. G. N. Kommer, A. Y. Pogromsky, B. Andrievsky and J. E. Rooda, Discrete-event implementation of observer-based feedback control of manu-

facturing system, in *Proc. 3rd IEEE Multi-conference on Systems and Control (MSC 2009)*, (IEEE, Saint Petersburg, Russia, July 8 – 10, 2009).

7. R. van den Berg, A. Y. Pogromsky, G. A. Leonov and J. E. Rooda, Design of convergent switched systems, in *Group Coordination and Cooperative Control. (Lecture Notes in Control and Information Sciences, Vol. 336, pp. 291-311)*, eds. K. Y. Pettersen and J. Y. Gravdahl (Springer, Berlin, 2006)
8. P. Hippe, *Windup in Control: Its Effects and Their Prevention* (Springer-Verlag, 2006).
9. A. Pavlov, A. Pogromsky, N. van de Wouw and H. Nijmeijer, *Systems & Control Letters* **52**, 257 (2004).
10. A. Pavlov, N. van de Wouw and H. Nijmeijer, *Uniform Output Regulation of Nonlinear Systems: A Convergent Dynamics Approach* (Birkhäuser, Boston, MA, 2005).
11. D. G. Luenberger, *IEEE Trans. Automat. Contr.* **AC16**, 596(December 1971).
12. A. Y. Pogromsky, B. Andrievsky, A. G. N. Kommer and J. E. Rooda, Decentralized feedback control of a line of manufacturing machines, in *Proc. 35th Annual Conference of the IEEE Industrial Electronics Society (IECON 2009)*, (IEEE, Porto, Portugal, November 3 – 5, 2009).

ANALYSIS AND SYNTHESIS OF CLOCK GENERATOR

E. V. KUDRYASHOVA[1], N. V. KUZNETSOV[1], G. A. LEONOV[2],
P. NEITTAANMÄKI[1] and S. M. SELEDZHI[2]

[1] *Department of Math. Information Technology*
University of Jyväskylä
Finland

[2] *Faculty of Mathematics and Mechanics*
St.Petersburg State University
Russia

This paper is devoted to analysis and synthesis of controllable clock generator. Here it is mathematically rigorously shown that RC-chain can be used as a delay line in the design of controllable clock generators. Bifurcation analysis of one-dimensional discrete model of controllable clock generator is done.

Keywords: Delay line, clock generator, DPLL, bifurcation.

1. Clock generator based on controllable delay line

In clocked circuits it is necessary that the delay was by one tact. For this purpose we need a special setting of parameters of delay lines, which will be described in details. The generators, constructed on logic elements and delay lines, are not high-stable with respect to frequency. Therefore, for their stabilization and synchronization by phase-locked loops it is necessary to introduce a controllable parameter in delay line. A class of such delay lines, the block-scheme of which is shown in Fig. 1, is considered.

We assume that the relation between the input u and the output x is described by the following standard equation of RC-chain

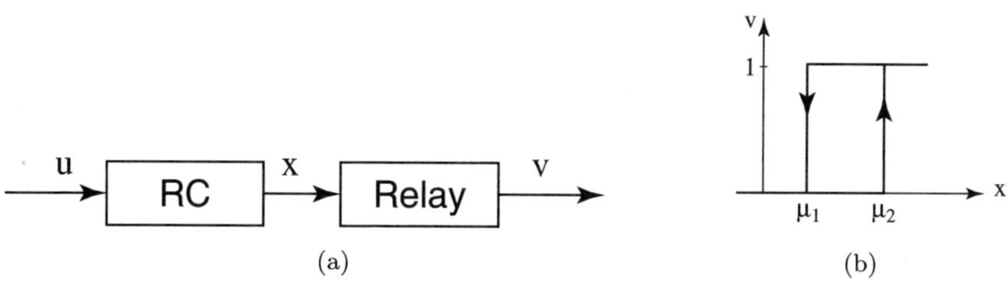

Fig. 1. (a) Delay line: RC-chain and relay. (b) Hysteresis.

$$RC\frac{dx}{dt} + x = u(t), \tag{1}$$

where R is a resistance, C is a circuit capacitance.

The relation between the input x and the output v is described by the graph of "relay with hysteresis" function (see Fig. 1). Here μ_1 and μ_2 are certain numbers from the interval $(0, 1)$. Consider function $u(t)$ which takes the values either 0 or 1.

Further it will be shown that the hysteretic effect is of great importance for synthesis of clock generators. This effect always occurs in real (non-ideal) logic elements. We can show here the analogy with a classical study of Watt's regulator by I.A.Vyshnegradskii [1]. Recall a main conclusion of Vyshnegradskii: "without friction — no regulation". But if a friction "is not sufficient", then it is possible to introduce a special correcting device, dashpot, which provides a stable operation of system. In the case being considered now the friction is replaced by hysteretic effect.

Consider the block-scheme in Fig. 2.

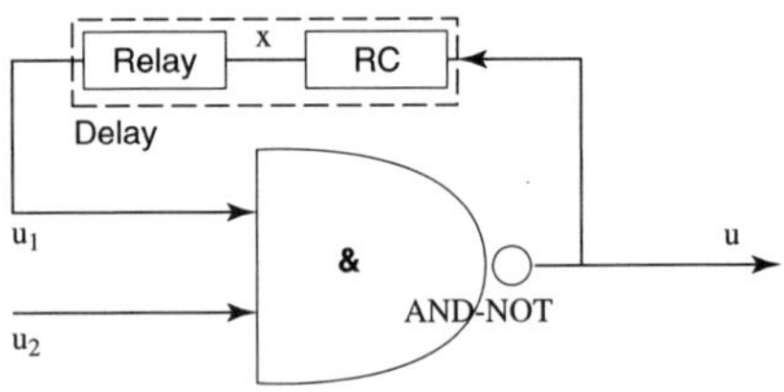

Fig. 2. Clock generator: block AND-NOT and delay line.

Let $u_2(t) = 0$ for $t < T$, $T > 0$. Then $u(t) = 1$ for $t < T$ and at the input there occurs (after a transient process) the signal $x(t) = 1$. Suppose, $x(t) = 1$ on $[0, T]$. Then $u_1(t) = 1$ on $[0, T]$ and the system is in equilibrium:

$$1 = u_1(t) = x(t) = u(t), \quad u_2(t) = 0.$$

Switch-on of clock generator is realized by the change of u_2 from the state 0 to the state 1: $u_2(t) = 1$, $\forall t > T$. Then on the certain interval (T, T_1) we have $u(t) = 0$. This implies that $u_1(t) = 1$ for $t \in (T, T_1)$, where

$$T_1 = T + RC \ln \frac{1}{\mu_1} \tag{2}$$

and $u_1(t) = 0$ on a certain interval (T_1, T_2).

Really, from equation (1) it follows that on (T, T_1) we have $x(t) = e^{-\alpha t}$, $\alpha = 1/RC$. In this case $u_1(t) = 1$ for $t \in (T, T_1)$, where T_1 is from relation (2), and $u_1(t) = 0$ for $t \in (T_1, T_2)$. From the latter relation it should be that $u(t) = 1$ for $t \in (T_1, T_2)$. This implies the following relation

$$T_2 = T_1 + RC \ln \tfrac{1-\mu_1}{1-\mu_2}, \quad x(T_2) = \mu_2.$$

In the case when $\mu_1 = 1 - \mu_2$, $\mu_2 \in (1/2, 1)$, we obtain

$$\tau = T_1 - T_0 = T_2 - T_1 = RC \ln \frac{\mu_2}{1-\mu_2}, \quad T_0 = T + RC \ln \frac{1}{\mu_2},$$

and 2τ-periodic sequence at the output u:

$$u(t) = 0, \quad \forall t \in [T_0, T_0 + \tau); \quad u(t) = 1, \quad \forall t \in [T_0 + \tau, T_0 + 2\tau).$$

Thus, the block-scheme in Fig. 2 is a clock generator with the frequency

$$\omega = \frac{1}{2\tau} = \left(2R \ln \frac{\mu_2}{1-\mu_2}\right)^{-1} C^{-1} \tag{3}$$

Here ω can be compared with the frequency of LC-oscillator: $1/\sqrt{LC}$.

At present it is developed different methods of frequency control of harmonic oscillators by means of a slow (with respect to the high frequency ω) change of parameter C. It is especially widely extended the phase-locked loops [2,3]. In the past decade similar constructions are actively developed and applied to the clock generators.

2. Bifurcations in discrete one-dimensional model of controllable clock generator

Let us consider a control of clock generator by DPLL. Consider a block-scheme of the simplest DPLL with analog input (Fig. 3).

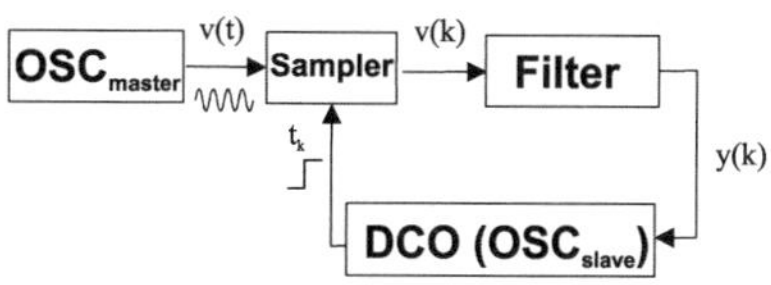

Fig. 3. Functional block-scheme of PLL.

Here the master oscillator OSC_{master} generates the sinusoidal signal

$$v(t) = A \sin(\omega_{in} t + \theta_0), \tag{4}$$

where A is an amplitude, θ_0 is a phase displacement, ω_{in} is a frequency, $\phi(t) = \omega_{in} t + \theta_0$ is a phase of input signal, $T_{in} = \dfrac{2\pi}{\omega_{in}}$ is a period of input signal.

The signal $v(t)$ enters the Sampler and is transformed in discrete signal $v(k) = v(t_k)$ at instants of time $\{t_k\}_{k=1,2...}$, defined by the pulses of controllable oscillator DCO. In the simplest case of filterless system (filter works as a gain) at the input of oscillator DCO we have

$$y(k) = Gv(k), \quad k = 1, 2, ... \tag{5}$$

Let T_{ctrl} be a nominal period of controllable pulse generator DCO. Then in open-loop system the interval of time $T(k)$ between sequential pulses (which occur at instants of time t_k and t_{k+1}) is constant and equal to T_{ctrl}.

Suppose in closed-loop system the interval of time $T(k)$ depends linearly on control signal $y(k)$

$$T(k) = t_{k+1} - t_k = T_{ctrl} - y(k). \tag{6}$$

Expressing t_k from (6), we obtain an equation of a tune, of instant of time t_k and the interval of time $T(k)$ between sequential impulses of controllable oscillator, to the signal $v(t)$

$$\begin{aligned} T(k) &= t_{k+1} - t_k = T_{ctrl} - GAv(t_k), \\ t_k &= t_1 + (k-1)T_{ctrl} - \sum_{i=1}^{k-1} GAv(t_i). \end{aligned} \tag{7}$$

Here arises the question of the existence (attainability in a definite time) and the study of stability of stationary modes of operation of system (7) in the case when after transient processes, beginning with certain $k = K$, the interval between the pulses of oscillator DCO and, accordingly, the control signal will be constant

$$T(k) = t_{k+1} - t_k = T_{ctrl} - y(k) = T_{sp},\ y(k) = GAv(t_k) = y_{sp}\ \forall k \geq K. \tag{8}$$

Let us introduce the denotation for the value of phase of input signal at instants of time of input signal sampling: $\phi(k) = \phi(t_k) = \omega_{in} t_k + \theta_0$ and for the nominal frequency of pulse sampling $\omega_{ctrl} = \frac{2\pi}{T_{ctrl}}$. Consider the case $\omega_{in} = \omega_{ctrl}$. Representing the first equation of (7) in terms of a phase of input signal and denoting by $\sigma(k)$ the value of phase $(\omega_{in} t_k + \theta_0)$ in $[-\pi, \pi]$, we obtain

$$\sigma(k+1) = f(\sigma(k)) = \sigma(k) - r \sin \sigma(k), \tag{9}$$

where r is a positive number.

One of the first works dedicated to analysis of system (9) belongs to Osborne. In [4], was considered the algorithm of investigation of periodic solutions, and it was shown that even in a simple discrete model of PLL, the bifurcation phenomenon of arising of new stable periodical solutions and changing of their period are observed. Later, in [5] for such systems, a model of transition to chaos through a cascade of period-doubling bifurcations was considered. Application and development of these ideas in the works [6] have allowed to construct bifurcation tree of transition to chaos through a cascade of period doubling (Fig. 2(a)). Also for the obtained bifurcation values of system (9) there is shown that, although in a system the violation of cascade of bifurcations of period doubling by the bifurcation of "splitting" of cycle occurs (Fig. 2(b)), in this system the effect of convergence similar to Feigenbaum effect is observed.

It was proved that system (9) is globally asymptotically stable for $r \in (0,2)$ and if $r \in (r_1 = 2, \beta)$, where β is the root of $\sqrt{r^2-1} = \pi + \arccos\frac{1}{r}$ then $\sigma(k) \in [-\pi, \pi]$ for all $k = 1, 2, \ldots$. For $r \in (r_2, r_3)$, where $r_2 = \pi$ and $r_3 = \sqrt{\pi^2+2} \approx 3.445229$, there are two asymptotically stable solutions with period 2. The second bifurcation value $r = r_2 = \pi$ value corresponds to bifurcation of splitting: the globally stable cycle of period 2 loses its stability, and two locally stable cycles of period 2 appear. For $r = r_3 = \sqrt{\pi^2+2} \approx 3.4452$, the third bifurcation occurs: two cycles of period 2 lose stability, and two 4-periodical cycles appear. Further transition to chaos through a cascade of period-doubling bifurcations takes place.

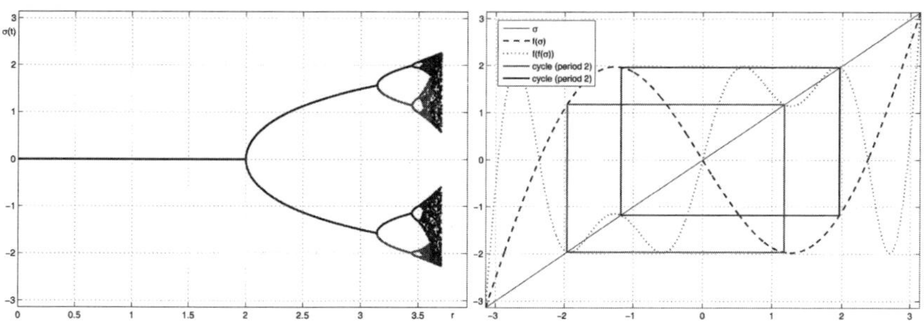

(a) (b)

Fig. 4. (a) Bifurcation tree. (b) Two cycles with period 2 ($r = 3.2$).

Note that the phenomenon of transition to chaos through a cascade of period-doubling bifurcations is well studied for class of maps of an interval into itself. It was found out that the convergence of $\delta_j = \frac{r_j - r_{j-1}}{r_{j+1} - r_j}$ is universal for one-dimensional one-parameter families of maps of an interval into itself. The value $\delta = \lim_{j \to +\infty} \delta_j = 4.6692...$ is the famous Feigenbaum's constant. The renorm-group theory explains this phenomenon for the class of unimodal maps (unique critical point and monotonous behavior on either side of the critical point) and for some special cases.

It is clear that $f(\sigma)$ is not unimodal, so for bifurcation analysis of (9) we apply numerical methods. Here the first 14 calculated bifurcation values of parameter r and Feigenbaum's numbers are shown.

$r_1 = 2$	$r_2 = \pi$	$r_3 = 3.44522922330131$
$r_4 = 3.512892457411257$	$r_5 = 3.527525366711579$	$r_6 = 3.53066537639108$
$r_7 = 3.531338162105000$	$r_8 = 3.531482265584890$	$r_9 = 3.53151312897655$
$r_{10} = 3.531519739097210$	$r_{11} = 3.531521154835959$	$r_{12} = 3.53152145808026$
$r_{13} = 3.531521523045159$	$r_{14} = 3.531521536968802$	

$\delta_2 = 3.759733732581654$	$\delta_3 = 4.487467584214882$	$\delta_4 = 4.62404520668058$
$\delta_5 = 4.660147831971297$	$\delta_6 = 4.667176508904449$	$\delta_7 = 4.66876798830324$
$\delta_8 = 4.669074658227896$	$\delta_9 = 4.669111696537520$	$\delta_{10} = 4.66902573654454$
$\delta_{11} = 4.668640891299296$	$\delta_{12} = 4.667817727564633$	$\delta_{13} = 4.66579740025095$

So in this system the effect of convergence similar to Feigenbaum's effect is observed.

Acknowledgments

This work was partly supported by projects of Ministry of education and science of RF (2.1.1/3889, NK-14P), RFBR and CONACYT.

References

1. G.A. Leonov, *Mathematical problems of control theory.* (World Scientific, Singapore, 2001).
2. G. Leonov, N. Kuznetsov and S. Seledzhi, *Nonlinear Analysis and Design of Phase-Locked Loops*, in the book Robotics, Automation and Control. (I-Tech Education and Publishing, 2009)
3. N.V. Kuznetsov, *Stability and Oscillations of Dynamical Systems: Theory and Applications.* (Jyväskylä Univ. Printing House, 2008).
4. H.C. Osborne, IEEE Transactions on Communications, 28(8) (1980).
5. Belykh V.N., Lebedeva L.V., J. of App. Math. and Mech., 46(5) (1983).
6. G.A. Leonov and S.M. Seledzhi, International Journal of Bifurcation and Chaos, 15(4) (2005).

CHAOS SYNCHRONIZATION WITH TIME-DELAYED COUPLINGS: THREE CONJECTURES

W. KINZEL and A. ENGLERT

Institute for Theoretical Physics
University of Wuerzburg, 97074 Wuerzburg, Germany

I. KANTER

Departement of Physics, Bar-Ilan University
Ramat-Gan, 52900 Israel

Networks of chaotic units with time delayed couplings and feedbacks are investigated analytically and numerically. Based on the results of simple models, three general conjectures are postulated, which need a rigorous proof (or counter example).

Keywords: Chaos synchronization, secure communication, delayed complex system.

Chaos synchronization is a counter intuitive phenomen. On one hand, a chaotic system is unpredictable. Two chaotic systems, starting from almost identical initial states, end in completely different trajectories. On the other hand, two identical chaotic units which are coupled to each other can synchronize to a common trajectory. The system is still chaotic, but after a transient the two chaotic trajectories are locked to each other [1,2]. This phenomenon is attracting a lot of research, since it has the potential to be applied for novel secure communication systems [3–5]. In addition, networks of chaotic units have been realized with electronic circuits and lasers and they are being discussed in the context of neural networks [6,7].
In this contribution, we report some results of our recent work on chaos synchronization with time-delayed couplings. From these results on simple models we postulate some conjectures, and we encourage the reader who is familiar with control theory and dynamical systems to prove these conjectures or to find counter examples.

We consider complete synchronization, only. This means, a network of chaotic units synchroinzes to a common identical trajectory, without any time shift and identical in amplitude and phase. For simplicity, we discuss the phenomena with a network of coupled maps. This model allows analytical and extensive numerical investigations. The dynamic is defined as

$$x_t^j = (1-\epsilon)f(x_{t-1}^j) + \epsilon\kappa f(x_{t-\tau_d}^j) + \epsilon(1-\kappa)\sum_k G_{jk} f(x_{t-\tau_c}^k) \qquad (1)$$

$x_t^j \in [0,1]$ are the dynamic variables, t is the discrete time and $j = 1, ..., N$ is the index of the N nodes. $\epsilon \in [0,1]$ is the strength of all delay terms while $\kappa \in [0,1]$ denotes the relative strength of the self-feedback term. G_{jk} is the coupling matrix and $f(x)$ is a map of the unit interval to itself. τ_c and τ_d are positive integers which model the time delay of the coupling and feedback, respectively.

In the following, we will use the Bernoulli shift $f(x) = (\alpha x)\text{mod}1$ with $\alpha > 1$, which gives a chaotic trajectory. We define the matrix G_{jk} such that the completely synchronized trajectory $x_t^1 = x_t^2 = ... = x_t^N = x_t$ is a solution of Eq.(1), $\sum_k G_{jk} = 0$. The stability of this synchronization manifold is determined by linearizing Eq.(1). With $d_t^j = x_t^j - x_t$ one obtains

$$d_t^j = (1-\epsilon)\alpha d_{t-1}^j + \epsilon\kappa\alpha d_{t-\tau_d}^j + \epsilon(1-\kappa)\alpha \sum_{k\neq j} G_{jk} d_{t-\tau_c}^k \qquad (2)$$

This equation can be analyzed in terms of the eigenvalues of the matrix G. In general, one obtains the master stability function which is analyzed numerically [8], but for the Bernoulli maps Eq.(2) gives a polynomial of degree τ_c (for $\tau_c \geq \tau_d$) which determines the spectrum of τ_c Lyapunov exponents for each eigenmode [9]. Usually, the eigenvalue zero describes perturbation tangential to the synchronization manifold whereas all other eigenvalues determine the transverse Lyapunov exponents.

Consider two units, $N = 2$ and $G_{12} = 1 = G_{21}$, with identical feedback and coupling times $\tau = \tau_c = \tau_d$. In the limit of infinite τ, one finds an analytic result of the phase diagram shown in Fig. 1 (left) [9,10].

In regions I and II the two units synchronize to a common chaotic trajectory $x_t^1 = x_t^2$. After a transient, synchronization occurs without any time shift although the transmitted signal is delayed by the time τ which can be arbitrarily long. Note that without self-feedback, $\kappa = 0$, the two units cannot synchronize. Zero lag synchronization is discussed in the context of public channel cryptography [5,11–14].

Complete synchronization is found in other networks, as well. For example, consider a ring of 4 units with $G_{ij} = G_{ji} = \frac{1}{2}$ for neighbours, as in Fig. 1 (right). The analytic stability analysis yields complete synchronization in region II of Fig. 1. In addition, however, we find a new kind of synchronization. In region III the unit A and C are synchronized and

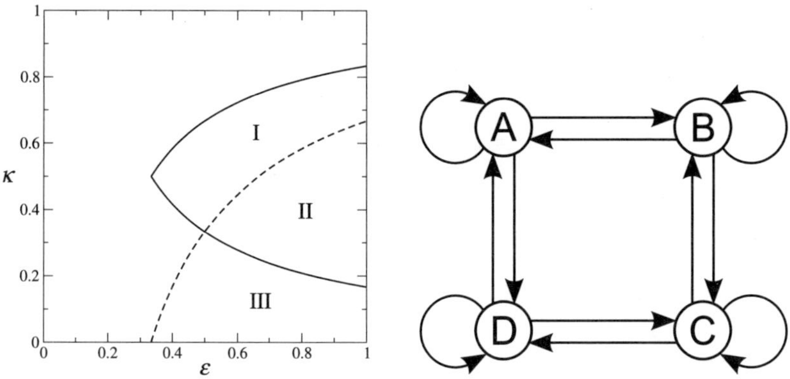

Fig. 1. Left: Phase diagram in the (ϵ, κ)-space (with ϵ and κ as coupling parameters) for the Bernoulli-shift parameter $\alpha = 3/2$ (analytical result). The areas I, II and III refer to different regimes of (sublattice-) synchronization. Right: Ring of four bidirectionally coupled units, each unit with self-feedback.

B and D are synchronized to a different trajectory. One finds sublattice synchronization, the network relaxes to the pattern $\begin{matrix} A & B \\ B & A \end{matrix}$. Note that other patterns are solutions of Eq. (1), as well. In this case, the patterns $\begin{matrix} A & A \\ B & B \end{matrix}$ and $\begin{matrix} A & B \\ A & B \end{matrix}$ are solutions, in agreement with the general classification of [15]. These configurations break the symmetry of the square, and we find that they are unstable [10]. We made similar observations in other networks, as well. Hence we conjecture:

For any networks, stable patterns of synchronized chaotic trajectories do not break the symmetry of the network.

This conjecture is based on some examples, but we do not have a proof. We encourage the reader to find a counter example.

As mentioned in the introduction, chaos synchronization is a counter intuitive phenomenon. A chaotic unit has a irregular unpredictable motion, and it is surprising that a set of these units can be tamed to a common trajectory by coupling them. However, a detailed analysis shows that this is only true when the transmission delay is not too large. We conjecture [16]:

A network of chaotic units cannot be synchronized if the coupling delay times are much larger than the characteristic time scales of the individual units.

This conjecture is based on the stability equation (2). The corresponding master stability equation for any Bernoulli network is

$$\zeta_t = \alpha\zeta_{t-1} + b\alpha\zeta_{t-\tau} \tag{3}$$

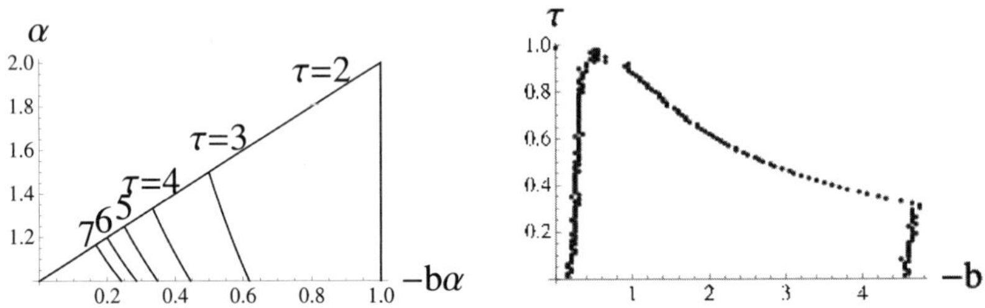

Fig. 2. Left: Regimes of stable complete synchronization. Inside the triangle the master stability function for a chaotic network is negative. α is the parameter of the Bernoulli shift with $\lambda_0 = \ln\alpha$ and b is the rescaled coupling constant. With increasing delay time τ the region of synchronization shrinks. Right: Regime of stable synchronization for a network of chaotic Roessler units. b is the rescaled coupling constant and τ is the delay time of the transmitted signals.

with $\tau_d = \tau_c = \tau$, and b contains the eigenvalues of the coupling matrix, including the feedback term. Fig. 2 shows the analytic result of the regions of synchronzation. The isolated unit without feedback is chaotic for $\alpha > 1$ with Lyapunov exponent $\lambda_0 = \ln\alpha$. The network can only be synchronized if τ is smaller than

$$\tau_{max} = \frac{e^{\lambda_0}}{e^{\lambda_0} - 1} \sim \frac{1}{\lambda_0} \tag{4}$$

For large λ_0, the maximal transmission delay is given by the Lyapunov time $1/\lambda_0$. Fig. 2 shows the corresponding stability regime for any network of Roessler units [16]. Again, synchronization of chaotic units is only possible for small values of coupling delay τ. In this example, τ is even smaller than the Lyapunov time $1/\lambda_0 \simeq 10$. These results are based on simple models. However, we believe that our conjectures is true for any network and for any kind of chaotic units. The stability of the synchronization manifold is determined by linear difference of differential equation, similar to (3) but with time dependent coefficients. The first term, without delay, leads to an exponential divergence. The question is, whether the second term with time delay is able to compensate the local term such that the solution decays to zero. We conjecture that this is only possible if the delay time is short, as in Figs. 2 and 2.

Fig. 3. Two bidirectional coupled units, each with a private filter.

As mentioned before bi-directional coupling may open the possibility to apply chaos synchronization to public channel cryptography [5]. This means, that two partners who want to send a secret message are not allowed to exchange secret keys before the transmission. any attacker who is recording any exchanged signal has complete knowledge about the details of the algorithms and equipments. But we assume that he cannot influence the two partners. For chaos synchronization, this poses the following question: Two chaotic systems A and B synchronize by exchanging signals. An attacker E can record the exchanged signals and has the same esquipment as A and B. Can E synchronize, as well? Recently, we have suggested a configuration, as in Fig. 3, which - to our present understanding - can realize such a public channel synchronization [13,14]. The method is based on several principles:

(1) Each partner selects a private secret filter through which all exchanged signals are transmitted.
(2) The communication is periodically switched on and off, and the filters are changed randomly during the off period.
(3) Integer values and nonlinearities are used for the transmitted signals.

The secret filter may be a convolution with a random kernel, for example for the iterated maps the transmitted signal is defined by

$$T_t = \sum_{s=0}^{N} K_s f(x_{t-s}) \tag{5}$$

The first ingredient ensures that an attacking unit which is driven by the two exchanged signals cannot synchronize. But an attacker E may be able to calculate the private secret filters of A and B. Therefore, the number of equations which E can use is limited by ingredient 2. Ingredient 3 relates this problem to the solution of equations with integers, which is proven to be in the complexity class of NP problems. Hence, we finally conjecture:

It is possible that two chaotic units synchronize whereas a third unit, being driven by the transmitted signals, cannot synchronize.

Reference [14] gives some arguments that this conjecture is true. On the other side, a lot of information is transmitted between the two units. The partner receives enough information to synchronize. An attacker receives the same information as the partner. Thus it is not obvious at all, that the conjectures is true, and we encourage the reader to put this conjecture on a sound mathematical basis.

References

1. L. M. Pecora and T. L. Carroll, **64**, 821 (1990).
2. A. Pikovsky, M. Rosenblum and J. Kurths, *Synchronization, a universal concept in nonlinear sciences* (Cambridge University Press, Cambridge, 2001).
3. A. Argyris, D. Syvridis, L. Larger, V. Annovazzi-Lodi, P. Colet, I. Fischer, J. García-Ojalvo, C. R. Mirasso, L. Pesquera and K. A. Shore, **438**, 343 (2005).
4. L. Kocarev and U. Parlitz, **74**, 5028 (1995).
5. F. review see W. Kinzel and I. Kanter, Secure communication with chaos synchronization, in *Handbook of Chaos Control*, eds. E. Schöll and H. Schuster (Wiley-VCH, Weinheim, 2008) Second edn.
6. S. Boccaletti, J. Kurths, G. Osipov, D. L. Valladares and C. S. Zhou, *Physics Reports* **366**, 1(August 2002).
7. S. Boccaletti, V. Latora, Y. Moreno, M. Chavez and D.-U. Hwang, *Physics Reports* **424**, 175(February 2006).
8. L. M. Pecora and T. L. Carroll, *Phys. Rev. Lett.* **80**, 2109(Mar 1998).
9. I. K. J. Kestler, W. Kinzel, *Phys. Rev. E* **76**, p. 035202 (2007).
10. I. K. W. K. J. Kestler, E. Kopelowitz, **77**, p. 046209 (2008).
11. E. Klein, N. Gross, M. Rosenbluh, W. Kinzel, L. Khaykovich and I. Kanter, **73**, p. 066214 (2006).
12. E. Klein, , R. Mislovaty, I. Kanter and W. Kinzel, **72**, p. 016214 (2005).
13. E. Klein, N. Gross, E. Kopelowitz, M. Rosenbluh, L. Khaykovich, W. Kinzel and I. Kanter, *Phys. Rev. E* **74**, p. 046201 (2006).
14. I. Kanter, E. Kopelowitz and W. Kinzel, *Physical Review Letters* **101**, p. 084102(August 2008).
15. I. S. M. Golubitsky, *Bull. Am. Math. Soc* **43**, p. 305 (2006).
16. W. Kinzel, A. Englert, G. Reents, M. Zigzag and I. Kanter, *Phys. Rev. E* **79**, p. 056207 (2009).

PART F

Microfluidics: theory, methods and applications

MOVING BOUNDARY PROBLEMS FOR THE BGK MODEL OF RAREFIED GAS DYNAMICS

G. RUSSO

Department of Mathematics and Computer Science
University of Catania
Catania, Italy
E-mail: russo@dmi.unict.it

A new semilagrangian method is presented for the numerical solution of the BGK model of the Boltzmann equation in a domain with moving boundary. The method is based on discretization of the equation on a fixed grid in space and velocity. The equation is discretized in characteristic form, and the distribution function is reconstructed at the foot of the characteristics by a third order piecewise Hermite interpolation. Reflecting moving boundary at the piston are suitably described by assigning the value of the distribution function at ghost cells. A comparison with Euler equation of gas dynamics for the piston problem has been performed in the case of small Knudsen number.

Keywords: Rarefied gas dynamics, semilagrangian methods, moving boundary problems.

1. Introduction

This work is motivated by the computation of rarefiled flow in MEMS (Micro Electro Mechanical Systems) [5]. The size of such devices is small enough that gas flow requires a kinetic treatment even at normal pressure and temperature conditions. Micro accelerators are often composed of several elements, each of which consists of a moving part, the *shuttle*, which is free to oscillate inside a fixed part, the *stator*. Although under certain conditions one can obtain an accurate description of the flow by quasi-static approximation [4], more general flow conditions inside the element require the treatment of a domain whose boundaries are not fixed. As a warm up problem, we consider the evolution of a gas in a one dimensional piston. Since we are interested in description of the moving boundary, we choose the simple BGK model to describe the gas [1], which is a simple relaxation approximation of the Boltzmann equation of rarefied gas dynamics. The numerical method that we use is a deterministic semilagrangian method on a fixed grid in space and velocity. Such a method is illustrated in detail in paper [7].

2. Description of the method

First let us assume that the integration domain in space is $[0, L]$, with a fixed L. The initial-boundary problem can be written as

$$\begin{aligned} \frac{\partial f}{\partial t} + v\frac{\partial f}{\partial x} &= \frac{1}{\tau}(M[f] - f), \\ f(t, x, v) &= f_0(x, v) \end{aligned} \tag{1}$$

where $v \in \mathbb{R}$, $x \in [0, L]$, and $t > 0$, and $M[f]$ represents the local Maxwellian that has the same conservative moments of f.

Suppose we want to integrate the equation up to a fixed time $t = t_f$. For simplicity we assume constant time step $\Delta t = t_f/N_t$ and uniform grid in physical and velocity space, with mesh spacing Δx and Δv, respectively, and denote the grid points by $t_n = n\Delta t$, $x_i = i\Delta x, i = 0, \ldots, N_x$, $v_j = j\Delta v, j = -N_v, \ldots, N_v$, where $N_x + 1$ and $2N_v + 1$ are the number of grid nodes in space and velocity, respectively. We assume that the distribution function is negligible for $|v| > v_{\max} = N_v \Delta v$.

Let f_{ij}^n denote the approximation of the solution $f(t_n, x_i, v_j)$ of the problem (1) at time t_n in each spatial and velocity node, and assume that it is given.

Integration of Eq. (1) along the characteritsics by implicit Euler scheme gives

$$\begin{aligned} f_{ij}^{n+1} &= \tilde{f}_{ij}^n + \frac{\Delta t}{\tau}(M_{ij}^{n+1} - f_{ij}^{n+1}), \\ x_i &= \tilde{x}_{ij} + v_j \Delta t, \quad i = 0, \ldots, N_x, \\ j &= -N_v, \ldots, N_v. \end{aligned} \tag{2}$$

The value of the function $\tilde{f}_{ij}^n$ is reconstructed at position $\tilde{x}_{ij} = x_i - v_j \Delta t$ by a suitable high order reconstruction. In particular, here we use a piecewise cubic polynomial, which is obtained by Hermite interpolation in each interval $[x_i, x_{i+1}]$. The first derivatives of the function at location x_i, $(\partial f_j/\partial x)_{x_i}$, are computed by second order central difference. The reconstruction is linear, without limiters. This guarantees that the scheme is conservative [3].

2.1. *Implicit calculation*

The implicit term can be explicitly computed by multiplying Eq. (2) by $1, v, |v|^2$ and summing over the velocities. This procedure allows the computation of the moments, because $M_{i,\cdot}^{n+1}$ and $f_{i,\cdot}^{n+1}$ have the same moments.

Therefore one obtains

$$\rho_i^{n+1} = \sum_j \tilde{f}_{ij}^n, \quad (\rho u)_i^{n+1} = \sum_j v_j \tilde{f}_{ij}^n, \quad E_i^{n+1} = \frac{1}{2}\sum_j |v_j|^2 \tilde{f}_{ij}^n. \tag{3}$$

Once the moments have been computed, the Maxwellian can be calculated from the moments, and the density function can be explicitly computed as

$$f_{ij}^{n+1} = \frac{\tau \tilde{f}_{ij}^n + \Delta t M_{ij}^{n+1}}{\tau + \Delta t}. \tag{4}$$

Notice that as $\tau \to 0$ the distribution function f_{ij}^{n+1} is projected onto the Maxwellian. Furthermore, in this limit the whole scheme becomes a relaxation scheme for the Euler equations. We say that the scheme is *Asymptotic Preserving* [6].

3. The piston problem

The system consists in a gas inside a one dimensional slab, which is driven by a moving piston (see Fig. 1). On the left boundary of the domain there is a fixed wall (the origin of our coordinate system), at the right end there is a piston, whose position is an assigned function of time $x_p : t \in \mathbb{R} \to x_p(t) \in [0, L]$. We assume that the gas inside the slab is governed by the BGK equation. The system is discretized on a uniform grid in the computational domain $[0, L]$ by $N_x + 1$ grid points of coordinates $x_i = ih$, $i = 0, \ldots, N_x$, $h = L/N_x$. As the piston moves, the domain occupied by the gas changes, while the position of the grid points remains fixed. As a consequence, only a certain number $N_x(t)$ of grid points is actually used (*active points*) while other points lie outside of the domain (*ghost points*).

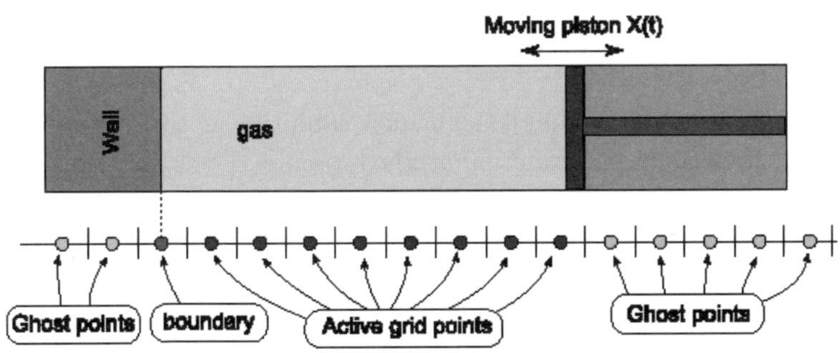

Fig. 1. Setup of the piston problem. The equations are solved for the values of the distribution function in the active grid points. The values outside of the computational domain (*ghost points*) are computed by making use of the boundary conditions.

The number of equations to be solved changes with time. We choose the time step in such a way that the piston can move by at most one grid point in one step, and denote by $u_p(t) \equiv \dot{x}_p(t)$ the assigned piston velocity.

Different boundary conditions may be assigned to the boundary. Here we consider the case of specular reflection.

3.1. *Specular reflection*

At the wall, at each time t, the distribution function, for positive velocities, is given by

$$f(t, 0, v) = f(t, 0, -v),$$

which is discretized as

$$f^n_{-i,j} = f^n_{i,-j}, \quad i \leq 0, j > 0,$$

keeping in mind that $v_j = j\Delta v$.

A similar condition can be used to treat reflecting boundary conditions near the piston:

$$f(t, x_p, v) = f(t, x_p, v^*), \quad v^* = 2u_p - v.$$

We convert the condition into an initial value for the ghost point using the following argument. We approximate the motion of the piston by a piecewise linear function of time, i.e. we assume that in time interval $[t_n, t_{n+1}]$ the velocity of the piston is unchanged. Then the value of the density function $f(t_n, \tilde{x}_{ij}, v_j)$, at the foot of the characteristics corresponding to the velocity $v_j < u_p$, is set to $f^n(x^*, v^*)$, where $x_{ij} + x^* = 2x_p(t_n)$ and $v_j + v^* = 2u_p(t_n)$ (see Fig. 2).

The simplest way to implement such condition is to precompute the values of the distribution function at ghost points $x_i > x_p(t_n)$, for $v_j < u_p$, as $f^n(x_i, v_j) = f^n(x^*, v^*)$, with $x_i + x^* = 2x_p(t_n)$ and $v_j + v^* = 2u_p(t_n)$, and then use the standard piecewise Hermite interpolation from grid points (active or ghost) at time level t_n. In general point (x^*, v^*) is not on a grid in phase space, therefore interpolation in x and v has to be used. In some cases, point (x^*, v^*) is in a cell whose values of the function is known at the vertices, and bilinear interpolation can be used. In other cases, the function at the vertices is itself not known, and an iterative procedure has to be used.

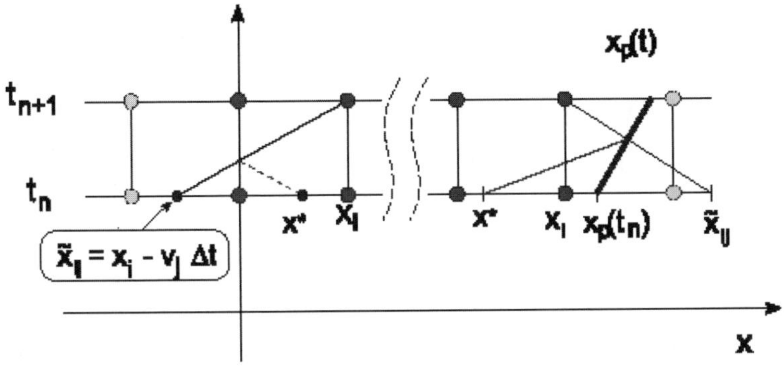

Fig. 2. Definition of the specular boundary conditions at the wall (left) and at the piston (right).

4. Numerical tests

As numerical test we solve the BGK equation with Maxwellian initial condition, and reflecting boundary conditions at the wall and at the piston. We impose the motion of the piston with a given velocity $u_p(t) = 0.25 \sin(t)$. The piston induces waves that move back and forth into the slab. For small Knudsen number the behavior of the gas should be well described by the Euler equations of gas dynamics. To validate this expectation, a comparison is performed between solution of the BGK equation and the solution of the Euler equations of gas dynamics. The latter is obtained by writing the equations in Lagrangian form, so that the domain in Lagrangian coordinates becomes fixed, and then applying a finite volume central scheme to solve the equations numerically (see [2] for details).

The pressure at the piston and at the wall for the BGK model and for the Euler equations are shown in Fig. 3. During the talk, the time evolution of the distribution function $f(x, v, t)$ is shown.

Implementation of Maxwell boundary conditions and extension to two space dimensions will allow a realistic simulation of the oscillation of the shuttle in MEMS.

References

1. P.L. Bhatnagar, E.P. Gross,and M. Krook, A model for collision processes in gases. Small amplitude processes in charged and neutral one-component systems, *Physical Reviews*, **94**, 511–525 (1954).
2. R. Fazio, and G. Russo, A Lagrangian central schemes and second order boundary conditions for 1D interface and piston problems, *Communucation*

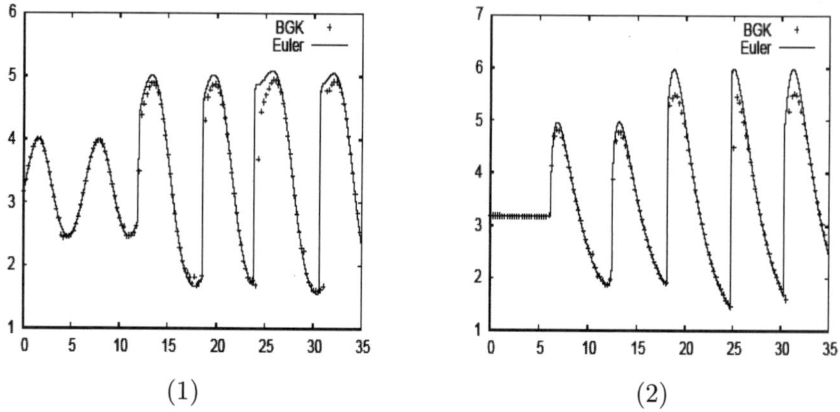

Fig. 3. $\tau = 10^{-3}$: the pressure at the boundary (1) $x = x_p(t)$ and (2) $x = L$ obtained by the semi-Lagrangian method for BGK equations and a Lagrangian scheme for Euler equations.

in Computational Physics, submitted.
3. F. Filbet, and E. Sonnendrucker, Comparison of Eulerian Vlasov Solvers, *Comput. Phys. Communications*, **150** 247–266 (2003).
4. A. Frangi, A. Frezzotti, and S. Lorenzani, On the application of the BGK kinetic model to the analysis of gas-structure interactions in MEMS, *Computers and Structures* **85**, 810–817 (2007).
5. M. Gad-el-Hak, editor, *The MEMS handbook* (CRC Press 2002).
6. S. Jin, Efficient asymptotic-preserving (AP) schemes for some multiscale kinetic equations, *SIAM J. Sci. Comput.* **21**, 441–454 (1999).
7. G. Russo, and F. Filbet, *Semilagrangian schemes applied to moving boundary problems for the BGK model of rarefied gas dynamics*, Kinetic and Related Models, **2**, 231–250 (2009).

EXPERIMENTAL INVESTIGATION ON PARAMETERS FOR THE CONTROL OF DROPLETS DYNAMICS

F. SAPUPPO*, F. SCHEMBRI and M. BUCOLO

Dipartimento di Ingegneria Elettrica Elettronica e dei Sistemi
University of Catania, Catania, 95125, Italy
*E-mail: *fsapuppo@diees.unict.it*

In this work temporal nonlinear dynamics in two-phase microfluidic phenomena are investigated. Time series representing the dynamics of two phase flow in an in vitro microfluidic snake mixer were captured using an ad hoc optical system in order to access fast bubble flow patterns in time domain. This work presents significant relationships between the input control parameters under investigation and nonlinear indicators extracted from experimental time series.

Keywords: Two phase flow, nonlinear time series analysis, micro-mixer, signal processing, temporal chaos.

1. Introduction

The two phase microfluidics is related to emulsions generation as dispersions of micrometric droplets in liquids. Such emulsions are widely used in a number of industrial domains, such as pharmaceutical [1], cosmetic, and food industries. In particular, droplets generated by immiscible fluids, such as oil and water, or bubbles generated as gases in liquids in microfluidic devices, are considered of particular interest because they become popular for generating nanovolume vessels for biochemical assays. Two-phase microfluidics uses discrete fluid packets as carriers to achieve various fluidic functions, bio-chemical reactions, and detections in the microscale, thus differing from continuous flow systems that deal with continuous liquid streams. Droplet manipulations can be performed inside microchannels [1], and such applications are called droplet-based microfluidics where droplets of micrometer size are mainly produced and individually manipulated [2]. Given the complexity of the microfluidic droplets dynamics [3], the usage of nonlinear analysis methods applied to experimental time series representative of the microfluidic processes results appropriate. The aim of this study is to investigate on the nonlinear dynamics of two phase flow in microfluidics and it represents the starting point toward a deeper study on the effect of the input flow rate and frequency to droplets dynamics.

2. Experimental system

A general purpose design of an the experimental non-invasive workbench for microfluidic systems study has been already realized [4]. Such system is made up of three main functional blocks: the microfluidic device, the sensing system and the monitoring-control system.

The microfluidic Device. The serpentine snake mixer channel (SMS0104, Thinxxs) has section of 640 μm and internal and external radius of curvatures of 280 μm and 920 μm. Air and water are pumped through the mixer's y-junction to obtain a two-phase flow (Fig. 1(a)). Piezoelectric twin pumps (TPS1304, Thinxxs) consisting in piezo-driven diaphragm micropumps are here used. They are controlled by an electronic pump control (EDP0704, Thinxxs) actuating fluid motion and input flow rates for both air and water through control frequencies (*fair*, *fwater*)(Figs. 1(b) and 1(c)).

Electro-optical system. The micro-scales of microfluidic devices requires an optical system for the extraction of parameters and variables involved in the microfluidic process. The optical system used here is based on discrete optical components (Fig. 2) and it provides image magnification (M=3X) suitable for acquisition and processing [4].

Photodiode sensing system. Two photodiodes, shown in Fig. 2(b) (SLD-70BG2A, Silonex) have been chosen to capture light variation due to the droplets passage in the microchannel. They are planar photodiodes and measure 3.6 mm x 3.6 mm, with an active area of 9.8 mm^2 and are placed on a axis parallel of the magnified image of the channel (Fig. 2(b)).

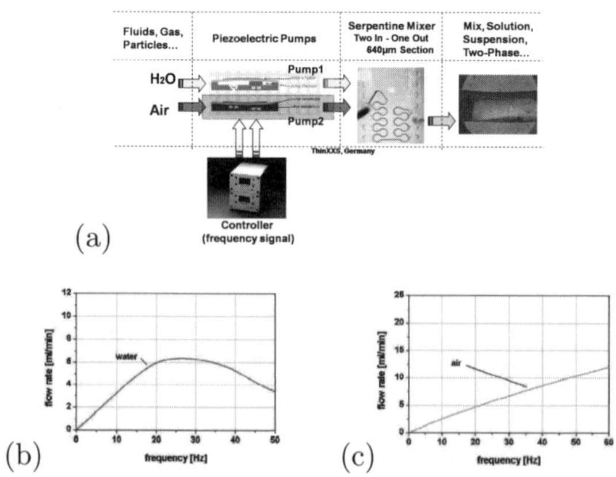

Fig. 1. (a) Microfluidic setup. Flow rate versus pump frequencies for (b) water (c) air.

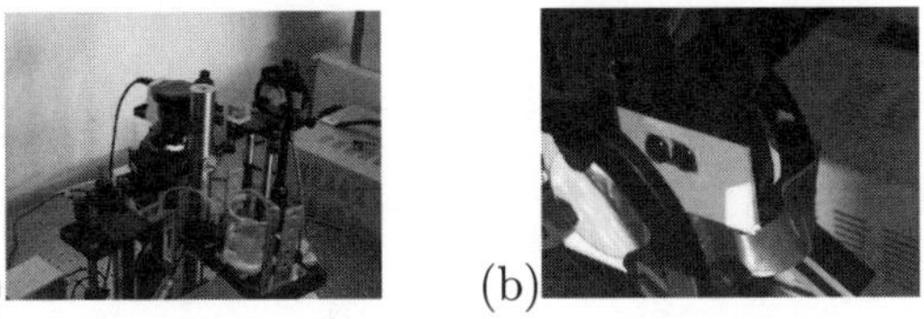

Fig. 2. Electro optical system. (a) Experimental setup. (b) Details of photodiodes.

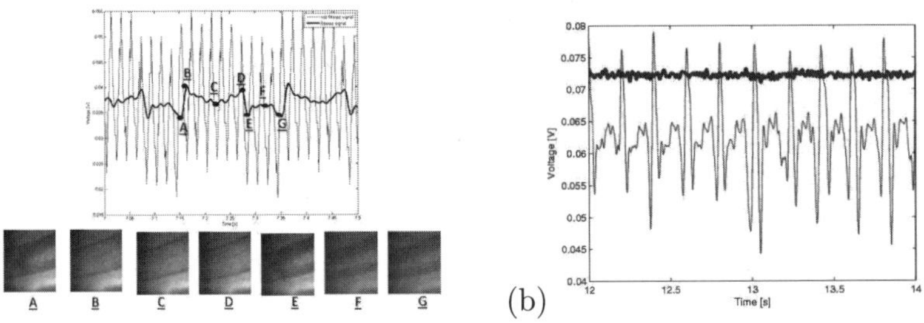

Fig. 3. (a) Filtered (bold line) and not filtered signal (*fwater*=*fair*=5 Hz - grey line) (b) Signal of droplets dynamics (*fwater*=25 Hz, *fair*=5 Hz - thin line) and reference signal of water (*fwater*=10 Hz, *fair*=0 Hz - bold line).

3. Droplets dynamics and signal processing

3.1. *Experimental campaign*

A campaign of 12 experiments has been carried out by varying the control frequencies of the pump control in a range between 5 Hz to 60 Hz with step of 5 Hz for water, with the air frequency fixed at 5 Hz. Such range of frequencies corresponds to a range of input flow rate (0.2-3.5 ml/min for water and a fixed component of 2.3 ml/min for air). During each experiment the pumps frequencies were kept constant for the acquisition time. Moreover a reference experiment (*fwater*=10 Hz, *fair*=0 Hz) was carried out.

3.2. *Photodiode signals and nonlinear time series analysis*

For each experiment, signals representing optical information related to the bubbles passage in the microchannel (Fig. 3(a)) were acquired using a photodiode for a period of 15 s at a sample frequency $Fs=1\ kHz$. Following a spectral analysis all the signals have been filtered with a low pass digital

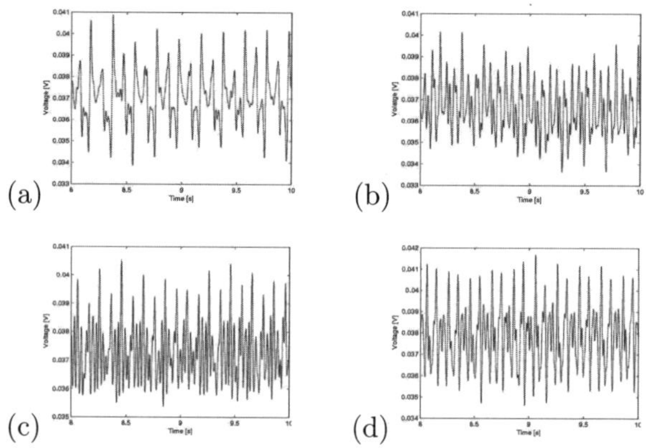

Fig. 4. Signals for *fair*=5 Hz and varying the water frequency: (a) *fwater*=5 Hz (b) *fwater*=10 Hz (c) *fwater*=15 Hz (d) *fwater*=20 Hz.

filter with cut off frequency of 60 Hz and a notch filter at 50 Hz.
Photodiode signals related to four experiments were, at first, compared to the reference signal (Fig. 3(b)), and then analyzed in a time windows of 2s (Fig. 4).
The nonlinear dynamic analysis has been carried out on the time series related to optical information [5] by means of the software TISEAN [6]. The embedding dimension m, time delay τ were used in order to calculate the trajectory divergence curve (d_j) and therefore the largest Lyapunov exponents $\lambda_{\max}$ and d-infinite (d_∞) [7] characterizing and quantifying the dynamics of a nonlinear time series. In particular the divergence curve, defined as d_j, was determined by averaging the distance between N couples of trajectories starting from two nearby points. The $\lambda_{\max}$ can be considered as proportional to the divergence curve initial slope, taking into account stretching phenomena, while the (d_∞), defined as d_j asymptotic value [7] takes into account both trajectories stretching and folding effects.

4. Results and comparison

Results related to nonlinear analysis for 4 experiments (1. *fwater*=5 Hz, 2. *fwater*=10 Hz, 3. *fwater*=15 Hz, 4. *fwater*=15 Hz, 5. *fwater*=20 Hz) are here presented in term of state space reconstruction (Fig. 5) and divergence curve (Fig. 6). In particular the divergence curves d_j (Fig. 6) are sensitive to the changes in the pulsate inputs. The initial slope, proportional to the $\lambda_{\max}$ and the asymptotic value d_∞ allows an intuitive comparison

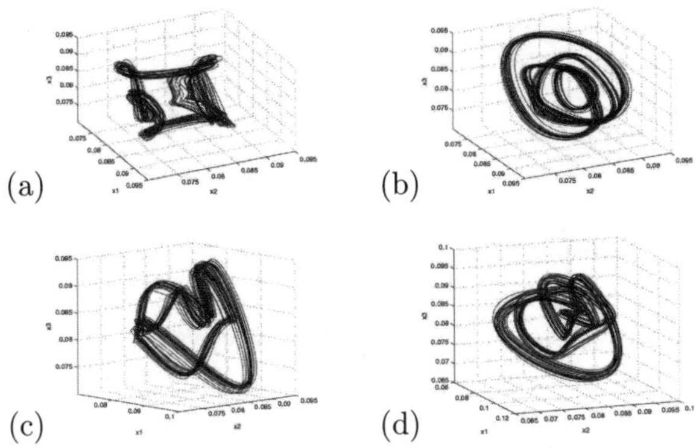

Fig. 5. Reconstructed phase space for *fair*=5 Hz and varying the *fwater* (a) *fwater*=5 Hz (b) *fwater*=10 Hz (c) *fwater*=15 Hz (d) *fwater*=20 Hz.

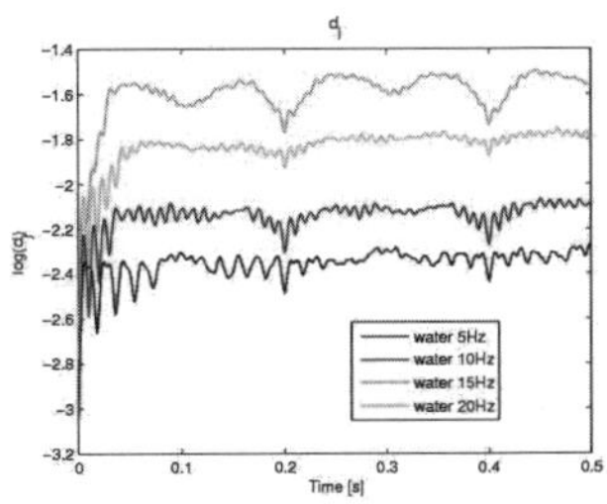

Fig. 6. Logarithmic divergence curve (d_j) for fixed *fair*=5 Hz and *fwater*=5 Hz (black line), *fwater*=10 Hz (red line), *fwater*=15 Hz (green line), *fwater*=20 Hz (cyan line).

among the different dynamics. Finally Fig. 7 shows the relationship of the largest Lyapunov exponent ($\lambda_{\max}$) and d-infinite (d_∞) with the water frequency. The decreasing trend as the carrier fluid (water) frequency increase indicates the reduction in the nonlinear features of the air bubbles flow. Moreover, it is worth noticing that all the time series have a positive $\lambda_{\max}$.

5. Conclusion

Experimental methods were used in this work to extract important two-phase flow (air - water) dynamic features for different input flow conditions (flow rate and frequency). A mathematical nonlinear analysis of experimental time series, obtained as optical information, was then performed varying the input flow rate and frequency of water underlining a relationship

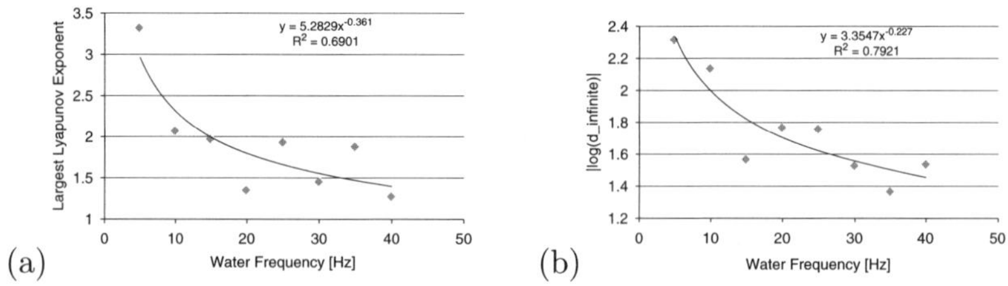

Fig. 7. (a) Largest Lyapunov exponent vs water frequencies, (b) absolute value of d-infinte vs water frequencies.

between the control parameters under investigation and the nonlinear parameters λ_{max} and d_∞. This leads to the concept that such parameters represent a sensitive control for the dynamics of the droplets in microfluidic phenomena.

References

1. M. He, J.S. Edgar, G.D.M., J. R.M. Lorenz., J.P. Shelby and D.T. Chiu, Selective encapsulation of single cells and subcellular organelles into picoliter- and femtoliter-volume droplets *Anal. Chem.*, **77**, pp. 1539-1544, (2005).
2. M. R. Bringer, C. J. Gerdts, H. Song, J. D. Tice and R. F. Ismagilov, Microfluidic systems for chemical kinetics that rely on chaotic mixing in droplets *Phylos. Trans. Royal Society* , **362**, pp. 1087–1104, (2004).
3. M. Bucolo, L. Fortuna, F. Sapuppo, and F. Schembri, Chaotic dynamics in microfluidics experiments. In *Proc. Mathematical Theory of Networks and Systems*, Blacksburg, Virginia, USA, July 28- Aug. 1. pp. 1–12, (2008).
4. F. Sapuppo, M. Bucolo, M. Intaglietta, L. Fortuna and P. Arena, Cellular nonlinear network: real-time technology for the analysis of microfluidic phenomena in blood vessels *Nanotechnology- Institute of Physics Publishing*, **17**, pp. S54–S63, (2006).
5. H. Kantz and T. Schreiber, *Nonlinear Time Series Analysis.* Cambridge University Press, Cambridge, (2004).
6. R. Hegger, H. Kantz and T. Schreiber, Practical implementation of nonlinear time series methods: The TISEAN package *CHAOS*, **9**, pp. 413–435, (1999).
7. A. Bonasera, M. Bucolo, L. Fortuna, M. Frasca and A. Rizzo, A New Characterization of Chaotic Dynamics: the d-infinite Parameter *Non linear Phenomena in Complex Systems*, **6** (3), pp. 779–786, (2003).

RAPID PROTOTYPING OF THIOLENE MICROFLUIDIC CHIPS

G. MISTURA*, D. FERRARO, A. FRANZOI,
D. FRIZZO, A. PACETTI and M. PIERNO

Dipartimento di Fisica G.Galilei and CNISM
Universit di Padova, via Marzolo 8, 35131 Padova, Italy
*E-mail: *mistura@padova.infm.it*

We present some microfluidic devices we have recently fabricated using a rapid prototyping process that relies on the rubber elasticity and adhesive power of partially cured thiolene optical adhesives: i) a chip provided with metal electrodes for electroporation experiments; ii) a device with a heater to study the thermophoresis of small particles.

Keywords: Microfabrication, photolithography, replica molding, micro-particle image velocimetry, microchannel flow, slip length, thermophoresis.

1. Introduction

Multilayer fabrication has recently become an important topic in microfluidics because of the ability to realize increasingly complex channel networks that more effectively can deal with demands posed by applications. As with single layer devices, the material most popular for multilayer microfluidics is polydimethylsiloxane (PDMS) [1], mainly because its easy processing permits rapid prototyping and its elastomeric nature allows fabrication of pneumatic valves and pumps. A drawback of PDMS is its low resistance to organic solvents that tend to swell PDMS, leading to channel obstruction and/or delamination of chips. Liquid photopolymerization of thiolenes [2] and other monomer formulations [3,4] across photolithographic masks has recently emerged as a popular rapid prototyping method to produce microfluidic chips with good solvent resistance useful for chemical microreactors [5]. To produce multilayer devices with these polymers, sacrificial materials have so far been necessary to fill channels in lower levels while fabricating channels in a upper level. Very recently Bartolo *et al.* [6] proposed a process that avoids this complication by photopolymerizing e.g. thiolenes in PDMS molds and then laminating the fabricated patterns to a substrate. However this process requires preliminary fabrication of photoresist "master" and PDMS "molds" for each layer. We have recently developed a much more rapid fabrication process that uses direct photolithographic patterning of liquid thiolene adhesives on thin polyethilene supporting sheets and their subsequent transfer to a substrate by lamination [7]. Such a method has been applied for the realization of a microfluidic bead array sensor for pH measurements [8]. Here we present other thiolene microfluidic devices we have recently fabricated: i) a chip provided with metal electrodes for electroporation experiments; ii) a chip for thermophoresis investigations.

2. Experimental methods

2.1. *Fabrication process*

The general fabrication process can be described as follows: a commercial liquid thiolene optical adhesive (NOA61 from Norland products) is sandwitched between two surfaces, one of which has suitably tailored anti-stick properties, while the other shows good adhesion to the thiolene glue. Typically, we use microscope glass slides as a supporting substrate, while for anti-stick surfaces we employ either a glass slab covered with a thin polyesther sheet and a thin water layer between the two to guarantee adhesion or a glass slide functionalized with an anti-stick self-assembled monolayer. Hereafter, we describe the various steps of our rapid prototyping method.

(i) The two surfaces are kept separated by spacers of known thickness. The adhesive is then selectively photopolymerized across a photomask in such a way that desired parts of the liquid solidify.
(ii) Unpolymerized parts remain liquid and can be easily rinsed away after separating the anti-stick surface. A patterned structure that has an elasticity typical of a rubber results and remains attached to the adhesive surface.
(iii) The partially cured thiolene pattern can be directly laminated to another glass slide with pre-drilled fluid inlet/outlet holes to obtain a complete single layer glass/thiolene/glass chip. By further UV curing ("hard-curing") the bond between the different parts are strengthened and the thiolene structure is hardened.
(iv) The chip is annealead to 50 C for about 10 hours to guarantee a good stability of the thiolene glue.

2.2. *Materials*

Microscope glass slides (Menzel Glaeser) were cleaned with water and soap, then rinsed with ultrapure water followed by etanol and acetone rinses.

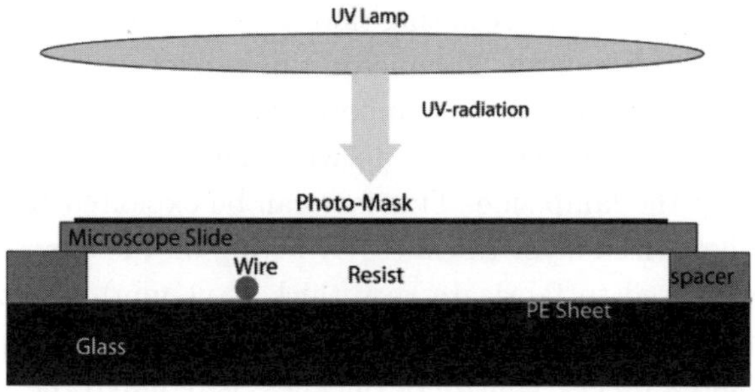

Fig. 1. Scheme of the lithographic process.

Immediately before use, the glass slides were cleaned in an UV-ozone cleaner (Jelight Inc. 42-220 model). In some cases, microscope cover glasses (170 micron thick) were used as top layers to seal chips and were cleaned by UV-ozone alone. As spacers, polymer sheets of known thickness, pieces of glass slides or silicon wafers were employed. Spacers in the range 25 μm - 1 mm were successfully tested, but thicker layers could be achieved if necessary, eventually by building a suitable container for liquid thiolene. Photomasks were printed at a resolution of 4000 dpi on photographic polyester films by a local photoplotting/image-setting service company. Thiolene based UV curable adhesives were obtained from Norland Products that commercialises them as optical adhesives under the acronym NOA. Different grades are available that differ in viscosity and sensivity and other properties. We extensively tested grade NOA61 that is reported to have viscosity 300 cps at 25°C and good adhesion to glass. NO61 reaches its full hardness and best adhesion to glass by ageing 1 week after UV hard curing. This material is moderately hydrophobic with a characteristic water contact angle of 70°. For rinsing uncured thiolene adhesive we used boiling water followed by ethanol, which has only a mechanical effect. Polyester sheets acting as temporary backing sheet for thiolene adhesive were cut from the transparent parts of photographic films used as photomasks. These films showed a good resistance to ethanol used during the rinsing step and a good adhesion towards thiolene glue. A 100 Watt mercury vapor UV flood lamp (Spectronics SB-100P), optimised for emission at 365 nm was used for photopolymerization (luminous flux 6000 μW/cm^2 at 15 cm distance). The lamp was held at a distance of 70 cm from the sample. At this distance, a reasonable collimation and a good homogeneity over the illuminated chip area is achieved. Exposure times depend on the thiolene layer thickness and the desired properties of the resulting thiolene structures but typically were between 7 min and 9 min for 200 micron thick spacers. To understand the curing behaviour of thiolene it is important to note that these materials polymerize by front-polymerization [9] with the liquid beginning to solidify starting from the lamp side. This fact can be exploited to continuously adjust the polymerized layer thickness by acting on the exposure dose and in our case was used to fabricate very thick (> 1 mm) structures for on-chip tube connectors. If spacers are used to confine thiolene liquid between two surfaces and a precise control of thickness is required the UV exposure dose should be at least as much as is required that the polymerization front arrives to the opposite surface (anti-stick surface). Thicker layers can be fabricated by using larger glass plates instead of microscope slides thereby improving capillary confinement. For fabrication of very thick on-chip tube connectors by dose controlled liquid front polymerization we found it useful instead to invert the sample geometry with respect to Fig. 1: a transparent glass container filled with liquid NOA glue was posed on top of a mask and illuminated from below through the mask.

3. Single layer microfluidic devices

Single layer devices produced by liquid photopolymerization of thiolene monomers are useful e.g. for chemical microreactors because of the good solvent resistance of thiolene resins [5]. In the single layer fabrication method reported in the literature [6] thiolene liquid is sandwitched between two adhesive surfaces (glass slides) and rinsing of unpolimerized liquid is done on the fabricated chip with the channel already sealed. Rinsing in this case is done by either applying pressure or a vacuum to the formed channel network. The narrow width of microfluidic channels and the viscosity of adhesives can make this rinsing step cumbersome and often delamination of the chip can result. In our approach on the other hand rinsing of the chip is performed while the channels are still open, greatly facilitating the rinsing step. In this way, narrower/longer channels and more complex interconnected channel patterns can be more easily produced. The chip is then "dry sealed" with a second glass slide by simply laminating this glass slide to the partially cured thiolene channel pattern. The elastomeric properties of partially cured thiolene resin and the tackiness of this rubber are exploited to obtain good conformal contact during lamination and can be performed even on non perfectly flat/smooth surfaces. To guarantee a better adhesion, an heavy weight (about 5 kg) with an opening the size of the glass slide is placed on top of the chip during the process of hard-curing.

Finally, connectors are glued to the chip above the drilled holes. Figure 2 shows the layout of a chip built for electroporation studies (A, B). It is formed by a single channel with a maximum width of 400 microns which decreases to 100 micron in the central region. The height of the channel is less than 100 microns. In the middle of the narrower channel, we have evaporated three gold stripes on the opposite sides of the glass walls of different widths. There are three electrodes to better evaluate which one works better. After the preliminary tests, the final chip will have only one pair of electrodes made out in ITO to allow direct optical access to the electroporation region. The length of the thin channel is much bigger than the entrance length to guarantee a well defined flow profile in proximity of the electrodes.

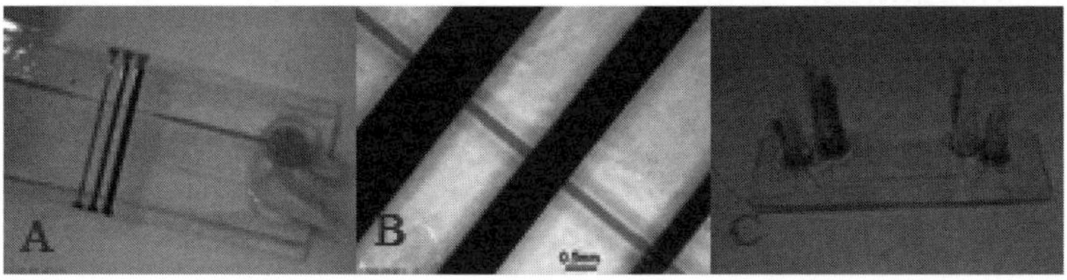

Fig. 2. Microdevices for electroporation (A, B) and thermophoresis (C) studies. The centre photo (B) shows an enlarged view of the microchannels underneath the gold electrodes.

The chip (C) in Fig. 2 is suitable for realizing a thermal gradient on the (micro)scale of the channel to induce migration of particles (microfluidic thermal filter). We have fabricated a simple chip in NOA whose layout is displayed in Fig. 4. A central rectangular channel 150 μm wide, 120 μm deep and 50 mm long, the *main channel*, is filled with the working fluid, like a colloidal solution. A Moleculoy alloy wire, with a diameter of 60 μm, is inserted at a distance of 1 mm from one side of the main channel and connected to a power supply, while a second wider channel, the *cooling channel*, is placed at the same distance at the other side of the main channel and is filled with running cold water. By regulating the heating power and water flow is is possible to vary the temperature gradient across the main channel.

4. Conclusions

We have presented a few examples of single-layer microfluidic chips fabricated by exploiting the adhesive and photoresist properties of NOA61, They show the good flexibility allowing a very rapid prototyping of chips that can easily include wires, electrodes, obstructions, during the fabrication process, as well as an easy sealing of the device. The required equipment, which includes a high-resolution laser printer and a UV lamp, is not expensive and can be easily found in most laboratories. Once the methodology has been optimized, it takes less than one day to fabricate the chip starting from the printing of the photomask. If one needs to realize channels having a typical width less than 100 μm or less, this technique is not longer reliable and it is more convenient to use the standard replica molding of PDMS. However, this requires the fabrication of a master by mechanical micromachining or by photolithography of SU8.

Acknowledgments

Partial support from Progetto di Eccellenza Cariparo 2007 "Mischa" and from CNISM-CNR Progetto d'Innesco della Ricerca Esplorativa 2007 "Viscous flows in microfluidic devices" are warmly acknowledged.

References

1. M. A. Unger, H. Chou, T. Thorsen, A. Scherrer, S. R. Quake, *Science* **288**, 113 (2000)
2. C. Harrison, J. T. Cabral, C. M. Stafford, A. Karim, and E. J. Amis, *J. Micromech. Microeng.* **14**, 153 (2004).
3. C. Khoury, G.A. Mensing and D. J. Beebe, *Lab Chip*, **2**, 50 (2002).
4. J. B. Hutchison, K. T. Haraldsson, B. T. Good, R.P. Sebra, Ning Luo, K.S. Anseth, and C.N. Bowman, *Lab Chip* **4**, 658 (2004)
5. Z. T. Cygan, J. T. Cabral, K.L. Beers, and E.J. Amis, *Langmuir* **21**, 3629 (2005).
6. D. Bartolo, G. Degre, P. Nghe and V. Studer, *Lab Chip* **8**, 274 (2008).
7. M. Natali, S. Begolo, T. Carofiglio and G. Mistura, *Lab Chip* **8**, 492 (2008).
8. L. Brigo, T. Carofiglio, C. Fregonese, F. Meneguzzi, G. Mistura, Giampaolo, M. Natali and U. Tonellato, *Sens. Actuators B* **130**, 477 (2008).
9. J. T. Cabral, S.D. Hudson, C. Harrison, and J.F. Douglas, *Langmuir* **20**, 10020 (2004).

DEVELOPMENT INTEGRATED INDUCTIVE SENSORS FOR MAGNETIC IMMUNOASSAY IN "LAB ON CHIP" DEVICES

B. ANDÒ, S. BAGLIO*, A. BENINATO and V. MARLETTA

DIEES, University of Catania, Catania, Italy
*E-mail: *salvatore.baglio@diees.unict.it*

G. FALLICA

STMicroelectronics, Catania, Italy
E-mail: giogio.fallica@st.com

In this paper an inductive integrated sensor to be adopted in sensing application on microfluidic systems for biomedical applications is presented. It is based on the use of magnetic particles that, suitably coated, act as markers of the bio-molecule to be detected. The device consists of a primary coil and two secondary coils arranged in a differential configuration. The sensing principle of the device is related to the output voltage variation on the secondary coils due to the presence of the magnetic beads only over one of the two coils. A general transduction mechanism is therefore presented together with analytical models and some discussion on novel integrated devices whose layout is proposed here. These sensors will be embedded into a more general "lab on a chip" device in which suitable microfluidics will be implemented both to bind markers with analyte and to drive the fluid into the sensing chamber.

Keywords: Magnetic immuno-assay, magnetic beads, inductive sensors, lab-on-chip.

1. Introduction

In the modern research the realization of biosensors for applications in different fields like public health, clinical analysis, water and air pollution, biotechnology holds an important position. For such devices a very high sensitivity and specificity, a short analysis time, low cost, easiness to handle, capability for portable applications are of fundamental importance. Biosensors with high sensitivity and specificity can be obtained by using Immunoassay techniques [1], [2], which is a biochemical test that measures the concentration of a substance in a biological liquid using the reaction of an antibody or antibodies to its antigen. Detection is made by coupling these molecules to suitable markers, such as enzymes, radioactive compounds, fluorophores, luminescence and magnetic ones. By comparison with the other, magnetic markers have potential advantages, which are related to their low price, very high stability and absence of toxicity. In addition, biomolecules fixed to magnetic nanoparticles can be easily localized and manipulated by suitable magnetic fields [2], [3], [4]. The problem of detecting biological agents is therefore shifted to the ability of sensing the presence of the chosen marker. In Fig. 1(a) the detection principle is shown.

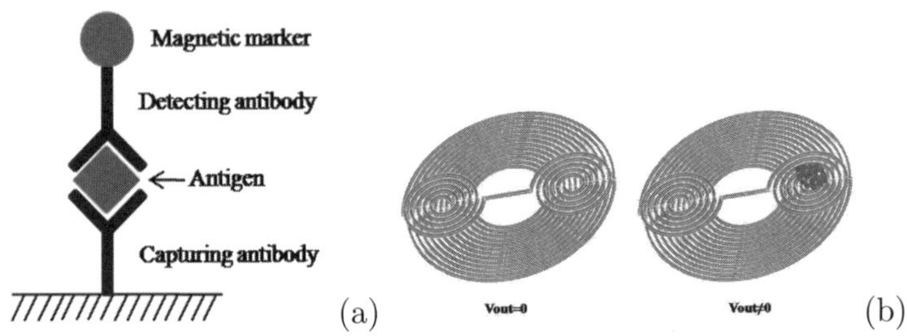

Fig. 1. (a) Bond antibody-antigen-antibody-marker in the magnetic immunoassay. (b) Working principle of the planar differential transformer.

Several different works have been presented in the scientific literature which deal with the problem of detecting the presence of magnetic markers into a selected area, [5], [6], [7], [8], [9]: in these sensors the inductance value changes as function of a certain density of magnetic particles in the active region of the sensor device. First the active region must be functionalized, very often silanizated, to allow the binding of the antibody over the sensor surface; then both the active region and the magnetic particles must be functionalized with a antibody specific to the analyte to be detected. The interaction antibody-analyte-antibody binds the magnetic particles to the sensors surface. If a magnetic field source is considered, the distribution of the magnetic flux lines will be affected by the presence of a number of magnetic particles, this in turn induces a change in the inductive impedance of the coil placed in the active area of the sensor. In comparison with other kinds of magnetic biosensors, inductive devices have several potential advantages, which are related to their higher simplicity, fully compatibility with standard Si technology materials, low cost and higher flexibility. In this work an inductive sensor will be presented together with its analytical model, then some technology related issues will be discussed and the layout of some novel inductive devices for magnetic immunoassay will be presented. These layout refer to more rich devices where an on-chip temperature control is planned. In fact both the heater and the temperature sensor will be embedded into the same device.

2. The inductive sensor

The device presented here is based on a planar coreless differential transformer configuration. A primary coil generates a magnetic flux which links with two secondary coils, having opposite winding sense, connected in a differential arrangement. The primary coil generates a magnetic flux that therefore induces voltages with equal values but opposite in sign in the secondary coils, due to their opposite winding sense and the symmetry properties of the device; therefore the resulting output voltage, which is the difference between the voltages across the secondary coils, is zero when no magnetic particles are present. On the other hand, the presence of magnetic particles in one of the secondary coils will cause an increase of the magnetic field density on a area close to the magnetic particles; in this way, the output voltage will be non-zero. This working principle is schematized in Fig. 1(b). In the proposed device only the secondary coil over which the magnetic particles are placed, acts as the "active" sensor, while the other one acts as "dummy", like in most differential sensing approaches. In particular here the differential configuration is used not to enhance sensitivity; in fact there are no opposite variations of inductance, but to lower the noise floor. In fact any unwanted external excitation will affect the two secondary coils at the same manner and therefore will be nulled by the differential arrangement, while only the signal produced by the magnetic markers placed over one of the two coils will be useful for a non zero output voltage. The primary coil is a source of excitation of the sensor. This approach allows a more flexible optimization of the device in terms of sensitivity; in fact in the case in which the primary coil is current driven and the secondary coil has an infinite impedance, the open circuit voltage at the secondary coil, if expressed in terms of the current applied to the primary winding, is proportional to the product of the number of turns of the primary and the secondary coils as reported in Equation (1)

$$S = N_1 N_2 \frac{2\Re_1 + \Re_2}{\left[\Re_1\left(2 - \frac{\Delta L}{L_2'}\right) + \Re_2\left(1 - \frac{\Delta L}{L_2'}\right)\right]^2} \frac{di_1}{dt} \tag{1}$$

where N_1 represents the number of turns of the primary coil; N_2 the number of turns of the secondary coil; $\Re_1$ and $\Re_2$ are the reluctances of primary and secondary coils; i_1 is the current in the primary coil; ΔL is the increase of inductance value of the secondary coil over which the magnetic beads are placed due to the increase in the magnetic flux; and L_2' is the inductance value of the secondary coil over which the magnetic beads are placed.

This differential transformer is realized in a dedicated technology based on STMicroelectronics fabrication facilities. It is made up of two metal layers (Metal1 and Metal2) separated by a layer of TEOS oxide. The primary winding has been realized in the Metal1 layer, while the Metal2 has been used to realize the secondary windings. The device was studied and optimized in function of the geometric parameters [10] such to determine the optimal dimension of both the primary and secondary coils and the optimal relative position. Particular efforts, related to this latter issue, have been payed to the top surface quality that must be suitable to the antibody binding necessary for realizing the immunoassay process, therefore planarization has been considered together with the deposition of a final thin layer of gold over the contact pad area in order to protect contacts from the further surface silanization. Some of these device layouts are shown in Fig. 2. In particular in Fig. 2(a) the basic device, made up of two primary coils superimposed with two secondary coils is shown. Only one side of the secondary windings is functionalized to bind the antibody. Moreover a twofold detection strategy is implemented: a larger coils is used to detect and quantify the amount of the magnetic particles while the smaller one will be used to detect the target presence (ideally it is aimed to sense the presence of the "single" magnetic bead). In Fig. 2(c), the differential mechanism has been implemented here by designing a couple of identical superimposed windings and then by connenting the secondary coils in a counterphased arrangment. In order to adequately perform the immunoassay protocol an accurate temperature control is required, therefore both a heater and a temperature sensor have been embedded into the proposed devices. This has required the introduction of a high resisitivithy polysilicon layer into the process to realize the resistive heater. The resistive heater has been designed to uniformely cover the region below the two secondary coils. In order to control the temperature suitable sensors have been considered. Given the materials avalable, two types of sensors have been designed: a RTD thermoresistor (Figs. 2(a)–2(c)) and a thermopile, series of 7 thermocouples, made by using the Metal1 and the Poly layers (Fig. 2(b)). In this latter case the RTD has been inserted only for comparison reasons.

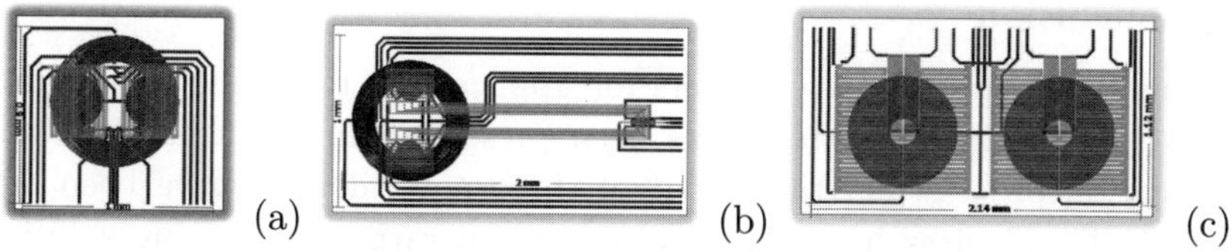

Fig. 2. Three optimized device layouts of the differential inductive sensor for immunoassay with magnetic beads. (a) Differential transformer with dual secondary winding for both "single bead" and "average" detection, resistive heater and RTD temperature monitoring; (b) Differential transformer with the thermopile as temperature sensor; (c) Dual transformer topology with differential connetion of the output terminals, resistive heater and RTD temperature sensor.

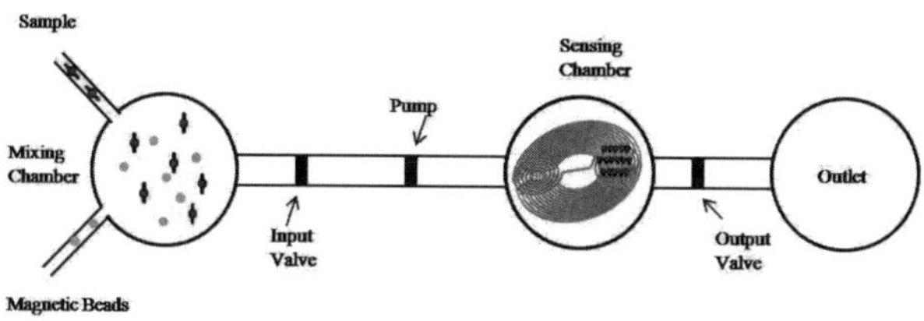

Fig. 3. The functional block scheme of the lab on chip system embedding the inductive sensor for magnetic beads.

3. The lab on chip device

This inductive sensor implements the sensing strategy of a more general device, that can be summarized as in Fig. 3. The functionalized magnetic beads and the fluid sample containing the analyte bounded to the antibodies, are driven into the "mixing chamber" to realize the antibody-antigene binding schematized in the upper section of the sandwich shown in Fig. 1(a). This agglomerate is driven to the "sensing chamber" by a valve-pump system. If the chambers are pre-loaded or, a positive pressure difference is applied, valves can be single-use because the sample goes from the "mixing chamber" to the "sensing chamber"; otherwise a volumetric pumping mechanism can be adopted to move the fluid to the "sensing chamber" without the need of valves. The inductive sensor will be placed on the bottom of the "sensing chamber" and here the immunoassay mechanism will be completed by the binding of the functionalized magnetic beads to the receptors over the active sensor surface to complete the sandwich shown in Fig. 1(a). In the "sensing chamber" the above described sensor reveals the amount of the molecule to detect.

4. Conclusion

In this paper an inductive sensors for applications to magnetic immunoassay in microfluidic systems has been presented. In particular here the fabrication process and the layout of three novel devices have been discussed. The devices here embed also a temperature monitoring solution that has been realized with the materials available in the dedicated fabrication process. The focus here is on the optimal design of the inductive transducers but the global goal, and work is in progress in this direction, is toward integrated microfluidic systems in lab-on-a-chip scenario.

References

1. K. Larsson, K. Kriz, D. Kriz, (1999). Magnetic Transducers in Biosensors and Bioassays. *Analusis*, **27**(7), pp. 617-621.
2. K. Kriz, J. Gehrke, D. Kriz (1998). Advance toward magneto immunoassays. *Biosensors and Bioelec.*, **13**, pp. 817-821.
3. D.R. Baselt, G.U. Lee, K.M. Handen, L.A. Chrisey and R.J. Colton (1997). High-sensitivity Micromachined Biosensor. *Proc. of the IEEE*, **85**, pp. 672.
4. R.L. Edelstein, C.R. Tamanaha, P.E. Sheehan, M.M. Miller, D.R. Baselt, L.J. Whitman and R.J. Colton (2000) The BARC biosensor applied to the detection of biological warfare agents. *Biosensors and Bioelectronics*, **14**, pp. 805.
5. S. Baglio, S. Castorina, N. Savalli (2005). Integrated inductive sensors for magnetic immuno assay applications. *IEEE Sensors Journal*, **5**(3), pp. 372-384.
6. C. Serre, S. Martnez, A.Pérez-Rodrìguez, J.R. Morante, J.Esteve, J.Montserrat (2006) Si technology based microinductive devices for biodetection applications. *Sensors and Actuators A*, **132**(3), pp. 499-505.
7. S. Baglio, A. Pérez-Rodrìguez, S. Martnez, C. Serre, J.R. Morante, J. Esteve, J. Montserrat (2007) Micro-inductive Signal Conditioning with Resonant Differential Filter: High Sensitivity Biodetection Applications. *IEEE transaction Instrumentation and Measurements*, **56**(5), pp. 1590-1595.
8. S.M. Azimi, M.R. Bahmanyar, M.Zolgharni, and W.Balachandran (2007). An Inductance-based Sensor for DNA Hybridization Detection. In *Proc. of the 2nd IEEE Int. Conference on Nano/Micro Engineered and Molecular Systems.*
9. A. Vilà, A. Romano-Rodrìguez, F. Hernàndez, S. Martìnez, C. Serre, A. Pèrez-Rodrguez, J.R. Morante (2005) Microcoils for biosensors fabricated by focused ion beam (FIB). *Electron Devices*, **56**, pp. 209-212.
10. B. Andò, S. Baglio, A. Beninato, G. Fallica, V. Marletta, N. Pitrone, (2009). Analysis and design of inductive biosensors for magnetic immunoassay. In *Proc. of the XIX IMEKO World Congress Fundamental and Applied Metrology*. Lisbon, Portugal.

PART G

Mathematical modelling of dynamic systems for volcano physics

WAVELET MULTI-RESOLUTION ANALYSIS FOR THE LOCAL SEPARATION OF MICROGRAVITY ANOMALIES AT ETNA VOLCANO

FILIPPO GRECO[(1)]*, GILDA CURRENTI[(1)], CIRO Del NEGRO[(1)], AGNESE Di STEFANO[(1,2)], ROSALBA NAPOLI[(1)], ANTONIO PISTORIO[(1,2)], DANILA SCANDURA[(1,3)], GENNARO BUDETTA[(1)] and MAURIZIO FEDI[(4)]

[(1)] *Istituto Nazionale di Geofisica e Vulcanologia*
Sezione di Catania, Italy
[(2)] *Dipartimento di Ingegneria Elettrica, Elettronica e dei Sistemi*
Universita' degli Studi di Catania, Italy
[(3)] *Dipartimento di Matematica e Informatica*
Universita' degli Studi di Catania, Italy
[(4)] *Dipartimento di Scienze Della Terra*
Universita' degli Studi di Napoli "Federico II", Italy
*E-mail: *greco@ct.ingv.it*
www.ct.ingv.it

A microgravity 14-year-long data set (October 1994-September 2007) recorded along a 24-kilometer East-West trending profile of 19 stations was analyzed to detect underground mass redistributions related to the volcanic activity involving the southern flank of Mt. Etna volcano (Italy). An important issue with the above data-set is the need of separating the useful signal (i.e. the volcano-related one) from unwanted components (instrumental, human-made, seasonal and other kinds of noise). To filter the gravity data-set from these last components we propose the wavelet multi-resolution analysis. This method provides a good separation of the long period component from the short period one, and allows exploring the local features of the signal with a detail matched to their characteristic scale. Once the useful signal has been suitably separated from the noise, the residual space-time-image evidences gravity anomalies correlated with the ensuing volcanic activity.

Keywords: Microgravity, volcano monitoring, Etna volcano.

1. Data analysis

Since 1994, 96 surveys were carried out along the East-West profile (Fig. 1), using the Scintrex CG-3M gravimeter (serial # 9310234). Figure 2 shows the gravity variations measured in the East-West profile stations (y axis) between October 1994 and September 2007 (x axis).

The space-time map shows significant gravity increase/decrease cycles, with different wavelengths, reaching the maximum amplitude of approximately 110 µGal. The main gravity cycles are concentrated along the central

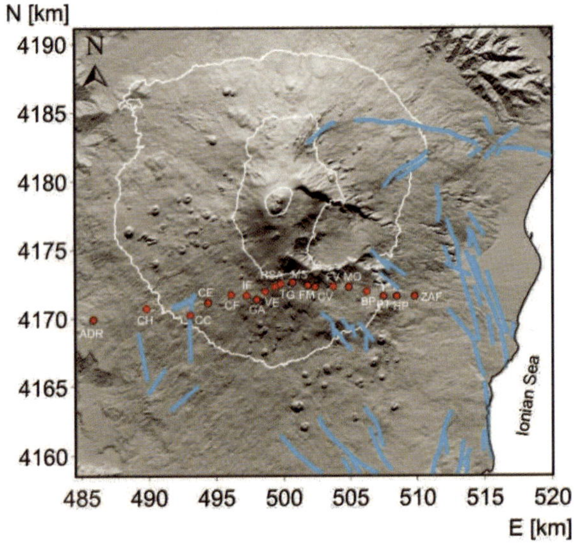

Fig. 1. Sketch map of Mt. Etna showing the gravity stations (red circles) that compose the East-West profile on the southern slope of the volcano which runs from Zafferana to Adrano across the Rifugio Sapienza. The blue lines show the major surface fault systems bordering the eastern and southern sectors of the volcano.

and eastern limb of the profile [1]. Besides useful signal (i.e. the volcano-related one), the gravity map includes unwanted components (instrumental, human-made, seasonal and other kinds of noise). In order to filter the gravity data-set from these last components the wavelet multi-resolution analysis was applied. Using the discrete wavelet transform (DWT), the gravity data are decomposed into a low-resolution approximation level and several detail levels [2,3]. In a two-dimensional multiresolution process three detail levels can be obtained along horizontal, vertical and diagonal directions at each scale [3]. We are particularly interested in the horizontal and vertical components that reflect temporal and spatial gravity variations, respectively. The space-time matrix consists of 50 x 100 grid data at about 500 m step and with a 1.5 month average sampling. The multiresolution analysis decomposes the data from the finest to the coarsest levels, corresponding to level ℓ = -(L-1) and $\ell = 0$, respectively, where L = log2(m) for discrete data of m values. In the case of a 2D multiresolution analysis m is the minimum size of the data matrix dimension. For the considered gravity data-set the finest level is at most ℓ = -5.

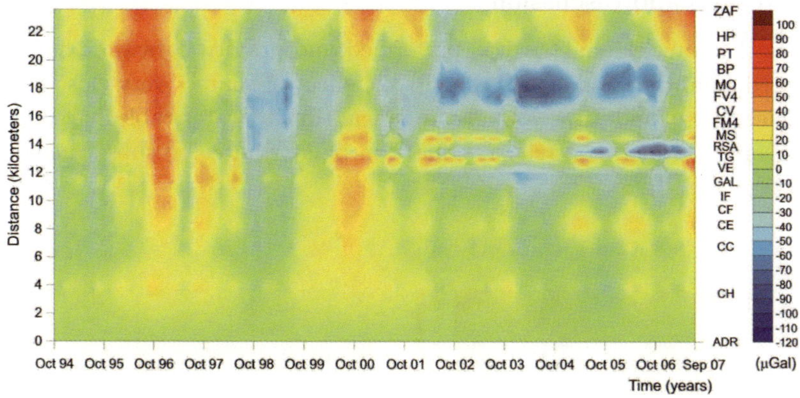

Fig. 2. Image showing the space/time gravity variations along a 24-kilometer East-West trending profile of 19 stations between October 1994 and September 2007. The error on temporal gravity differences along the East-West profile is 10 μGal.

Using the "Minimum Entropy Criterion" [2,4], the Shannon's entropy was calculated for several wavelet basis over the wavelet coefficients from level $\ell = 0$ to $\ell = -5$. The minimum value was obtained for the Symlet 4 basis (Table 1). The wavelet decomposition of gravity data set by Symlet 4 gives one low-resolution approximation map and five total detail maps with the most energetic levels from $\ell = -1$ to $\ell = -5$ (Fig. 3). The D_{-5} detail map (Fig. 3) evidences the presence of local noise at VE and TG stations situated at the central part of the profile (Fig. 1). The high resolution map of the D_{-4} detail shows up short-wavelength time-space gravity variations (Fig. 3). Most of the signal in the horizontal components of the D_{-4} detail map can be associated to quite shallow sources which produce localized gravity anomaly. Besides the presence of high frequency components in the horizontal details, seasonal variations come up in the vertical details of the D_{-4} and D_{-3} maps (Fig. 3). Gravity changes with amplitude ranges between ± 40 μGal peak-to-peak, and consistent with water-table fluctuations, was observed in the easternmost stations of the profile. Accordingly, yearly rainfall data show high precipitation value on the eastern flank with a precipitation ratio of about two with respect to the westernmost flank [5]. In addition, the stations located in the westernmost tip of the profile are very close to the extended outcrops of the sedimentary substrate [6]. Thus, only restricted aquifers are located near the westernmost stations [7], where any gravity change due to water-table fluctuations can be then considered meaningless.

2. Results and discussion

Once the wavelet analysis was performed and gravity data were suitably filtered from the (i) noise (D_{-5}), (ii) the effect of sources very close to the surface (D_{-4}), and (iii) seasonal components due to water-table fluctuations (D_{-3} and D_{-4}), the gravity residual can be attributed to subsurface bulk density (Fig. 4).

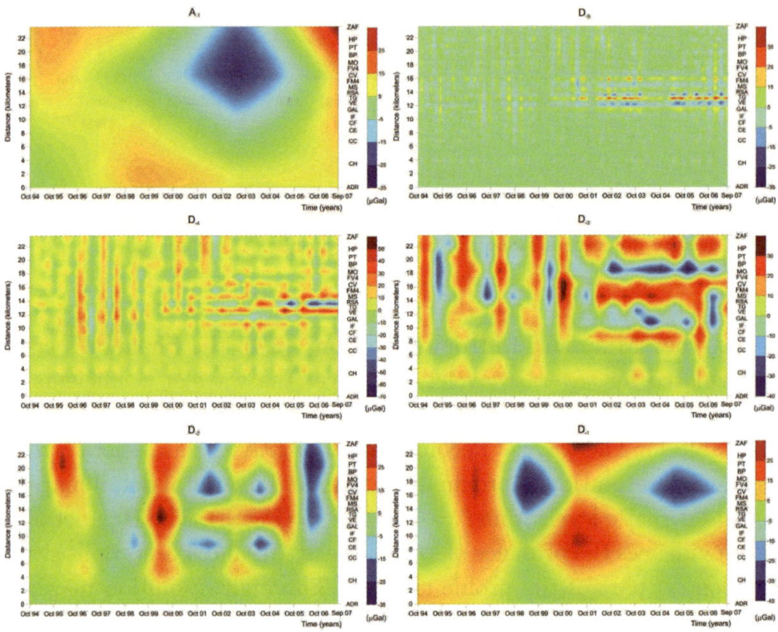

Fig. 3. Using a Symlet 4 wavelet base, five total detail maps and a low resolution map (A_1) were obtained. The meaningful gravity anomalies are related to the total detail levels from D_{-3} to D_{-1}.

The residual gravity map shows long period gravity increase and decrease cycles with duration ranges between a few years and several years, and with a wavelength of order of 10–12 km (Fig. 4). The gravity increase/decrease cycles affect mainly the central and eastern stations of the profile, whereas the gravity changes at stations closest to the reference station (ADR) remain within 20 μGal during the entire period (Fig. 4). A first complete gravity increase/decrease cycle attains a maximum amplitude of approximately 90 μGal, which started during the early months of

1995 (gravity increase), culminated at the end of 1996, and continued until late-1998 (gravity decrease), when the mean value of gravity at each station reached a level lower than it was in 1994-95, before the increase took place (Fig. 4). The considerable increase/decrease of gravity field recorded in this period (late-1995 to late-1998) is among the largest ever recorded along the profile until now.

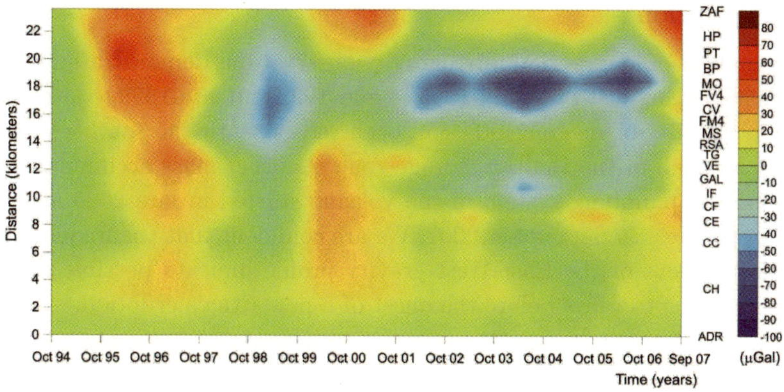

Fig. 4. The 14-year-long gravity data-set after filtering the contributions identified in the D_{-5} and D_{-4} total maps and the D_{-3} vertical detail.

A second increase/decrease gravity cycle, affecting the same stations of the previous one, but mostly evident in the central portion of the profile, was observed between mid-1999 (gravity increase), culminating at the mid-2000 (maximum amplitude about 80 μGal) and continuing until early-2004. The 1999-2000 gravity increase was interrupted by a progressive gravity decrease of about 80 μGal, which started between early-2001, continued very slowly until early-2004, and partially compensated for the previous gravity increase. After the 2001 no significant gravity increase/decrease cycles, in terms of both amplitude and wavelength, occurred. Moreover, the gravity field on the easternmost stations shows a persistent negative gravity anomaly during the period 2001–2006, whereas the gravity in the western flank is almost unchanged. After about 5 years of absence of significant gravity cycles, a new semi-cycle with characteristics similar to the previous ones seems to start at the end of 2006 and continues during the survey carried out in September 2007.

Table 1. Shannon's entropy values relative to wavelet basis.

Coiflet 2	Coiflet 3	Coiflet 4	Daubechies 1
3.24	3.97	4.50	4.38
Daubechies 2	Daubechies 3	Daubechies 4	Symmlet 1
3.60	4.02	3.20	4.38
Symmlet 2	Symmlet 3	**Symmlet 4**	Symmlet 5
3.60	4.02	**3.14**	3.55

3. Conclusion

The analysis of long-period microgravity observations demonstrated that in the last 14 years just two (mid-1995 to mid-1998 and late-1999 to late-2001) major episodes of magma intrusion occurred beneath the southern sector of the volcano in the shallow storage zone. Another mass accumulation in the shallow plumbing system of the volcano started in late 2006 and was still in progress in September 2007. We are confident that the frequent and regular surveys of the East-West gravity profile indicate possible magma rising months to years before the onset of a new Etna's eruption.

References

1. F. Greco, G. Currenti, C. Del Negro, R. Napoli, D. Scandura, G. Budetta, M. Fedi, E. Boschi, *Space-time gravity variations to look deep into the southern flank of Etna volcano* (J. Geophys. Res., submitted, 2009).
2. M. Fedi, T. Quarta, *Wavelet analysis for the regional-residual and local separation of potential field anomalies* (Geoph. Prospect. 46, pp. 507–525, 1998).
3. M. Fedi, G. Florio, *Decorrugation and removal of directional trends of magnetic fields by the wavelet transform: application to archaeological areas* (Geoph. Prospect. 51, pp. 261–272, 2003).
4. Coifman, RR and M.V. Wickerhauser, *Wavelets and adapted waveform analysis. A toolkit for signal processing and numerical analysis* (Proceedings of Symposia in Applied Mathematics 47, pp. 119–145, 1993).
5. Regione Siciliana, *Regione Siciliana - SIAS - Servizio Informativo Agrometeorologico Siciliano.*
6. R. Romano, *Geological map of Mt. Etna* (Progetto Finalizzato Geodinamica, Istituto Internazionale di Vulcanologia, 1:50.000 scale, Catania, 1982).
7. L. Ogniben, *Lineamenti idrogeologici dell'Etna* (Rivista Mineraria Siciliana 100-102, pp. 1–24, 1966).

INTEGRATED INVERSION OF NUMERICAL GEOPHYSICAL MODELS USING ARTIFICIAL NEURAL NETWORKS

A. Di STEFANO*, G. CURRENTI and C. Del NEGRO

Istituto Nazionale di Geofisica e Vulcanologia
Sezione di Catania Catania, Italy
*E-mail: * distefano-a@ct.ingv.it*
www.ct.ingv.it

L. FORTUNA and G. NUNNARI

Dipartimento di Ingegneria Elettrica, Elettronica e dei Sistemi
University of Catania Catania, Italy
www.ing.unict.it

A unified modelling procedure is proposed to jointly interpret the variations observed in geophysical data and to properly take into account the relationship between the intrusive processes and the geophysical variations expected at the ground surface. We focus on the joint inversion of geophysical data by a procedure based on Artificial Neural Network (ANN) for the estimation of the volcanic source parameters. As forward model, we develop a 3D numerical model based on Finite Element Method (FEM) for computing ground deformation, magnetic and gravity changes caused by magmatic overpressure sources, with the aim to consider a more realistic description of Etna volcano, including the effects of topography and medium heterogeneities.

Keywords: Identification, modeling, numerical methods.

1. Introduction

Geodetic and potential fields investigations have been playing an increasingly important role in Mt. Etna eruptive processes [1–3]. The amount of available data collected represents a valuable database, but limited efforts have been made for an effective integration of different data. When the cause of their variations can be ascribed to the same volcanic source, a joint inversion would be advisable in order to identify the source parameters with a greater degree of confidence [4]. The rationale of the inversion modelling approach requires: (i) solution of forward models, (ii) numerical inversion procedure. The forward problem consists in deriving a relationship between sources and observations. Over the last decades, straightforward analytical solutions for simplified geometric sources have been devised under the assumption of homogeneous elastic half-space medium [5–7]. To overcome this intrinsic limitation and provide more realistic models, which consider various geometries as well as complicated distribution of medium properties and real topography, we developed a numerical procedure based on Finite Element Method (FEM) to evaluate geophysical changes caused by overpressure source at Mt. Etna in a 3D formulation [8]. We investigate the ability of an inversion procedure based on Artificial Neural Networks (ANNs) for the estimation of the volcanic source parameters from magnetic, gravity and ground deformation data. Artificial Neural networks have been used to invert geophysical models based on analytical models, because once the network is trained, the inversion process requires a very short time, while in traditional optimization algorithm the whole search procedure has to be reiterated. Numerical models are not often used to train neural network for inversion because of the high computational time to obtain the solution of forward model. We performed a hybrid inversion in which neural networks are trained with numerical patterns, obtained using Finite Element Method. Synthetic modelling is performed in order to provide a data set large enough to represent the training and testing sets of the possible models within the model space. Multilayer perceptrons (MPLs), once correctly trained, can solve the inversion problem very fast and with an appreciable degree of accuracy.

2. Forward numerical model

As forward models, we developed a numerical procedure based on Finite Element Method to evaluate geophysical changes caused by spherical overpressure sources that are quite appropriate for modelling inflation/deflation of magma reservoirs. We used the software Pylith to solve the elastostatic problem for the elastic deformation field. A computational domain of a 100x100x50 km is considered for the deformation calculation. As for boundary conditions, horizontal and vertical displacements are fixed to zero at the lateral and bottom boundaries respectively, representing the vanish displacement at the infinity. The upper boundary is stress free and represents the ground surface. To test the accuracy of the numerical solutionl, benchmark tests are carried out comparing the numerical solutions of deformation field in a homogeneous half-space with the analytical ones [5]. Then we used numerical model including medium heterogeneities and irregular topography to model a more realistic description of Mt. Etna. As magnetic effect generated by a pressure source, we analyzed the thermomagnetic effect, that is related to thermal demagnetization or remagnetization due to temperature changes of rocks. When temperature exceeds Curie point, rocks lose their magnetization and then modify the intensity of superficial magnetic field. The magnetic change is expressed as follows:

$$\begin{aligned} \Delta T &= \tfrac{\Delta M_m}{4\pi r^2} \left\{1 - 3(\tfrac{x}{r}\cos I - \tfrac{z}{r}\sin I)^2\right\} \\ \Delta M_m &= \tfrac{4}{3}\pi r^3 m \end{aligned} \tag{1}$$

Where ΔM_m is the magnetic moment, R is the radius of the sphere, m is the magnetization, I is the magnetic inclination. Gravity field changes are due to additional mass input at some depth. Migration of magmatic mass generates a density variation that can be observed at the surface through gravity field measures. The gravity change due to input of new mass in a spherical source is expressed as follows:

$$\begin{aligned} \Delta g &= G \Delta M_g \tfrac{z}{(x^2+y^2+z^2)^{\frac{3}{2}}} \\ \Delta M_g &= \tfrac{4}{3}\pi r^3 \Delta\rho \end{aligned} \tag{2}$$

where ΔM_g is the mass change, ρ is the density contrast, G is the gravitational constant. Using the forward model described above, we generated a synthetic set of deformation, magnetic and gravity data for training the neural network. The procedure of pattern generation is divided in several step and is executed automatically. First the computational domain of Mt. Etna is meshed with the software LaGrit to finally obtain 598948 isoparametric and arbitrarily distorted tetrahedral elements connected by 103424 nodes, with a spatial resolution that reaches 300m in the area where the topographic relief is sharper and in the area where the sources are located. Then the parameters of the sources are generated with random distribution in the ranges reported in Table 1. The volume where the pressure sources are located contains the position of all the pressure sources active during the last decades [1].

Table 1. Ranges of the random generated parameters of the source.

Source Parameters	Minimum	Maximum
$X_C[km]$	496	502
$Y_C[km]$	4175	4183
$Z_C[km]$	-9	-1
$\Delta V[m^3]$	$5\ 10^6$	$10\ 10^6$
$\Delta M_g[kg]$	$3\ 10^9$	$150\ 10^9$
$\Delta M_m\ [Am]$	$1\ 10^9$	$4\ 10^9$

Once all the patterns are defined, the meshed sources are iteratively introduced in the domain mesh to finally obtain 1050 complete meshes. Every mesh is characterized by the source in a different random position that is contained in the range reported in Table 1. The numerical solutions are calculated applying a pressure of 100 MPa to the source wall and then rescaling them with the random values of ΔV, assuming valid the hypothesis of point source. At the end solutions have been interpolated at the coordinates of the stations of deformation, magnetic and gravity networks on Mt. Etna, whose maps are reported in Fig. 1. The rectangle corresponds to the projection on surface of the volume where the sources are located.

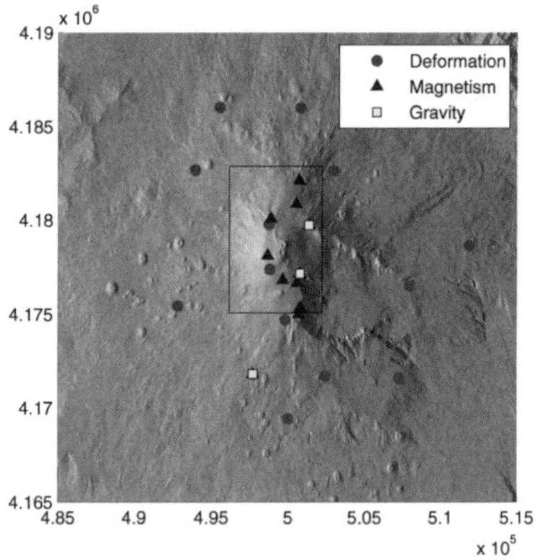

Fig. 1. Deformation, gravity and magnetic monitoring network on Mt. Etna.

3. ANN based inverse model

The problem of inversion is approached numerically with the Artificial Neural Networks: the forward model $f()$ is used to generate synthetic data d_i corresponding to model parameters x_i, (randomly chosen in the space of parameters). At the same time the pairs (d_i, x_i) are used to approximate the inverse model $f^{-1}(.)$ by using an approximating function of the form:

$$f_a(x) = \sum_{j=1}^{NH} c_j \phi(w_j^T x + t_j) + c_0 \tag{3}$$

where $\phi(.)$ represents the sigmoid function, x is the MLP input vector, w_j is a vector of coefficients (weight of the connections), and c_j, t_j are additional adjustable coefficients. Once the neural network has been trained, it is able to approximate the f^{-1} function and to identify, for a set of geophysical observations, the source parameters that better reproduce these variations. The neural network used for the inversion is a three layered network, with 29 inputs (the three component of deformation field, thermomagnetic field and gravity filed at the network stations), 20 hidden neurons, 6 outputs (position of the source, volume change, mass change, magnetic momentum). The deformation inputs are rescaled computing the quantities ux/uz, uy/uz uz instead of ux, uy, uz, so that the function that relates the first two quantities with the source parameters becomes easier to be interpolated. The outputs are normalized linearly to finally obtain values in the range [-1;1]. The ANN is trained with numerical patterns, obtained using Finite Element Method. Then the inverse function is tested with a data set that was not used previously for training. The performance index used to test the results is the root mean square error, which expression is $RMSE = [1/N \sum(P_i - O_i)^2]^{1/2}$, where P_i is the calculated value, O_i is the observed value, N is the dimension of data set. We performed the inversion of deformation data for the analytical and numerical solution, to compare the accuracy of the two results. Then we inverted deformation, magnetic and gravity data together to investigate if the integrated approach allows for a more accurate solution. The performance index of the three cases are reported in Table 2.

Table 2. Performance index RMSE for the inversion of analytic and numeric deformation model and for the integrated numeric model.

	Deformation (Analytic)	Deformation (Numeric)	Integrated (Numeric)
$X_C[m]$	239.44	648.78	224.24
$Y_C[m]$	301.05	807.42	620.62
$Z_C[m]$	329.99	672.24	335.80
$\Delta V[m^3]$	0.1 10^6	0.18 10^6	0.25 10^6
$\Delta M_g[kg]$	-	-	0.31 10^9
ΔM_m $[Am]$	-	-	0.07 10^9

Numerical and analytical inversion of deformation data provide similar results, demonstrating that also the numerical problem can be inverted with good accuracy, even if the inverse function to interpolate is more complicated because of complex distribution of heterogeneity and irregular topography. The mean square errors related to the integrated inversion are smaller than those obtained inverting only deformation data, evidencing the advantages of the integrated approach. This approach, involving geophysical data of different kinds, allows for a more accurate solution than when ground deformation data alone is considered.

4. Conclusions

A hybrid approach is proposed using neural networks and numerical FEM models for geophysical modelling. This method permits to interpret geophysical data avoiding the intrinsic limitation of analytical solutions and providing a more realistic description of volcanic processes. ANN is used in inverse scheme to identify the source parameters from geophysical observations. The results show that, notwithstanding the high nonlinearity of the considered inverse problems, it can be solved with acceptable accuracy. The work highlights the usefulness of integrated version of geophysical data, that permits to constrain the solution better than when single data type are considered.

References

1. A. Bonforte, A. Bonaccorso, F. Guglielmino, D. Palano, G. Puglisi, *Journal Geophys. Res.* **113**, B05406 (2008), doi:10.1029/2007JB005334.
2. R. Napoli, G. Currenti, C. Del Negro, F. Greco, D. Scandura, *Geophys. Res. Lett.* **35**, L22301 (2008), doi:10.1029/2008GL035350.
3. D. Carbone, G. Currenti, C. Del Negro, *Bull. Volcanol.* **69**, 553 (2007), doi:10.1007/s00445-006-0090-5.
4. G. Nunnari, L. Bertucco, F. Ferrucci, *IEEE Transaction on Geosciences and Remote Sensing* **39**, 736 (2001).
5. K. Mogi, *Bull. Earthq. Res. Inst., Univ. Tokyo* **36**, 99 (1958).
6. Y. Sasai, *Bull. Earthq. Res. Inst., Univ. Tokyo* **66**, 585 (1991).
7. Y. Hagiwara, *Bull. Earthq. Res. Inst., Univ. Tokyo* **52**, 301 (1977).
8. G. Currenti, C. Del Negro, G. Ganci, *Geophys. J. Int.* **169**, 775 (2007), doi: 10.1111/j.1365-246X.2007.03380.x.

SPH MODELING OF LAVA FLOWS WITH GPU IMPLEMENTATION

A. HÉRAULT*, C. DEL NEGRO and A. VICARI

Istituto Nazionale di Geofisica e Vulcanologia, Sezione di Catania piazza Roma 2, 95123 CATANIA, Italy
*E-mail: *herault@ct.ingv.it*

G. BILOTTA† and G. RUSSO

Dipartimento di Matematica e Informatica, Universitá di Catania viale A. Doria 6, 95125 CATANIA, Italy
E-mail: †bilotta@dmi.unict.it

We describe the implementation of the Smoothed Particle Hydrodynamics (SPH) method on graphical processing units (GPU) using the Compute Unified Device Architecture (CUDA) developed by NVIDIA. The entire algorithm is executed on the GPU, fully exploiting its computational power. The code faces all three main components of an SPH simulation: neighbor list constructions, force computation, integration of the equation of motion. The simulation speed achieved is one to two orders of magnitude higher than the equivalent CPU code. Applications are shown for simulating the paths of lava flows during volcano eruptions. Both static problems with purely thermal effects (such as lava lake solidification) and dynamic problems with a complete lava flow were simulated.

Keywords: SPH, complexity, nonlinear systems, CFD.

1. Introduction

A complete modeling of lava flow is challenging from the modelistic, numerical and computational points of view. The described phenomenon has at its core a complex fluid dynamic problem with free boundaries: the natural topography irregularities, the dynamic free boundaries and phenomena such as solidification, presence of floating solid bodies or other obstacles and their eventual fragmentation make the problem difficult to solve using traditional numerical methods (finite volumes, finite elements).

Recent developments by researchers at INGV have led to the creation of increasingly sophisticated models with ever more detailed representations of the mechanical and thermal aspects of lava flows. Simple models based on the concept of maximum slope and stochastic perturbation of topography (DOWNFLOW [1]) were integrated by models that included a more complete physical description, based on Cellular Nonlinear Networks and Cellular Automata (MAGFLOW [2–4]), able to describe the spatial and temporal evolution of lava flow on the basis of given eruptive parameters. These models have been applied with success in collaboration with the Dipartimento di Protezione Civile for the creation of scenarios during the Mt Etna eruptions over the last years.

Although the latter models include a detailed and accurate physical description of the lava rheology, including thermal effects such as radiation, solidification and the dependency of the physical parameters on the temperature, they are inadequate for the description of more sophisticated thermal-based phenomena such as crust and lava tube formation and their rupture with consequent ephemeral vent opening.

Traditional methods such as finite volumes or finite elements also meet significant challenges in the simulation of lava flows, being tied to spatial discretization with fixed or adaptive meshes. The need to refine the discretization grid in correspondence of high gradients, when possible, is computationally expensive and with an often inadequate control of the error; for real-world applications, moreover, the information needed by the grid refinement may not be available (e.g. because the Digital Elevation Models are too coarse). Eulerian discretization has an additional problem with boundary tracking, which for complex fluids such as lava is further complicated by the presence of internal boundaries given by fluid inhomogeneity and presence of solidification fronts. Another problem is given by the need to solve the implicit system of equations needed to determine the pressure at every time-step for every grid-point. Lagrangian methods such as finite elements, instead, are challenged by the problems related to the continuous and deformable nature of the lava flow, which inevitably leads to significant deformations in the finite element structure, with consequent loss of accuracy and the need for remeshing.

An alternative approach is offered instead by mesh-free particle methods [5] that solve, in a natural way, most of the problems connected to the dynamics of complex fluids. Particle methods discretize the fluid using nodes which are not forced on a given topological structure: boundary treatment is therefore implicit and automatic; the movement freedom of the particles also permits the treatment of deformations without incurring in any significant penalty; finally, the accuracy is easily controlled by the insertion of new particles where needed.

2. The Smoothed Particle Hydrodynamics (SPH) method

To this purpose, a new model has been developed, based on the Smoothed Particle Hydrodynamics (SPH) meshless method. Formulated by at the end of the '70s by Gingold and Monaghan [6] and Lucy [7] for astrophysics problem, the SPH method has recently seen a growing involvement for fluid dynamic applications [8,9]. As a particle method, SPH doesn't suffer from the limitations traditional mesh-based numerical methods (finite differences, finite volumes, finite elements) encounter when describing a complex, free-surface fluid flow; in comparison to other particle methods, SPH also provides additional benefits such as the automatic preservation of mass and the direct computation of most physical quantities (e.g. pressure) without resorting to large, sparse implicit systems.

The underside of the SPH method is that it is necessary to employ a number of particles higher than the number of nodes in grid methods to achieve simulations of comparable resolution, thus increasing the computational requirements. However, since most calculations in SPH algorithms are direct, this method can be parallelized to a much higher degree than most traditional mesh methods, a characteristic that makes the SPH method particularly favorable to implementation on highly parallel computational hardware such as modern video cards.

3. Scientific computing on graphic cards

Since the introduction of 3D rendering on computers, video cards have evolved from simple devices dedicated to video output into powerful parallel computing devices. The graphical processing units (GPUs) on modern video cards often surpasses the computational power of the CPU that drives them. Until recently, however, such computational power has been limited to the rendering of complex 3D scenes, satisfying the needs of computer gamers and professional designers.

The increasing computational power of GPUs has led to a growing interest in their usage for computation beyond video rendering; their computational power these days allows turning a desktop computer into a teraflop high-performance computer able to match the most expensive clusters in terms of performance, but at a fraction of cost, both in terms of initial price and total cost of ownership. However, full exploitation of their capabilities requires appropriate tools and problems that are computational rather than data-intensive.

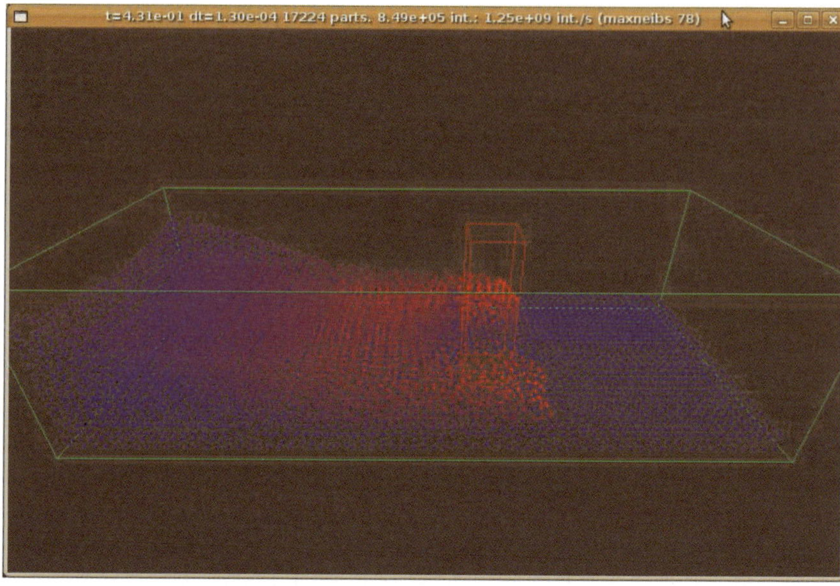

Fig. 1. GPU SPH simulation running on an NVIDIA GTX 280, with an average 1.2×10^9 interactions per second.

Previously, General Programming for the GPU (GPGPU) has relied mostly on the OpenGL standard, an architecture designed to standardize programming of the most common operations needed to render detailed static and dynamic 3D scenes. Its usage for more generic numerical computations requires an intimate knowledge of computer graphics and a number of programming tricks to convert mathematical operations into equivalent graphical rendering operations and, conversely, to interpret the rendered scene as mathematical results of the operations. These transformations exact a significant coding cost and impose a number of constraints on the operations that can be performed.

4. SPH on CUDA

The CUDA architecture and programming language, introduced by NVIDIA in the spring of 2007, works around these limitations by allowing GPU programming using the C language extended to handle the specific needs of their GPU and its interfacing with the CPU host. While traditional GPGPU programming uses the GPU as a coprocessor for the CPU, performing only the most expensive computations on the graphics card while keeping much of the algorithm structure on the CPU host, CUDA encourages porting nearly all computations to the GPU.

Our lava simulation model uses the SPH method with a GPU implementation in CUDA to achieve a high computational performance. Additionally, the peculiar nature of GPUs provides us with hardware support for special functions such as bi- and trilinear interpolation, which we can exploit to compute the normal to the Digital Elevation Model (DEM) of the terrain, essential to the implementation of boundary conditions for the flow.

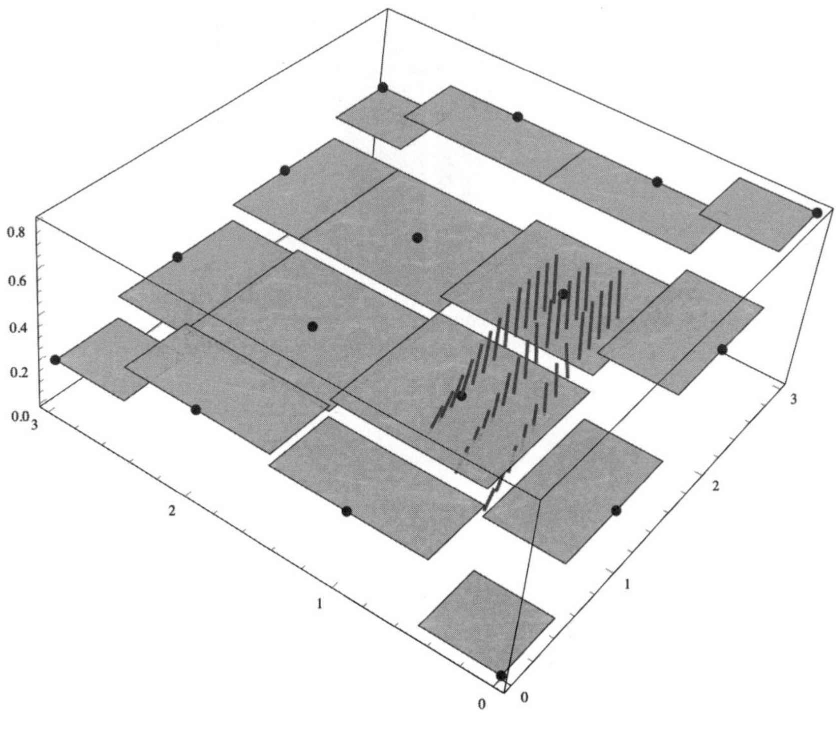

Fig. 2. DEM normals computed by exploiting the interpolation hardware in GPUs.

Both static problems with purely thermal effects (such as lava lake solidification) and dynamic problems with a complete lava flow can be simulated.

A direct comparison between SPH and finite elements for the lava lake solidification shows the superiority of the SPH method, that guarantees a significantly improved accuracy in proximity of the contact area of two or more solidification fronts. Although purely static problems are less of a direct interest for hazard assessment, they are still an important tool for risk management, when used in conjunction with dynamic problems, for example when evaluating the effects of barrier formation with respect to lava flow diversion or halting, and the consequent formation of lava lakes.

For the dynamic part of the model, the SPH algorithms are based on the ones of the SPHysics simulator, with the addition of thermal effects and the treatment of non-Newtonian fluids. Following recent developments in the physical modeling of lava flow rheology, both Bingham and power-law fluids can be simulated by our code.

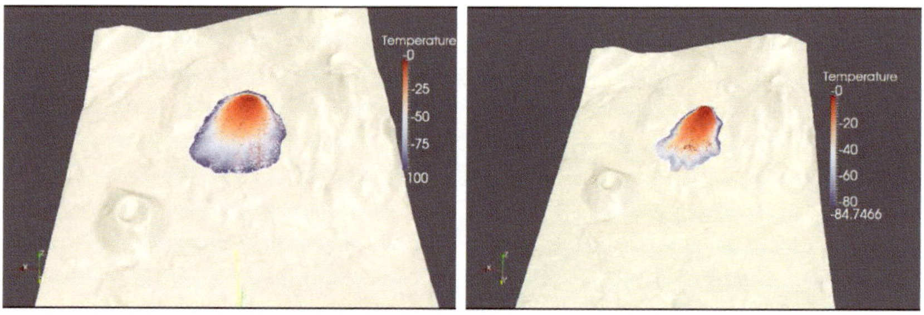

Fig. 3. Side-by-side comparison of constant vs temperature-dependent viscosity in a lava flow. The fluid is color-coded according to temperature delta from the eruption (1363 K).

References

1. M. Favalli, M. Pareschi, A. Neri and I. Isola, *Geophys. Res. Lett.* **32** (2005).
2. A. Vicari, A. Herault, C. Del Negro, M. Coltelli, M. Marsella and C. Proietti, *Environmental Modelling & Software* **22**, 1465 (2007).
3. C. Del Negro, L. Fortuna, A. Herault and A. Vicari, *Bull. Volcanol.* **70**, 805 (2008).
4. A. Herault, A. Vicari, A. Ciraudo and C. Del Negro, *Computer & Geosciences* **35** (2008).
5. R. W. Hockney and J. W. Eastwood, *Computer simulation using particles* (Hilger, Bristol U.K, 1988).
6. R. A. Gingold and J. Monaghan, *Mon. Not. R. Astr. Soc.* **181**, 375 (1977).
7. L. Lucy, *Journal Astronomical* **82**, 1013 (1977).
8. J. Monaghan, *Jour. Comp. Phys.* **110**, 399 (1994).
9. R. Dalrymple, M. Gómez-Gesteira, B. Rogers, A. Panizzo, S. Zou, A. Crespo, G. Cuomo, and M. Narayanaswamy, Smoothed particle hydrodynamics for water waves, in *Advances in Numerical Simulation of Nonlinear Waves*, ed. Q.Ma (World Scientific Press, 2009)

3D DYNAMIC MODEL FOR CHANNELED MODEL FLOWS WITH NONLINEAR RHEOLOGY

MARILENA FILIPPUCCI* and ANDREA TALLARICO†

Dipartimento di Geologia e Geofisica
Università di Bari, I-70125 Bari, Italy
*E-mail: *marilena.filippucci@geo.uniba.it*
†tallarico@geo.uniba.it
www.uniba.it

MICHELE DRAGONI

Dipartimento di Fisica, Università di Bologna
I-40127 Bologna, Italy
E-mail: michele.dragoni@unibo.it
www.unibo.it

In this work we studied the effect of a power-law rheology on a gravity driven lava flow. We consider a viscous fluid flowing in the x direction in an inclined rectangular conduct. The flow is assumed steady, laminar and subjected to the gravity force. The fluid is assumed isothermal, isotropic and incompressible, with constant density and power law rheology. Analytical solution for the equation of the motion does not seem to be possible so an approximated solution was found. We used the finite volume method to obtain the discretized equation that was solved with a classical iterative technique. The convergence, the stability and the order of approximation were tested for the test case $n = 1$ comparing the numerical solution with the analytical solution available for the Newtonian rheology. The results indicate that, for a lava flow with constant volume flow rate, the use of a power-law rheology produces important differences on the height of the channel and on the average velocity of the flow with respect to the Newtonian case.

Keywords: Power-law rheology, lava flow, finite volume.

1. Introduction

Viscosity measurements are related to temperature, composition, crystallinity, and vesicularity of lava samples. Several authors agree in approximating the behavior of the basaltic lava as a Newtonian fluid at temperature above 1100° C [4] or for crystal concentration below 25% [9]. Other studies have demonstrated that the rheological behavior of subliquidus basalts may be analyzed in terms of the pseudoplastic and of the Bingham models (e.g. [3]). Among these authors, Shaw *et al.* [10] showed that for shear rates greater than about $2s^{-1}$, the rheological response of the sample can be approximated to the Bingham model, while at lower shear rates evidence of pseudoplasticity becomes apparent. Pseudoplasticity is a power law rheology, according to which the viscosity depends on the strain rate according to a power law with exponent n. If n is lower than 1, the fluid is pseudoplastic and it thins with an increase in stress. If n is greater than 1, the fluid is said dilatant and it thickens with an increase in stress.

Since the lava is a mix of silicate liquid, crystals, and gas bubbles, and during the evolution of the flow crystal and bubble fractions increase [6], many efforts have been devoted in the attempt to understand the role of crystals and bubbles in lava behaviour. Several authors agree in attributing to lava melts a pseudoplastic behaviour with n depending on temperature [12], on crystal concentration [2] or on bubble concentration [1] and without any observable yield strength [7]. On the contrary, Smith [11] concluded that, by increasing crystal concentration, rheology of crystal-poor magmas must progressively give way to dilatant rheology.

Regarding the modeling of lava flows, several models have been proposed assuming for lava the Newtonian rheology [13] and the Bingham rheology [14]. In this paper we present the modelling of lava flowing in an inclined rectangular channel under the action of the gravity force assuming the power-law rheology. We describe the mathematical problem and the numerical solution and show the tests on the numerical solution in order to understand the stability and the accuracy of the solution.

2. Model description

We consider a viscous fluid flowing in the x direction in an inclined rectangular channel, with the cross section parallel to the yz plane. The width of the channel is α and the thickness is h; the slope of the inclined plane is α. The channel and the coordinate system are shown in Fig. 1. The flow is assumed steady, laminar and subjected to the gravity force. The fluid is assumed isothermal, isotropic and incompressible, with constant density ρ. The constitutive equation for a power law fluid is the following:

$$\sigma_{ij} = 2k\dot{e}^{n-1}\dot{e}_{ij} \tag{1}$$

where:

$$\dot{e} = 2\sqrt{|I_2|} \tag{2}$$

where I_2 is the second invariant of the strain rate tensor. The apparent viscosity of the fluid is:

$$\eta_{app} = k\dot{e}^{n-1} \tag{3}$$

where k is the fluid consistency, which is a measure of the resistance to shear and n is the power-law exponent.

The flow velocity is purely longitudinal and varies with the coordinates y and z. The equation of motion reduces to the form:

$$\rho g_x + \frac{\partial}{\partial y}\left(\eta_{app}\frac{\partial v}{\partial y}\right) + \frac{\partial}{\partial z}\left(\eta_{app}\frac{\partial v}{\partial z}\right) = 0 \tag{4}$$

where v and g_x are the x components of the velocity and of the acceleration of gravity respectively. The boundary conditions are the no-slip at the wall and the symmetry at the centre of the channel.

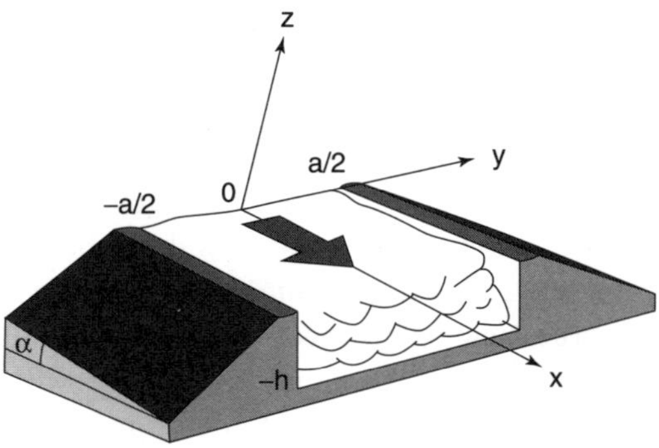

Fig. 1. System of coordinates and geometrical parameters.

3. Numerical solution

We solved the mathematical problem using the Finite Volume method [8]. We employed a uniform rectangular grid with half-thickness control volumes (CV) at the boundaries of the domain. The differential equation is integrated over each CV using linear interpolation functions between the two nearest grid points. This is a second-order approximation scheme and corresponds to the Central-Difference approximation of the first derivative in the Finite Difference methods [5]. The discretization procedure consists in integrating Eq. (4) over each CV in order to obtain an algebraic equation system that can be implemented at and solved by the calculator, by using an appropriate solver. The algebraic equations were solved iteratively

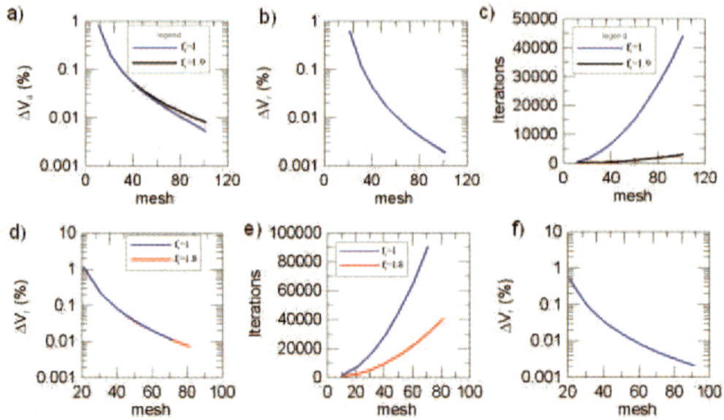

Fig. 2. (a)–(c) Test case $n = 1$; (d) and (e) test case $n = 0.5$; (f): test case $n = 1.5$.

using a classical point by point Gauss-Seidel method in an accelerated version that uses a relaxation factor f_r. The convergence criterion is that the maximum residual computed over all the CVs drops below 10^{-4}.

We tested the numerical solution comparing it with the analytical solution available for the Newtonian rheology [13]. Results of the test indicate that the error on the average numerical velocity with respect to the analytical $(\Delta\bar{V})_a$ (Fig. 2(a)) and respect to the mesh refinement $(\Delta\bar{V})_r$ (Fig. 2(b)) decrease by increasing the number of CVs of the mesh. As we expected, f_r strongly affects the number of iterations necessary for convergence, improving the computational time costs (Fig. 2(c)). Since for non Newtonian fluids the analytical solution for the velocity field is not available, the only way to test the numerical solution is the evaluation of the effect of the grid refinement $(\Delta\bar{V})_r$. For this test, we set $n = 0.5$, that corresponds to a pseudoplastic model (Fig. 2(c)). Also in this case, the use of the over-relaxation factor f_r reduces the number of iterations necessary for convergence (Fig. 2(d)). The test case for the dilatant fluid was obtained by setting $n = 1.5$ and, as for the pseudoplastic test case, we evaluated the error due to the mesh refinement $(\Delta\bar{V})_r$ (Fig. 2(e)). In this case the use of the over-relaxation produces numerical instability and convergence cannot be reached. From Figs. 2(a), 2(c) and 2(e) it can be observed that $(\Delta\bar{V})_r$ is about 0.01% for a mesh 50×50. This grid seems a god compromise between numerical precision and time cost and will be used for the following results.

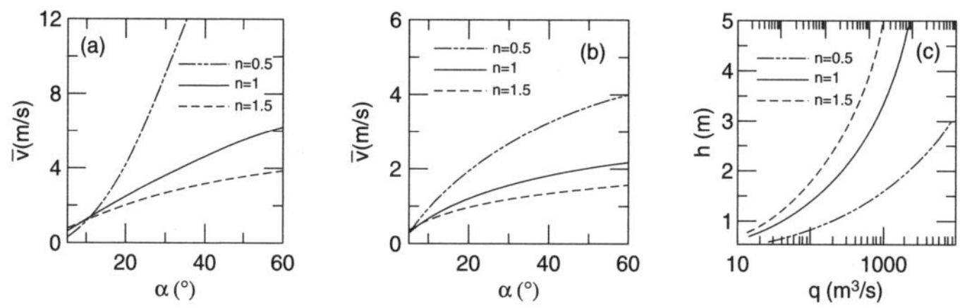

Fig. 3. (a) Plot of the average velocity with respect to the channel slope for a fixed geometry ($a = 10$ m, $h = 4$ m, $\rho = 2650$ kg/m^3, $k = 10^4$ Pa s^n); (b) Plot of the average velocity with respect to the channel slope for a fixed flow rate ($a = 10$ m, $q = 500$ m$^3/s$, $\rho = 2650$ kg/m^3, $k = 10^4$ Pa s^n); (c) Plot of the channel thickness with respect to the volume flow rate ($a = 10$ m, $\alpha = 20°$, $\rho = 2650$ kg/m^3, $k = 10^4$ Pa s^n).

4. Results

The effect of the slope on the average velocity and on the flow rate can be remarkably different for a Newtonian and a non-Newtonian fluid flowing in a rectangular channel down slope. In Fig. 3(a) we plotted the effect of the slope α on $\bar{v}$ varying n and fixing k and the geometrical properties a and h of the channel. Differences between the linear and the non-linear rheology are enhanced by increasing α. The effect of the slope has been evaluated also for constant volume flow rate q (Fig. 3(b)). The effect of non-linearity on $\bar{v}$ varies with α and with k. Finally, we evaluated the dependence of the volume flow rate q variation on h for a fixed value of a and α (Fig. 3(c)).

5. Conclusions

Results showed that velocity profiles and average velocities can vary significantly depending on the assumption of linear or non-linear rheology. This involves two considerations: first, if we measure lava flow velocity, and retrieve flow parameters such as viscosity, the assumption of the rheological model can bring to significant errors, since velocity profiles strongly vary with the degree of non-linearity; second, if we wish to calculate the velocity of a flow, the assumption on rheology is important, since the average velocity can vary over two orders of magnitude under the effect of non-linearity.

The effect of the power law rheology with respect to the Newtonian rheology is small for small slopes, where the effect of the gravity is not important. On the contrary, for steeper slopes, where the gravity effect is higher, the effect of the power law rheology becomes not negligible and the flow behaviour, in terms of the average velocity or in terms of the lava thickness, can be considerably different from the Newtonian case. This means that the assumption of the Newtonian rheology, not supported by laboratory analyses can lead to significant errors in flow parameters.

References

1. Bagdassarov, N., and H. Pinkerton (2004), Transient phenomena in vesicular lava flows based on laboratory experiments with analogue materials, J. Volcanol. Geotherm. Res., 132, 115-136.og
2. Champallier R., M. Bystricky, and L. Arbaret (2008), Experimental investigation of magma rheology at 300 MPa: From pure hydrous melt to 75 vol. % of crystals. Earth and Planetary Science Letters, 267, 3-4, 571-583.
3. Chester, D.K., C.R.J. Kilburn, and J.E. Guest (1985), Mount Etna: the anatomy of a volcano, Stanford Univ. Pr.
4. Dragoni, M., and A. Tallarico (1994), The effect of crystallization on the rheology and dynamics of lava flows. J Volcanol Geotherm Res, 59,241-252.
5. Ferziger, J.H., and M. Peric (2002), Computational methods for fluid dynamics, Berlin: Springer.
6. Griffiths, R.W. (2000), The dynamics of lava flows, Annu. Rev. Fluid Mech., 32, 477-518.
7. Lavallèe Y., K.U. Hess, B. Cordonnier, and D.B. Dingwell (2007), Non-Newtonian rheological law for highly crystalline dome lavas, Geology, September 2007, 35, 9, 843-846; doi: 10.1130/G23594A.1.
8. Patankar, S.V. (1980), Numerical heat transfer and fluid flow, Series in Computational Methods in Mechanics and Thermal Sciences, McGraw-Hill.
9. Pinkerton, H., and R. Stevenson (1992), Methods of determining the rheological properties of magmas at sub-solidus temperatures, J. Volcanol. Geotherm. Res., 53, 47-66.
10. Shaw, H.R., T.L. Wright, D.L. Peck, and R. Okamura (1968), The viscosity of basaltic magma: an analysis of field measurements in Makaopuhi lava lake, Hawaii. Am. J. Sci., 266, 255-264.
11. Smith, J.V. (2000), Textural evidence for dilatant (shear thickening) rheology of magma at high crystal concentrations, J. Volcanol. Geotherm. Res., 99, 1-7.
12. Sonder, I., B. Zimanowski, and R. Bttner (2006), Non-Newtonian viscosity of basaltic magma, Geoph. Res. Let., 33, L02303, doi:10.1029/2005GL024240.
13. Tallarico, A., and M. Dragoni (1999), Viscous Newtonian laminar flow in a rectangular channel: application to Etna lava flows, Bull. Volcanol., 61, 40 - 47.
14. Tallarico, A., M. Dragoni, and G. Zito (2006), Evaluation of lava effusion rate and viscosity from other flow parameters, J. Geophys. Res., 111, B11205, doi:10.1029/2005JB003762.

CLUSTERING OF INFRASONIC EVENTS AS TOOL TO DETECT AND LOCATE EXPLOSIVE ACTIVITY AT MT. ETNA VOLCANO

P. MONTALTO* and G. NUNNARI

Dipartimento di Ingegneria Elettrica, Elettronica e dei Sistemi
Universitá degli studi di Catania, Catania, 95125, Italy
*E-mail: *montalto@ct.ingv.it*
http://www.diees.unict.it

A. CANNATA†, E. PRIVITERA and D. PARTANÉ

Istituto Nazionale di Geofisica e Vulcanologia section of Catania
Catania, 95125, Italy
E-mail: †cannata@ct.ingv.it
http://www.ct.ingv.it

Active volcanoes characterized by open conduit conditions effectively generate sonic and infrasonic signals, whose investigation provides useful information for both monitoring purposes and study of the dynamics of explosive phenomena. At Mt. Etna volcano (Italy) a clustering algorithm based on spectral features and amplitude of the infrasonic events was developed. It allows to recognize the active vent with no location algorithm and by using only one station. Moreover, a waveform inversion procedure was coded, based on genetic algorithm, that enables us to quantitatively investigate the infrasound source parameters.

Keywords: Infrasound, volcanoes, clustering, Self-organizing map, K-means, source modelling.

1. Introduction

During the last decade, new insights into explosive volcanic processes have been achieved by studying infrasonic signals [1]. In fact, infrasonic activity on volcanoes is generally evidence of open conduit conditions and can provide important indications on the dynamics of the explosive processes. Unlike the seismic signal whose wavefield can be strongly affected by topography [2] and path effects [3], the infrasonic signal maintains almost unchanged its features during the propagation. In most of cases the infrasonic signals are related to the internal magma dynamics, as the acoustic resonance of fluids in the conduit, triggered by explosive sources; this implies propagation of sound waves in both magma and atmosphere [4]. Other studies relate the source of sound to the sudden uncorking of the volcano [5], the local coalescence within a magma foam [6] and the Strombolian bubble vibration [7]. In this paper we illustrate an unsupervised clustering of infrasound events, able to recognise the active vent without location algorithm. Moreover, a method to quantitatively investigate the source mechanism is shown.

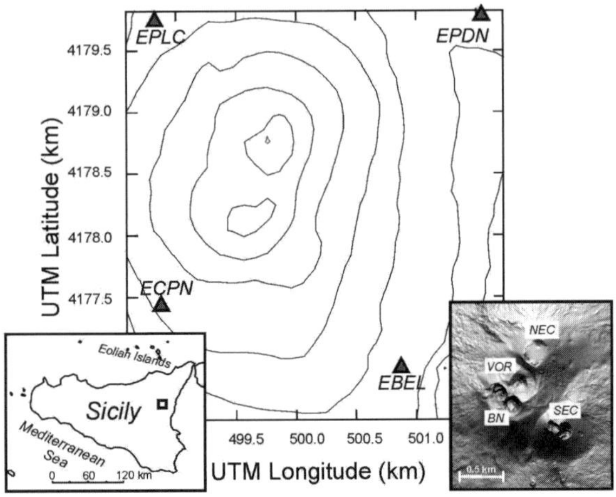

Fig. 1. Map of the summit area of Mt. Etna with the location of the four infrasonic sensors (triangles), composing the permanent infrasound network. The digital elevation model in the lower right corner shows the distribution of the four summit craters (VOR=Voragine, BN=Bocca Nuova, SEC=South-East Crater, NEC=North-East Crater).

2. Infrasonic features clustering

The time period September-November 2007 was characterised at Mt. Etna volcano (Italy) by explosive activity and intense degassing. During this time interval infrasonic signals were recorded by an infrasonic network, composed of 4 sensors azimuthally distributed around the summit area (Fig. 1). By a triggering procedure, about 1000 infrasonic events were found, consisting in amplitude transients with short duration (from 1 to over 10 s), impulsive compression onsets and peaked spectra with most of energy in the frequency range 1-5 Hz (Fig. 2).

Generally, in order to reduce the size of certain information (signals, data, etc.), definition of peculiar features or properties by a "feature extractor" may be useful. In our case, we can use both the spectral characteristics,

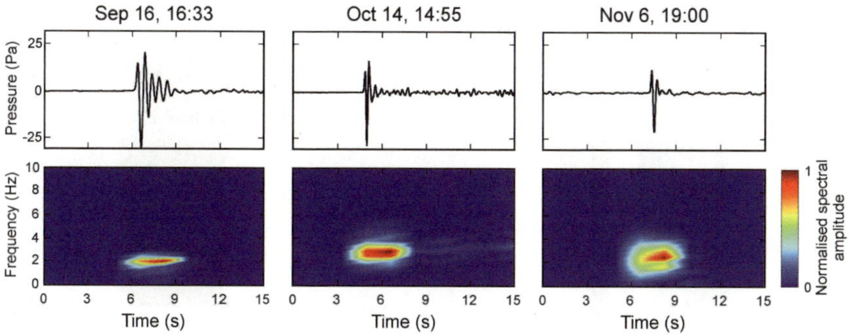

Fig. 2. Infrasonic events recorded by EBEL station and corresponding Short Time Fourier Transform.

computed by Sompi method [8] and consisting in dominant frequency and quality factor, and the peak-to-peak amplitude, as features to describe the infrasonic events. Then, in order to investigate prospective similarities or differences among the features extracted from the infrasonic signals, we plotted the frequency, the quality factor and the peak-to-peak amplitude, in the x-axis, y-axis and z-axis, respectively, and obtained the so-called 'feature space'. In order to discover cluster in the feature space, the Self-Organizing Map (SOM) was chosen. SOM is a neural network based on unsupervised learning useful in data visualization and exploration [9]. The SOM maps high-dimensional input vectors onto two-dimensional grid of prototype vectors that are easier to visualize and explore than the original data. After a learning process the clustering structure can be visualized using a common tool for visual inspection called U-matrix (unified distance matrix). It visualizes distances between neighbouring map units, and thus shows the cluster structure of the map: high values of the U-matrix indicate a cluster border while uniform areas of low values indicate clusters themselves. In Fig. 3(a) the SOM U-matrix after training algorithm is presented. Each group of neurons constitutes a cluster. In the obtained U-matrix we can see three dark blue regions, that correspond to low values in the U-matrix, and hence to clusters in the data. These regions are separated by lighter colours. Thus through the visual inspection of the U-matrix we can recognise three clusters in the feature space. By studying the final U-matrix map, and the underlying features planes of the map, a number of cluster can be identified by K-means algorithm [10]. The best clustering structure, which was obtained by the K-means algorithm, is selected using Davies-Bouldin index [11].

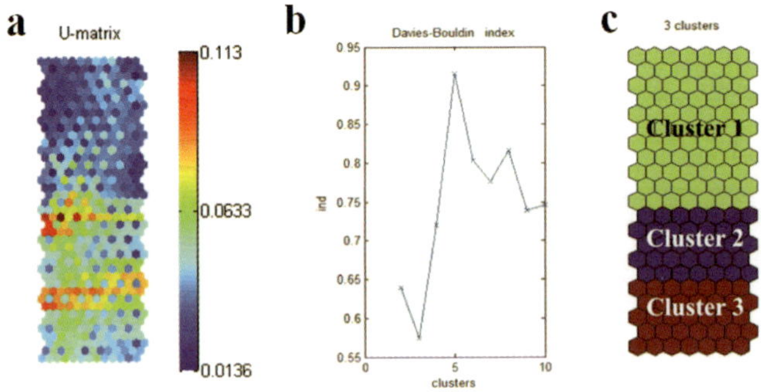

Fig. 3. (a) U-matrix, (b) Davies-Bouldin index and (c) best clustering structures.

This index uses the within-cluster distance and the between-clusters distance. The Davies-Bouldin index is suitable for evaluation of K-means partitioning because it gives low values indicating good clustering results. Fig. 3(b) shows the Davies-Bouldin index where the best clustering corresponds to the number of three clusters and then it has been projected onto the SOM (Fig. 3(c)). According to [12], a cluster (called cluster 1) was related to the degassing activity of the North East Crater, while the other two (called clusters 2 and 3) to two different explosive activities of the South East Crater (Figs. 3(c) and 4). The analyses were performed using SOM Toolbox [13].

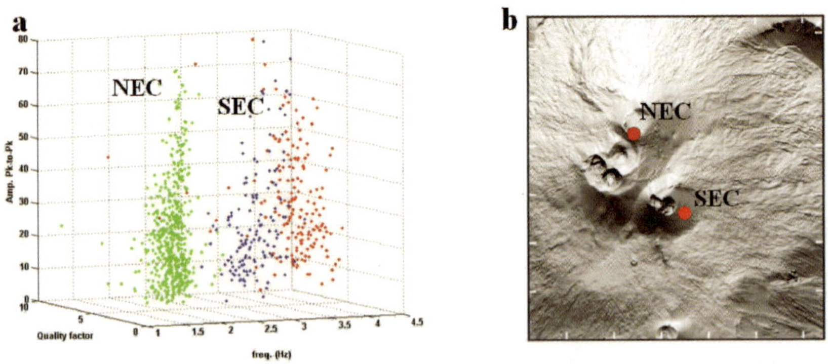

Fig. 4. (a) Infrasonic feature space (green, blue and red dots indicate clusters 1, 2 and 3 respectively). (b) Location of the vents, source of infrasound.

3. Infrasonic source modelling

Once the events are characterised and the active vents located, the source processes can be studied. The detected infrasonic events are similar to the signals described in [7] and [14], and explained as generated by the vibration of a large gas bubble, before it bursts. Therefore, in order to quantitatively investigate the source mechanism of the infrasonic events, a waveform inversion procedure was developed. Using the equations reported in [7], we were able to calculate synthetic waveforms. Then, by optimization algorithms, we can constrain the values of the three unknown parameters, radius, length of the bubble and initial overpressure (indicated by R, L and ΔP, respectively) that allow finding the best fit between synthetic and measured waveforms. Optimization method chosen to look for the best fit between observed and synthetic signals was Genetic Algorithm. Examples of waveform inversion are reported in Fig. 5.

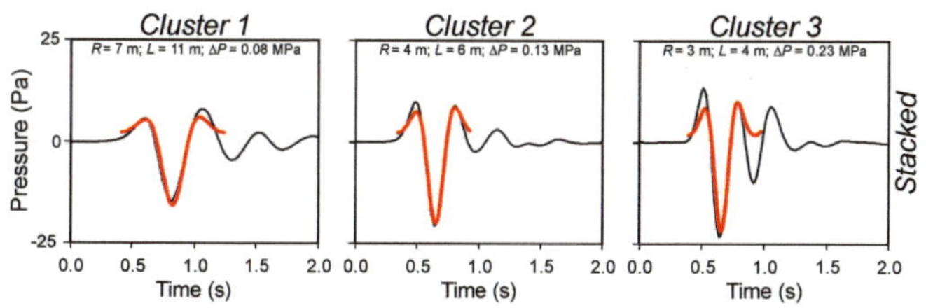

Fig. 5. Comparison between the stacked waveforms of the three clusters of infrasonic events (black) and the synthetic ones (red). The source parameters obtained by the waveform inversion are reported at top of the plots.

4. Conclusions

At active volcanoes the detection and location of explosive activity is generally obtained by videocameras and thermal sensors. However, the efficiency is strongly reduced or inhibited in case of poor visibility caused by clouds or gas plumes. In these cases the detection and characterization of explosive activity by infrasounds is very useful [12] and some techniques, based on infrasound signals recorded by arrays or networks, were developed to locate the source of this signal and then the active vent [14]. All these techniques require that most of the stations properly work and the noise is low. At Mt. Etna the events at a single vent for a certain type of activity maintain stable their features [15]. Therefore, once the link between event characteristics and vent is known we can understand which crater is active and which volcanic activity is going on by simply extracting the features of the infrasonic signal at a single station. In the light of it, a clustering based on spectral features and amplitude of the infrasonic events was developed. It allows to recognize the active vent with no location algorithm and by using only one station. Moreover, a waveform inversion procedure was coded, based on genetic algorithm, that enables us to quantitatively investigate the infrasound source parameters.

References

1. B. G. Vergniolle, S., *Geophys. Res. Lett.* **21**, 1959 (1994).
2. P. T. euberg, J., *Geophys. J. Int.* **143**, 239 (2000).
3. E. Gordeev, *J. Geophys. Res.* **98**, 19687 (1993).
4. M. S. Garces, M.A., *J. Volc. Geotherm. Res.* .
5. L. J. Johnson, J.B., *J. Volcanol. Geotherm. Res.* .
6. C.-A. J. Vergniolle, S., *J. Volcanol. Geotherm. Res.* **137**, 135 (2004).
7. B.-M. C.-A. J. Vergniolle, S., *J. Volcanol. Geotherm. Res.* **137**, 109 (2004).
8. F. Y. F. M. Y. M. Kumazawa M., Imanishi Y., *Geophys. J. Int.* **101**, 613 (1990).
9. K. T., *Self-Organizing Maps* (Springer, Berlin, 1995).
10. J. A. Dubes, R., *Pattern Recognition* **8**, 247 (1976).
11. B. D. Davies, D.L., *IEEE Transaction on Pattern Analysis and Machine Intelligence* **PAMI-1**, 224 (1979).
12. M. P. P. E. R. G. Cannata, A., *J. Geophys. Res.* **114** (2009).
13. A. E. P. J. Vesanto J, Himberg J, Self-organizing map in matlab: the som toolbox., in *Proceedings of Matlab-DSP Conference*, (Espoo, Finland, 1999).
14. M. E. Ripepe, M., *Geophys. Res. Lett.* **29** (2002).
15. M. P. P. E. R. G. Cannata, A., *Geophys. Res. Lett.* **36** (2009).

LAVA FLOW SUSCEPTIBILITY MAP OF MT. ETNA BASED ON NUMERICAL SIMULATIONS

A. CAPPELLO*, C. DEL NEGRO and A. VICARI

Istituto Nazionale di Geofisica e Vucanologia
Sezione di Catania, Italy
*E-mail: * cappello@ct.ingv.it*

We constructed maps of probability of lava inundation using computer simulations considering the past eruptive behaviour of the Mt. Etna volcano and data deriving from monitoring networks. The basic a priori assumption is that new volcanoes will not form far from existing ones and that such a distribution can be performed using a Cauchy kernel. Geophysical data are useful to update or fine tune the initial Cauchy kernel to better reflect the distribution of future volcanism. In order to obtain a final susceptibility map, a statistical analysis permits a classification of Etna's flank eruptions into twelve types. The simulation method consists of creating a probability surface of the location of future eruption vents and segmenting the region according to the most likely historical eruption on which to base the simulation. The paths of lava flows were calculated using the MAGFLOW Cellular Automata (CA) model, allowing us to simulate the discharge rate dependent spread of lava as a function of time.

Keywords: Volcanic hazard, cellular automata model, lava flow.

1. Introduction

Mt Etna will undoubtedly erupt again. When it does, the first critical question that must be answered is: which areas are threatened with inundation? What people, property, and facilities are at risk? These questions can be answered by estimating the areas most likely to be affected by eruptions on various parts of the volcano.Knowledge of the likely path and rate of advance of the front of a lava flow is of potential value for organizing both evacuations and mitigation efforts during the period of flow. Knowledge of the probability of a particular site being overrun by a lava flow is useful for long-term planning purposes. The aim of this work is to compute hazard maps that can be used as a general guide to assist emergency managers during an eruption, to plan emergency response activities, and to identify communities and infrastructure at risk.

2. Methodology

We have defined a methodology for the compilation of a new kind of map showing the hazard related to lava invasion in predefined study areas. A hazard map provides the probability that given areas will be affected by potential destructive volcanic processes. Its evaluation is generally based on the past eruptive behaviour of volcanoes and on data obtained by the monitoring networks. In order to compute hazard maps, the following steps should be computed:

(1) Definition of the source area;
(2) Computation of susceptibility map (that provides the spatial probability of vent opening);
(3) Characterization of the expected eruptions;
(4) Evaluation of the temporal probability for the occurrence of the hazard during the considered time interval;
(5) Numerical simulations of eruptive process and
(6) Construction of the hazard map.

Mt Etna (Sicily, Italy) was chosen as the study area since it as one of the most dangerous in terms of possible fracture reactivation. Having defined the study area, we have developed a methodology for estimating the long-term future spatial and temporal patterns in the Mt Etna. First a mid/long-term susceptibility map is generated, and then geophysical data are incorporated to develop the final susceptibility map. For the aim of this study, we took advantage of a database containing the main volcanological data of all eruptions at Mt Etna since 1600. Mid/long-term susceptibility maps of Etna have been generated using the following data: location of vents, fractures and vent alignments. All available datasets have been converted into a probability density function (PDF). Each of the PDFs should be given a relevance value (which measures its importance) and a reliability value (which measures the quality of the dataset) with respect to the evaluation of the susceptibility. Finally, the PDFs and their relative values are combined through a Poisson process. In the first stage a local intensity function λ_{xy} is computed using the kernel technique. A kernel function is a density function used to obtain the intensity of volcanic events at a sampling point x_p, y_p, calculated as a function of the distance to nearby volcanoes and a smoothing constant h. In estimating local volcanic densities in volcanic fields, the kernel function used is the Cauchy kernel. For the 2-D Cauchy kernel, the calculation of λ_{xy} at the grid point x_p, y_p is

$$\lambda_{xy} = \frac{1}{\pi h^2 N} \sum_{i=1}^{N} \left\{ \frac{l_i}{1 + \left(\frac{d_i}{h}\right)^2} \right\} \tag{1}$$

where d_i is the distance between x_p, y_p and the ith volcanic event, N is the number of volcanic events considered in the calculation and l_i is a factor for weighting eruption volume of the corresponding ith volcanic event. Figure 1 shows an example of PDF generated considering only past vent locations, a 1 km grid spacing and a smoothing coefficient of 2 km. Probability estimates for each grid point x_p, y_p are computed by using a

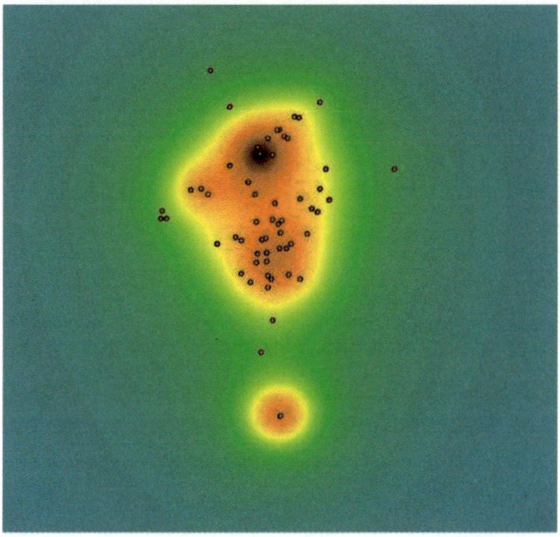

Fig. 1. An example of PDF based on past vent locations.

Poisson distribution where λ_{xy} represents the intensity function normalized to unity across the entire area:

$$P_{xy}\left\{N(t) \geq 1\right\} = 1 - \exp\left(-t\lambda_t\lambda_{xy}\Delta x\Delta y\right) \tag{2}$$

where $N(t)$ represents the number of future volcanic events that occur within time t and area $\Delta x\Delta y$. Having generated the PDF based on past, the next step is to condition it with additional data deriving from geophysical data. This step represents the short-term analysis performed using data provided by the monitoring networks. The corresponding PDF should be evaluated using the same procedure. Once all the PDFs has been calculated, they are combined (with an assigned relevance) in order to obtain the final susceptibility map. For the characterization of the volcanic events we based on the knowledge of past eruptions to fix four different volumes of total fluxes: 6, 20, 50 and 200 m^3. Then we established short, medium and large times of eruptions, setting respectively 15, 30 and 60 days of simulation. Combining these values with random distribution, we obtained twelve possible functions representing the variation of flux rate in relation to the time of eruption The shape of the curves has been considered as a kind of

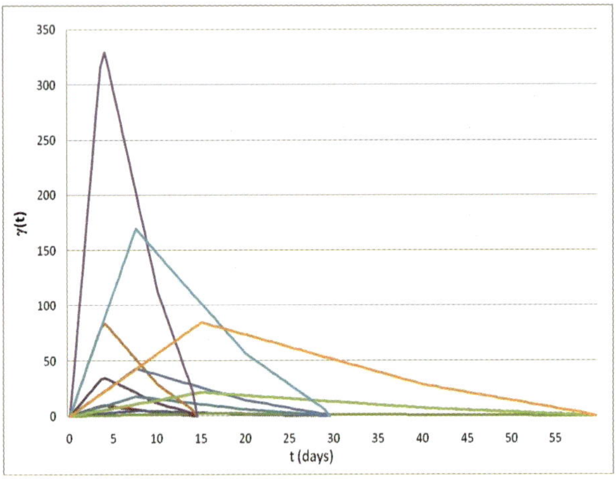

Fig. 2. Values of flux rate-days of eruption used for the simulations.

bell, in which the eruption starts from a low value of flux rate, reaching its maximum value after a 1/4 of the entire time of simulation. After 2/3 of the maximum time of simulation reaches 1/3 of the maximum value and, finally, gradually decrease until the end of the eruption is reached (Fig. 2). The susceptibility map assigns a probability of activation to each vent in the grid. Next step is the calculation of maps that assign a probability to each type of considered effusion rate and duration (event probability), devised on the basis of the emission behavior analysis of the study area. For every type i of eruption, the positional data y_i are been used to estimate the type i intensity function $\lambda_i(x)$ for each point x of the grid using the kernel estimator described by [3]. When j is an index of events in a category, i is a type of category, n is the total number of events in a category, then

$$\lambda_i(x; h_i) = h_i^{-2} \sum_{j=1}^{n_i} G\left\{\frac{x - y_{ij}}{h_i}\right\} \tag{3}$$

where h_i is a smoothing parameter that controls the size of the zone to which each data point contributes an increased intensity, and G is the truncated normal kernel

$$G(x) = \begin{cases} 0.5\pi e^{-0.5x'x} & x'x < 1 \\ 0 & x'x > 1 \end{cases} \tag{4}$$

The intensity function $\lambda_i(x)$ is then rescaled through the entire area. The choice of the parameter h_i is determinant for the calculation of the estimator. Its value for each type of event is chosen using the existing data. First of all, it is necessary to examine the strength of spatial autocorrelation in the data at different scales. This is tested using the adjusted semi-variogram

$$2\bar{\gamma}(h) = \frac{\left\{ \frac{1}{|N(h)|} \sum_{N(h)} \sqrt{|z(P_i) - z(P_j)|} \right\}^4}{\left(0.457 + \frac{0.494}{|N(h)|} \right)} \tag{5}$$

where $z(P_i)$ is the observed value at the point P_i, h is a specified distance, and $N(h)$ is the set of pairs separated by a distance near to h. For each data set in turn, the set of pairs of points is divided into suitable equal sized subsets. Each subset is used to calculate the statistic. The semi-variogram grows with increasing h until it reaches a constant level. The h-value at which it reaches this sill correspond to the distance at which autocorrelation disappears. Once the activation and the event probabilities are developed, numerical simulations could be computed. We used the MAGFLOW model for lava flow simulations based on Cellular Automatons [2,7]. The simulations were performed using the typical parameters of Etnean lava flows. A grid of vents is defined in the study area, and a prefixed number of simulations is executed for each of them, each one characterized by its own effusion rate and duration. Finally, the resulting hazard map is thus compiled by taking into account both information on lava flows overlapping, and their occurrence probability. This map is obtained by evaluating the hazard at each point in the study area as follows:

(1) for each simulation, the hazard related to a generic point in the study area is computed as the product of the defined probabilities of occurrence (conditioned probability) if it is affected by the simulated lava flow, zero otherwise;
(2) for each point, the conditioned probabilities are added over all the performed simulations.

The accuracy of the results strictly depends on the reliability of the simulation model, on the quality of input data and on the hypotheses on assigning the different probabilities of occurrence.

3. Conclusion

This work represents a preliminary methodology of applying computer simulation techniques to the assessment of hazard from lava flows. Using this method, MAGFLOW can be used to provide a series of simulation of lava flows and to produce most probable eruption scenarios. Any number of simulations can be created by sampling the probability surface and parameter of simulator. Much improvement can be done, as for example separation of different type of eruptions, or a better specification of probability surface of the location of future vents. Another improvement can be the introduction of a screening distance value (SDV) corresponding to the maximum distance from the source to the site at which the phenomenon could be a hazard. SDVs offer a way of arriving at realistic assessment of hazards based on a conservative worst-case scenario for the potential impact of each type of hazard.

References

1. M. Coltelli, C. Proietti, S. Branca, M. Marsella, D. Andronico, L. Lodato, *JGR* 112, F02029, doi:10.1029/2006JF000598, (2007).
2. C. Del Negro, L. Fortuna, A. Herault, A. Vicari, *Bull. Volcanol.*, doi:10.1007/s00445-007-0168-8, (2006).
3. P.J. Diggle, *J. R. Stat. Soc.*, Ser. A, 153, 349-362, (1900).
4. A. Felpeto, J. Marti, R. Ortiz, *Journal of Volcanology and Geothermal Research* 166, 106-116, (2007).
5. A.J. Martin, K. Umeda, C.B. Connor, J.N. Weller, D. Zhao, M. Takahashi, *JGR* 109, B10208, (2004).
6. McBirney, A. R., Serva, L., Guerra, M., Connor, C. B., *Journal of Volcanology and Geothermal Research* 126, 11-30, (2003).
7. A. Vicari, A. Herault, C. Del Negro, M. Coltelli, M. Marsella, C. Proietti, *Environmental Modelling & Software* 22, 1465-1471, (2007).
8. G. Wadge, P.A.V. Young, I.J. McKendrick, *JGR* 99 (B1), 489-504, (1994).

AUTOMATED DETECTION AND ANALYSIS OF VOLCANIC THERMAL ANOMALIES THROUGH THE COMBINED USE OF SEVIRI AND MODIS

G. GANCI* and L. FORTUNA

Dipartimento di Ingegneria Elettrica Elettronica e dei Sistemi, Universitá di Catania, Italy
*E-mail: *ganci@ct.ingv.it*

A. VICARI and C. DEL NEGRO

Istituto Nazionale di Geofisica e Vulcanologia Sezione di Catania, Italy

Multispectral infrared observations carried out by the spacecrafts have shown that spaceborne remote sensing of high-temperature volcanic features is feasible and robust enough to turn into volcano monitoring. Especially meteorological satellites have proven to be a powerful instrument to detect and monitor dynamic phenomena, such as volcanic processes, allowing very high temporal resolution despite their low spatial resolution. An automated system that uses both EOS-MODIS and MSG-SEVIRI thermal satellite data was developed at the Istituto Nazionale di Geofisica e Vulcanologia of Catania for early hot spot detection and for estimating the temporal evolution of the average effusion rate during eruptive events. The advantage of the use of both SEVIRI and MODIS data in increasing temporal coverage to improve satellite monitoring of active volcanoes was also confirmed on Etna volcano during the early phase of 2008 eruption.

Keywords: Remote sensing, Etna volcano, multi-platform.

1. Introduction

The timely and synoptic view afforded by satellite-based sensors provides an excellent means to monitor the thermal activity of a volcano and to estimate the effusion rate throughout an eruption [2,4,9]. In the past, data provided by Landsat with high spatial resolution and low temporal resolution (16 days) have been employed for the analysis of thermal analysis of active lava flows, lava domes, lava lakes and fumarole fields. Recently images with lower spatial but higher temporal resolution from meteorological satellites have been proved to be an ideal instrument for continuous monitoring of volcanic activity, even though the relevant volcanic characteristics are much smaller than the nominal pixel size (1–3 km^2). Indeed, despite of the fact that the volcanic features of interest are usually much smaller than the nominal pixel size of the sensors, meteorological satellites, such as the MODerate Resolution Imaging Spectro-radiometer (MODIS) and even the Spinning Enhanced Visible and Infrared Imager (SEVIRI), can detect emitted radiance in the shortwave infrared (SWIR) part of the electromagnetic spectrum, a region in which active lava flows, vents and domes emit copious amounts of energy. MODIS instrument offers up to 10 wavebands suitable for hot-spot detection, but provide data at least 4 times a day for any subaerial volcano with a spatial resolution of about 1 km. In particular band 21, also known as the "fire channel", was designed to have a much higher saturation temperature of about 500 K. For these reasons MODIS data were supported as basis for automated systems to detect and monitor volcanic eruptions for the entire globe. The launch of MSG SEVIRI on August 2002 provides a unique opportunity for a volcanic eruption detection system in real-time by providing images at 15 minutes interval. In spite of the low spatial resolution (3 km at nadir), the frequency of observations afforded by the MSG SEVIRI was recently applied both for fire detection [7] and for the monitoring of effusive volcanoes in Europe and Africa [6]. Since these two satellites show significantly different characteristics in spatial, spectral and temporal resolution, the aim of this work is to integrate the information coming from MODIS and SEVIRI data in order to obtain a better comprehension of the volcanic phenomenon. If near-real-time volcano monitoring is to be achieved using satellite data, images must be routinely received and analyzed rapidly. To this end, a multi-platform tool for satellite image analysis and volcanic processes characterization was developed. Computing routines were designed to allow for the joint exploitation of radiometers MODIS and SEVIRI in operational monitoring, in response to the need for fast and robust determination of hot spot detection and effusion rate estimation at active volcanoes. In this work, this monitoring system have been used to identify pixels covered by lava flow during the last Mt Etna eruption began in May 2008; then the hot spot pixels were processed in order to compute the time-varying discharge rates.

2. The satellite technique

The first step in the automatic system is the hotspot identification to locate possible thermal anomalies over satellite data. To this purpose, the contextual approach of Harris was implemented (i.e. the VAST code of [5]. As for MODIS images, this technique uses the difference between brightness temperature in channel 21 (or 22) and channel 31 (ΔT) and sets a ΔT

threshold obtained from within the image, to define whether a pixel is hot. Due to the low $NE\Delta T$ of channel 22, we use channel 22 data if they are unsaturated. If channel 22 is saturated, we use channel 21. The algorithm first defines a "nonvolcanic" portion of the image and uses the maximum ΔT from that portion to set a threshold. Pixels belonging to the volcanic area are then scanned and all the pixels that are greater than the threshold are classified as hot. An example of the hot spot detection results is given in Fig. 1, where the observed flow field is superimposed to satellite image of 22 July showing good correspondence between the pixels flagged as hot and the actual lava extent. The same procedure was applied to SEVIRI data channel 4 and 9. A hotspot image during 15 May 2008 is showed in Fig. 2. The thermal methodology for obtaining effusion rates is based on [1,8]. It was adapted to satellite thermal data by [3], who proposed a method of estimating time-averaged discharge rates using measurements of the total thermal flux of active surface flows obtained using data from satellite based sensors [10]. Thermal analysis is carried out for each hot pixel located to give an estimate of the area and radiated heat flux of the thermal feature contained within the pixel. To achieve this we assume a three component thermal surface within the pixel and use the approach of [5], where reasonable bounds are placed on the temperature of the lava crust, cracks and ambient background, to obtain lava flow crust and crack area. We use three bands to solve the following nonlinear three-equation system:

$$\begin{cases} R_x = p_b L_{\lambda_x, T_b} + p_c L_{\lambda_x, T_c} + p_h L_{\lambda_x, T_h} \\ R_y = p_b L_{\lambda_y, T_b} + p_c L_{\lambda_y, T_c} + p_h L_{\lambda_y, T_h} \\ R_z = p_b L_{\lambda_z, T_b} + p_c L_{\lambda_z, T_c} + p_h L_{\lambda z, T_h} \end{cases} \tag{1}$$

Where R_x, R_y, R_z are the corrected measured radiances in the channel x, y, z (respectively channel 4 [3.9μm], channel 9 [10.8μm], channel 10 [12μm] as for SEVIRI and channel 21/22 [3.9μm], channel 31[10.78–11.28μm], channel 32 [11.77–12.27μm] as for MODIS); $L_{\lambda,T}$ is the spectral radiance of the surface according to the Planck's law, at temperature T and wavelength λ; pb is the portion of pixel at background temperature Tb, pc is the portion of pixel at crust temperature Tc and ph is the portion of pixel with active lava at temperature Th. The unknown variables in this system are: pb, pc and Tc since we fix Th to 1363 K the extrusion temperature for Etna lavas and estimate Tb from the near lava-free pixels. Once we found the feasible solutions of the nonlinear system, we can compute the thermal flux for the high temperature component according to Stefan-Boltzmann law.

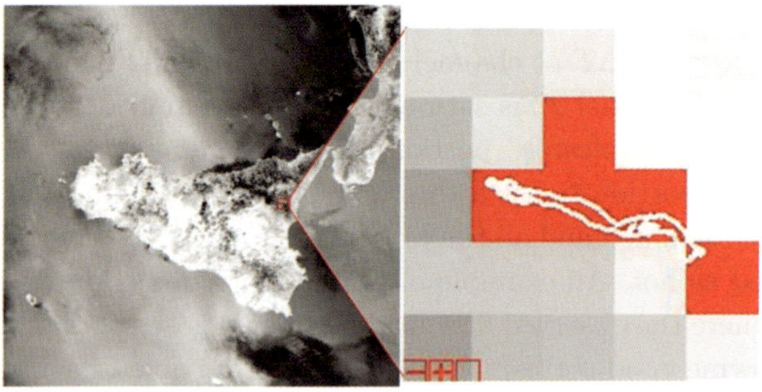

Fig. 1. HotSpot detection for MODIS data.

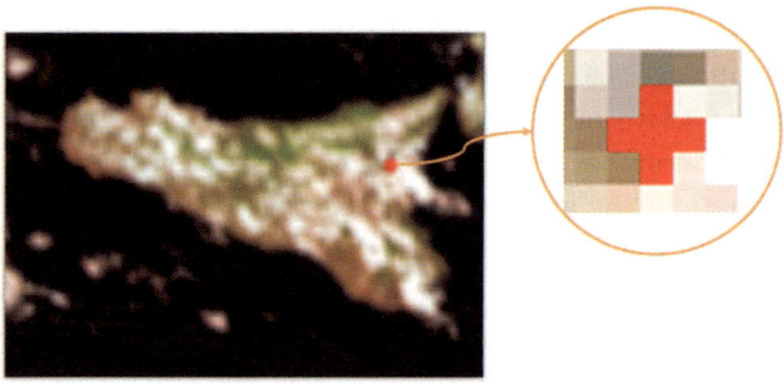

Fig. 2. HotSpot detection for SEVIRI data.

[3] and [10] showed that the total thermal flux measured using satellite infrared data can be converted to time-averaged discharge rate (E) at the time the image data were collected, by using:

$$E = \frac{Q_{TOT}}{\rho(C_P \Delta T + C_L \Delta\phi)} \tag{2}$$

where Q_{TOT} is the total thermal flux, ρ is the lava density, C_P is the specific heat capacity, ΔT is the eruption temperature of minus solidus temperature, C_L is the latent heat of crystallization, and $\Delta\phi$ is the volume percent of crystals that form while cooling through ΔT.

3. A case study: Etna 2008 eruption

Etna's 2008 eruption provided the opportunity to verify our model's ability to predict the path of lava flows while the event was ongoing and to produce different scenarios as eruptive conditions changed. A new eruption started on the morning of 13 May 2008 and it still going on. After a seismic swarm of more than 200 earthquakes and significant ground deformation a fissure erupted in the summit area immediately to the east of Etna's summit craters. On the afternoon of the same day, a new eruptive fissure opened with a number of vents displaying Strombolian activity and emission of lava flows toward the Valle del Bove (a wide depression that cuts the eastern flank of the volcanic edifice). An helicopter survey carried out on 14 May at 13:00 showed the two eruptive fissures: a first one opened on the east of the summit craters (3000 m asl) spreading along North-South direction and a second fissure started from the east flank of South-East Crater summit cone of Mt Etna (2900–2500 m asl) spreading with ENE-WSW orientation toward the Valle del Bove. During the following 24 hours the lava traveled approximately 6 km to the east, but thereafter its advance slowed and stopped, the most distant lava fronts stagnating about 3 km from the nearest village, Milo. Between 16 and 18 May ash emissions became more frequent and produced small but spectacular clouds, whereas the rate of lava emission showed a gradual diminution. During late May and the first week of June, the activity continued at low levels, with lava flows advancing only a few hundred meters from the vents as of 4 June. Four days later, on 8 June, there was a considerable increase in the vigor of Strombolian activity and lava output rate. During the following week, lava flows advanced up to 5 km from the source vents. Volcanic thermal anomalies have been observed almost continuously over the same Mt Etna flank in accordance with the occurrence of the lava effusion. A significant increase in the number of hotspots detected, with an evident increase even in their relative intensity, was instead recognized in the late evening of 13 May in accordance with the opening of the new eruptive fissure on NE side of SE Crater. Starting from 14 May several hotspots of high intensity have been flagged over the target area, indicating the clear presence of a lava flow. The time-varying effusion rate during 14 May-16 July was estimated from MODIS and SEVIRI data allowing a detailed chronology of lava flow emplacement (refresh rates: 15 minutes).The lava discharge reached a peak during the first days of the eruption; on 23 May the discharge rate decreased keeping medium-low values of about $1 - 2\mathrm{m}^3\mathrm{s}^{-1}$ until 8 June. After 8 June, the effusion rate increased showing scattered values; anyway this behavior could be due to an increasing of strombolian activity registered during this period. By integrating minimum and maximum effusion rate values we computed the cumulative curves of erupted lava, finding that the erupted volume is constrained between 8 and 20 million cubic meters. The estimated time-varying effusion rate was also tested, with good results, giving it as input to the numerical simulation of the lava flow path. The good agreement between simulated and mapped flow areas indicates that model-based inundation predictions, driven by time-varying discharge rate data, provide an excellent means for assessing the hazard posed by on-going effusive eruptions.

References

1. Crisp, J., Baloga, S.,*Journal of Geophysical Research*, **95**:1255-1270,(1990).
2. Harris, A.J.L., Dehn, J., Calvari, S., *Bulletin of Volcanology*, doi:10.1007/s00445-007-0120-y,(2007).
3. Harris, A. J. L., Butterworth, A. L., Carlton, R. W., Downley, I., Miller, P., Navarro, P., and Rothery, D. A., *Bulletin of Volcanology*, **59**, pp 49-64,(1997).
4. Herault, A., Vicari, A., Ciraudo, A., and Del Negro, C., *Computers & Geosciences*, doi:10.1016/j.cageo.2007.10.008,(2007).
5. Higgins, J., Harris, A.J.L., *Computers & Geosciences*, **23**(6):627-645,(1997).
6. Hirn, B., Di Bartola, C., Laneve, G., Cadau, E. and Ferrucci, F.,*IEEE International Geoscience & Remote Sensing Symposium July 6-11*, Boston, Massachusetts, U.S.A.,(2008).
7. Laneve, G., Cadau, E. G.,*Geoscience and Remote Sensing Symposium, IGARSS*, doi 10.1109/IGARSS.2007.4423337, pp 2447-2450,(2007).
8. Pieri, D.C., Baloga, S.M.,*J Volcanol Geotherm Res*, **30**:29-45,(1986).
9. Vicari, A., Ganci, G., Ciraudo, A., Herault, A., Corviello, I., Lacava, T., Marchese, F., Del Negro, C., Pergola, N., Tramutoli, V.,*Use of Remote Sensing Techniques for Monitoring Volcanoes and Seismogenic Areas, USEReST*,(2008).
10. Wright, R., Blake, S., Harris, A., Rothery, D.,*Earth and Planetary Science Letters*, **192**:223-233,(2001).

MODELING VOLCANOMAGNETIC DYNAMICS BY RECURRENT LEAST-SQUARE SUPPORT VECTOR MACHINES

STANISLAW JANKOWSKI* and ZBIGNIEW SZYMANSKI

Warsaw University of Technology, Poland
*E-mail: *sjank@ise.pw.edu.pl*

GILDA CURRENTI, ROSALBA NAPOLI and CIRO Del NEGRO

Istituto Nazionale di Geofisica e Vulcanologia
Sezione di Catania, Italy

LUIGI FORTUNA

Dipartimento Di Ingegneria Elettrica Elettronica e dei Sistemi Università di Catania
Italy

Nonlinear dynamic systems can be described by means of statistical learning theory: neural networks and kernel machines. In this work the recurrent least-squares support vector machines are chosen as learning system. The unknown dynamic system is a mapping of past states into the future. The recurrent system is implemented by special data preparation in the learning phase. The next iterations can be calculated but the convergence is usually not guaranteed. Due to the fact that the predicted trajectory can diverge from the real trajectory the semi-directed mode can be applied, i.e. after several prediction steps the system is updated by using the current values of the considered process as new initial conditions.The idea was applied to real magnetic data acquired at Etna volcano.

Keywords: Recurrent ls-svm, volcanomagnetic dynamics.

1. Recurrent least-squares support vector machine

Least-squares support vector machine (LS-SVM) originates by changing the inequality constraints in the support vector machine (SVM) [1] formulation to equality constraints with objective function in the least squares sense [2,3].

The learning set for the regression task consists of pairs: input vectors and the target function value $D = \{(\mathbf{x}_i, t_i)\}\mathbf{x}_i \in X \subset R^d, \quad t_i \in R$. The model is expressed as: $f(\mathbf{x}) = \mathbf{w}\varphi(\mathbf{x}) + b$.

The LS-SVM can be formulated as the following constraint optimization problem:

$$L = \frac{1}{2}||\mathbf{w}||^2 + \gamma \sum_{i=1}^{l} [t_i - \mathbf{w} f(\mathbf{x}_i) - b]^2 \tag{1}$$

Kernel function:

$$K(\mathbf{x}, \mathbf{x}') = \varphi(\mathbf{x}) \cdot \varphi(\mathbf{x}') \tag{2}$$

The solution can be expressed as the linear combination of kernels weighted by the Lagrange multipliers:

$$f(\mathbf{x}) = \sum_{i=1}^{l} \alpha_i K(\mathbf{x}_i, \mathbf{x}) + b \tag{3}$$

Hence, the learning of this system is performed by solving the system of linear equations

$$\begin{bmatrix} \mathbf{K} + \gamma^{-1}\mathbf{I} & \mathbf{1} \\ \mathbf{1}^T & 0 \end{bmatrix} \begin{bmatrix} \alpha \\ b \end{bmatrix} = \begin{bmatrix} \mathbf{t} \\ 0 \end{bmatrix} \tag{4}$$

We use the RBF kernel: $K(\mathbf{x}, \mathbf{x}') = \exp\{-\eta||\mathbf{x} - \mathbf{x}'||^2\}$

The global minimizer is obtained in LS-SVMs by solving the set of linear equations (instead of quadratic programming in case of SVM). However the sparseness of the support vectors is lost. In SVM, most of the Lagrangian multipliers α_i are zeros while in LS-SVM the Lagrangian multipliers α_i are proportional to the errors e_i.

In order to obtain the recurrent model of a given dynamic data we adopt the LS-SVM system for learning of dynamical feedback systems.

The learning algorithm consists of 2 phases. In the phase 1 the model state inputs are delayed measured output values of the process for the input-output representation. Based on the learning data set

$$\begin{aligned} &\{\mathbf{x}_k^{(0)}, y_k^p\}_{k=1}^N, \\ &\mathbf{x}_k^{(0)} = [y_{k-1}^p, y_{k-2}^p, y_{k-3}^p] \end{aligned} \tag{5}$$

the LS-SVM model is prepared

$$\{\alpha_k^{(0)}, \mathbf{x}_k^{(0)}\}_{k=1}^N, \quad N = 1000. \tag{6}$$

In the learning phase 2 the measured output values are replaced by the estimated output values of the predictor before performing the new learning phase. The procedure is described by following pseudo-code:

for n=1 to 2 do
  Estimate output values:
  $y_k^{(n)} = simLSSVM(\{\alpha_i^{(n-1)}, \mathbf{x}_i^{(n-1)}\}_{i=1}^{N},$
  $\mathbf{x}_k^{(n-1)}), k = 1..N$
  Prepare new learning set:
  $\{\mathbf{x}_k^{(n)}, y_k^{(n)}\}_{k=1}^{N},$
  $\mathbf{x}_k^{(n)} = [y_{k-1}^{(n)}, y_{k-2}^{(n)}, y_{k-3}^{(n)}]$
  Create new LS-SVM model:
  $\{\alpha_k^{(n)}, \mathbf{x}_k^{(n)}\}_{k=1}^{N},$

The organization of the data preparation is explained in Tables 1 and 2.

Table 1. The structure of learning data for the recurrent LS-SVM – learning phase I.

X input of LS-SVM			Y output of LS-SVM
$Y(0)$	$Y(1)$	$Y(2)$	$Y(3)$
$Y(1)$	$Y(2)$	$Y(3)$	$Y(4)$
...	...	...	...
$Y(i)$	$Y(i+1)$	$Y(i+2)$	$Y(i+3)$

Table 2. The structure of learning data for the recurrent LS-SVM – learning phase II.

X input of LS-SVM			Y output of LS-SVM
$Y'(3)$	$Y'(4)$	$Y'(5)$	$Y'(6)$
$Y'(4)$	$Y'(5)$	$Y'(6)$	$Y'(7)$
...	...	...	...
$Y'(i)$	$Y'(i+1)$	$Y'(i+2)$	$Y'(i+3)$

$Y(i)$ – function value (measured)
$Y'(i)$ – function value (predicted from recurrent LS-SVM)

Our approach differs from that presented in [3]. The organization of the learning phase follows the suggestion presented in [6].

2. Magnetic data analysis on Etna volcano

Over the last decades different analyses have been devoted to reveal the presence of the chaotic motion in geomagnetic time series in volcanic areas [4,5]. The geomagnetic time series from the magnetic network on Etna volcano are analyzed to investigate the dynamical behavior of magnetic anomalies. The predictability of the geomagnetic time series was evaluated to establish a possible low-dimensional deterministic dynamics.

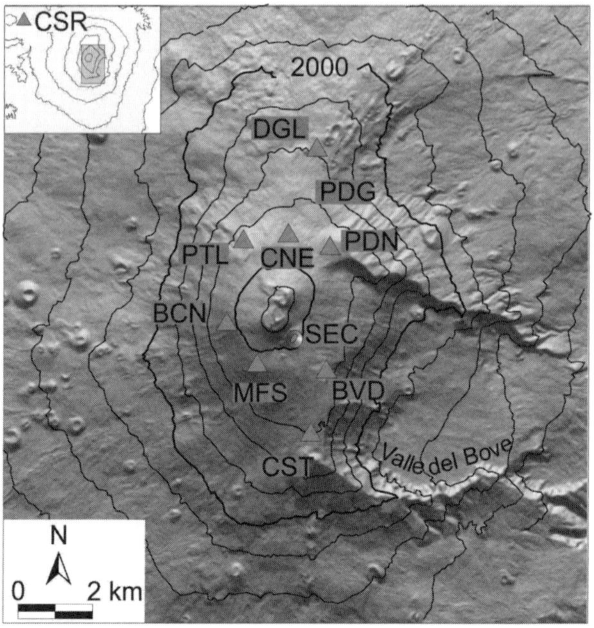

Fig. 1. Magnetic monitoring network on Etna volcano.

The analysis of the 10-minutes differences at Pizzi Deneri (PDN) monitoring station with respect to the Cesarò reference station CSR (located far away from the volcano edifice - see Fig. 1) shows prominent peaks centered around diurnal components at the period of 8, 12 and 24 h. After having removed the dominant periodic components, the filtered differences appear to be aperiodic and broadband.

Therefore, we look at the recurrent least-squares support vector machine approach as a way to provide evidence on the mechanism generating the time dependent variations.

The data from PDN station was firstly normalized to the range $[-1, 1]$. We used a learning data set from 7th to 14th January 2008. The testing data set spans from 15th to 21st January 2008.

The recurrent LS-SVM parameters are as follows: kernel type 'RBF' kernel, regularization parameter $\gamma = 2.13$, kernel width $\sigma = 0.264$. The current process value is calculated based on 3 previous measured values.

The model was tested on the data acquired between 15th and 21st January 2008. The goodness of the model can be evaluated using Absolute

Percentage Error (APE) defined as:

$$\text{APE} = \frac{|y_{\text{actual}} - y_{\text{predicted}}|}{y_{\text{actual}}} \cdot 100\% \tag{7}$$

The APE values for the test data were within range (−1.2%,0%). The Mean Absolute Percentage Error (MAPE) equals −0.13%, while the Root Mean Squared Error (RMSE) is 0.796.

The same recurrent LS-SVM was applied for the simulation of the data from 7th to 21th May 2008 (Fig. 2). On 13th May 2008 significant local magnetic field changes occurred which marked the resumption of the eruptive activity.

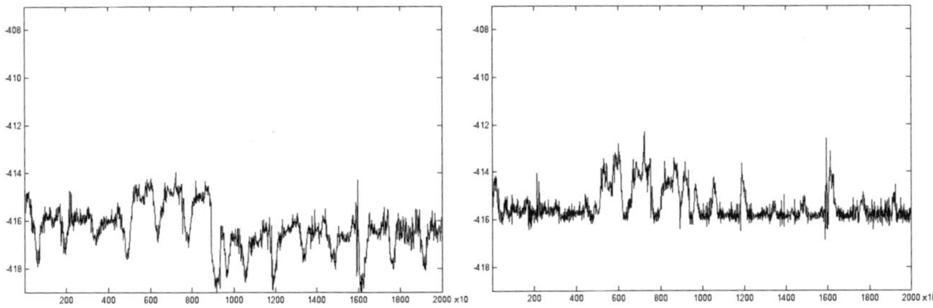

Fig. 2. a) Testing data set from PDN station (7th to 21st May 2008). b) Results of simulation on the testing data set (7th to 21st May 2008).

We assumed that the performance of the recurrent LS-SVM trained on the data where no significant magnetic changes occurred would be lower when applied to the data where such changes occurred. The APE values were within range (−1.5%,0%). The MAPE and RMS values (Fig. 3) were calculated for the 25 hour periods. Increased error rate is clearly visible at the point corresponding to occurrence of significant magnetic changes.

3. Conclusion

In this paper, identification methods are dedicated to understanding and describing the temporal dynamics of a geomagnetic time series gathered on Etna volcano. The results could have important implications on the study of the dynamical behavior of the volcanomagnetic signals. They underline that volcanomagnetic signals are the result of complex processes that cannot be easily predicted. The application of recurrent LS-SVM forecasting techniques has not provided strong evidence of nonlinear deterministic dynamics in volcanomagnetic data.

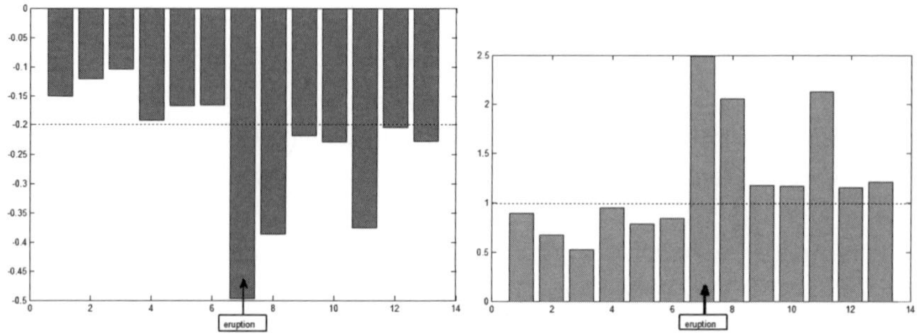

Fig. 3. (a) The Mean Absolute Percentage Error on the testing data (each bar represents 25 hour period). The data spans from 7th to 21st May 2008. (b) The Root Mean Square Error on the testing data (each bar represents 25 hour period). The data spans from 7th to 21st May 2008.

References

1. V.N. Vapnik: *Statistical Learning Theory*, Wiley Interscience, New York 1998
2. J.A.K. Suykens and J. Vandewalle: Least squares support vector machine classifier, *Neural Processing Letters*, **9**, 1999, 293-300
3. J.A.K Suykens, J. Vandewalle. Recurrent least squares support vector machines. *IEEE Transaction on Circuits and Systems*-I, **47** (2000), 1109–1114.
4. G. Currenti, C. Del Negro, L. Fortuna, A. Vicari, Nonlinear Identification of Complex Geomagnetic Models: An Innovative Approach, *Nonlinear Phenomena in Complex Systems*, 6,1 (2003) 524–533.
5. G. Currenti, C. Del Negro, L. Fortuna, R. Napoli, and A. Vicari, Non-linear analysis of geomagnetic time series from Etna volcano. *Nonlinear Processes in Geophysics* (2004) 11, 119-125.
6. M. Lucea: *Modélisation dynamique par réseaux de neurones et machines à vecteurs supports: contribution à la maîtrise des émissions polluantes de véhicules automobiles*, Ph. D. Thesis, Université Paris 6, 2006
7. H.N. Qu, Y. Oussar, G. Dreyfus, W. Xu: Regularized Recurrent Least Squares Support Vector Machines, *Proc. 2009 Int. Joint Conf. Bioinformatics, Systems Biology and Intelligent Computing*, Shanghai, China 2009 (preprint)

PART H

Geometric control for quantum and classical models

TIME-REVERSAL AND STRONG H-THEOREM FOR QUANTUM DISCRETE-TIME MARKOV CHANNELS

F. TICOZZI

Dipartimento di Ingegneria dell'Informazione, Università di Padova
via Gradenigo 6/B, 35131 Padova, Italy
E-mail: ticozzi@dei.unipd.it
www.dei.unipd.it

M. PAVON

Dipartimento di Matematica Pura ed Applicata, Università di Padova
via Trieste 63, 35131 Padova, Italy
E-mail: pavon@math.unipd.it

The time reversal of a completely-positive, non-equilibrium discrete-time quantum Markov evolution is derived via a suitable adjointness relation. Space-time harmonic processes are introduced for the forward and reverse-time transition mechanisms, and their role for relative entropy dynamics is discussed.

Keywords: Time reversal, space-time harmonic process, H-theorem.

1. Introduction

In this paper, we employ a mathematical framework for discrete-time Markovian processes, originating from Nelson's kinematics of diffusion processes [8,9], to derive the *time-reversal* of a Markovian evolution. The latter entails the Lagrange adjoint with respect to the (semi-definite) inner product induced by the flow of probability distributions. We show that this also holds for finite-dimensional, discrete-time quantum Markov evolutions. This result, and the properties of *space-time harmonic processes*, are used to derive monotonicity of an operator form of relative entropy and of the Belavkin-Staszewski's relative entropy. This paper is a shortened version of [10], to which we refer for the proofs and for further discussions.

2. Kinematics of Markov chains and time-reversal

Let $\{X(t), t \in \mathbb{Z}\}$ be a Markov chain taking values in the finite set $\mathcal{X} = \{x_1, x_2, \ldots, x_n\}$ which we identify from here on with the set of the indexes $\{1, 2, \ldots, n\}$. We denote by π_t the probability distribution of $X(t)$ over $\mathcal{X}$. In the following, π_t is always intended as a column vector, with i-th component $\pi_t(i) = \mathbb{P}(X(t) = i)$. Let $P(t)$ denote the *transition matrix* with elements $p_{ij}(t) = \mathbb{P}(X(t+1) = j|X(t) = i), i, j = 1, \dots, n$.

Let us agree that troughout the paper $\dagger$ indicates adjoint with respect to the natural inner product. The evolution is then given by the *forward* equation $\pi_{t+1} = P^\dagger(t)\pi_t$. For x and y n-dimensional column vectors, we define the semi-definite form:

$$\langle x, y\rangle_{\pi_t} = x^\dagger D_{\pi_t} y, \tag{1}$$

which is an inner product if $D_\pi = \operatorname{diag}(\pi_t(1), \pi_t(2), \dots, \pi_t(n))$ is positive definite. It represents the expectation of the random variable Z defined on $(\mathcal{X}, \pi_t)$ by $Z(i) = x_i y_i$. Let $\mathcal{F}_t^-, t \in \mathbb{Z}$ be the σ-algebra generated by $\{X(s), s \leq t\}$ and $\mathcal{F}_t^+$ to be the σ-algebra generated by $\{X(s), s \geq t\}$. Let $f : \mathbb{Z} \times \mathcal{X} \to \mathbb{R}$. Let us introduce the *forward conditional difference* $\Delta^+ f(t, X(t))$ with respect to the family $\{\mathcal{F}_t^-\}, t \geq 0$:

$$\begin{aligned} \Delta^+ f(t, X(t))_{|X(t)=i} : \quad &= \mathbb{E}(f(t+1, X(t+1)) - f(t, X(t))|X(t) = i) \\ &= \sum_j f(t+1, j) p_{ij}(t) - f(t, i). \end{aligned} \tag{2}$$

Henceforth, we denote by f_t and $\Delta^+ f_t$ the column vectors with i-th component $f(t, i)$ and $\Delta^+ f(t, X(t))_{|X(t)=i}$, respectively. We can then rewrite (2) in the compact form

$$\Delta^+ f_t = P(t) f_{t+1} - f_t. \tag{3}$$

Consider now the vector space $\mathcal{K} = \{f : \mathbb{Z} \times \mathcal{X} \to \mathbb{R} \,|\, \exists t_0, t_1, t_0 \leq t_1 \text{ s. t. } f(t, i) = 0, \forall i, t \notin [t_0, t_1]\}$, namely the set of functions with finite support. For $f, g \in \mathcal{K}$, we define the semi-definite *space-time inner product*

$$\begin{aligned} \langle f, g\rangle_\pi := \sum_{t=-\infty}^{\infty} \langle f_t, g_t\rangle_{\pi_t} \quad &= \quad \sum_{t=-\infty}^{\infty} f_t^\dagger D_{\pi_t} g_t \\ &= \quad \sum_{t=-\infty}^{\infty} \mathbb{E}(f(t, X(t))\, g(t, X(t))), \end{aligned}$$

where $\pi \sim \{\pi_t, t \in \mathbb{Z}\}$ denotes the family of the one-time chain distributions. By employing a discrete "integration by parts" formula, we get:

Proposition 2.1. *Let $f, g \in \mathcal{K}$. Then*

$$\langle \Delta^+ f, g\rangle_\pi = \langle f, \Delta^- g\rangle_\pi \tag{4}$$

In view of relation (4), we call Δ^- a $\langle\cdot,\cdot\rangle_\pi$-adjoint of Δ^+. Hence, the two conditional differences are *adjoint* with respect to the semi-definite *space-time inner product.* On the other hand, by using (3) and some straightforward calculations[a], we get

$$\sum_{t=-\infty}^{\infty} \mathbb{E}(\Delta^+ f(t,X(t))\, g(t,X(t))) = \sum_{t=-\infty}^{\infty} \langle f_{t+1}, D_{\pi_{t+1}}^{-1} P^\dagger(t) D_{\pi_t} g_t - g_{t+1}\rangle_{\pi_{t+1}}$$

Let $\pi_t(i) > 0$ for all t, i. In this case, $\langle f, g\rangle_\pi$ is an inner product and the corresponding adjoint is unique. We conclude that $\Delta^- g_{t+1} = D_{\pi_{t+1}}^{-1} P^\dagger(t) D_{\pi_t} g_t - g_{t+1}$. More explicitly, the matrices $Q(t) = D_{\pi_{t+1}}^{-1} P^\dagger(t) D_{\pi_t}$, are simply the matrices of the reverse-time transition probabilities. When $\pi_{t+1}(j) = 0$, $q_{ji}(t)$ may be defined arbitrarily to be any number between zero and one, provided it satisfies the normalization condition $\sum_i q_{ji}(t) = 1$. Notice that (5) leads to the correct form of the time-reversal even if the distributions $\{\pi_t\}$ are only non-negative. The space-time adjointness relation (4) for Markov chains admits an equivalent, compact formulation.

Proposition 2.2. *The space-time adjointness relation* (4) *holds if and only if for any t*

$$\langle P(t)x, y\rangle_{\pi_t} = \langle x, Q(t)y\rangle_{\pi_{t+1}}, \quad x, y \in \mathbb{R}^n. \tag{5}$$

3. Time-reversal for quantum Markov channels

Consider an n-level quantum system with associated Hilbert space $\mathcal{H}$ isomorphic to $\mathbb{C}^n$. Any linear, Trace Preserving and Completely Positive (TPCP) dynamical map $\mathcal{E}^\dagger$ acting on density operators can be represented by a Kraus operator-sum [6], i.e.: $\rho_{t+1} = \mathcal{E}^\dagger(\rho_t) = \sum_j M_j \rho_t M_j^\dagger$, $\sum_j M_j^\dagger M_j = I$. Following a quite standard quantum information terminology, we refer to linear, completely-positive trace-non-increasing Kraus maps as *quantum operations.* For observables, the *dual* dynamics is given by the *identity-preserving* quantum operation $\mathcal{E}(X) = \sum_j M_j^\dagger X M_j$. In the remaining of the paper, we consider the discrete-time quantum Markov evolutions associated to an initial density matrix ρ_0 and a sequence of TPCP maps $\{\mathcal{E}_t^\dagger\}_{t\geq 0}$.

In order to find the time-reversal of a given Markovian evolution, we rewrite the probability-weighted inner product (1) in symmetrized form

[a]In what follows, whenever a matrix M is not invertible, M^{-1} is to be understood as the generalized (Moore-Penrose) inverse $M^{\#}$.

$\langle x,y\rangle_\pi = \mathrm{trace}(D_x^{\frac{1}{2}} D_\pi^{\frac{1}{2}} D_y D_\pi^{\frac{1}{2}} D_x^{\frac{1}{2}})$. Dropping commutativity and allowing for a general density operator ρ and observables X, Y, we define:

$$\langle X,Y\rangle_\rho = \mathrm{trace}(X^{\frac{1}{2}}\rho^{\frac{1}{2}} Y \rho^{\frac{1}{2}} X^{\frac{1}{2}}).$$

This is a symmetric, real, semi-definite sesquilinear form on Hermitian operators. By analogy with the classical case, we then define the quantum operation $\mathcal{R}_{\mathcal{E},\rho_t}$ as the space-time $\{\rho_t\}$-adjoint of a quantum operation $\mathcal{E}$ using the quantum version of (5), which reads $\langle \mathcal{E}(X), Y\rangle_{\rho_t} = \langle X, \mathcal{R}_{\mathcal{E},\rho_t}(Y)\rangle_{\rho_{t+1}}$. In [10], it is shown that $\mathcal{R}_{\mathcal{E},\rho_t}$ admits an operator-sum representation with Kraus operators

$$R_j(\mathcal{E},\rho_t) = \rho_{t+1}^{-\frac{1}{2}} M_j \rho_t^{\frac{1}{2}}. \tag{6}$$

It is now natural to define a transformation between Kraus operators. Let $\mathcal{E}^\dagger$ be a quantum operation represented by Kraus operators $\{F_k\}$. For any ρ, define the map $\mathcal{T}_\rho$ from quantum operations to quantum operations

$$\mathcal{T}_\rho : \mathcal{E}^\dagger \mapsto \mathcal{T}_\rho(\mathcal{E}^\dagger), \tag{7}$$

where $\mathcal{T}_\rho(\mathcal{E}^\dagger)$ has Kraus operators $\{\rho^{\frac{1}{2}} F_k^\dagger (\mathcal{E}(\rho))^{-\frac{1}{2}}\}$. The results of [2] show that the action of $\mathcal{T}_\rho$ is independent of the particular Kraus representation of $\mathcal{E}^\dagger$. With this definition, we have that $\mathcal{T}_{\rho_t}(\mathcal{E}^\dagger) = \mathcal{R}^\dagger_{\mathcal{E},\rho_t}$. We are now ready for the main result of this section.

Theorem 3.1 (Time Reversal of TPCP maps, [10]). *Let $\mathcal{E}^\dagger$ be a TPCP map. If $\rho_{t+1} = \mathcal{E}^\dagger(\rho_t)$, then for any $\rho_t \in \mathfrak{D}(\mathcal{H})$, $\mathcal{R}^\dagger_{\mathcal{E},\rho_t} = \mathcal{T}_{\rho_t}(\mathcal{E}^\dagger)$ defined as in* (6) *is the* time-reversal *of $\mathcal{E}^\dagger$ for ρ_t, that is, it satisfies both:*

$$\rho_t = \mathcal{R}^\dagger_{\mathcal{E},\rho_t}(\rho_{t+1}) \tag{8}$$

$$\mathcal{T}_{\rho_{t+1}}(\mathcal{R}^\dagger_{\mathcal{E},\rho_t})(\sigma_t) = \mathcal{E}^\dagger(\sigma_t), \tag{9}$$

for all $\sigma_t \in \mathfrak{D}(\mathcal{H})$ such that $\mathrm{supp}(\sigma_t) \subseteq \mathrm{supp}(\rho_t)$. *Morover, it can be augmented to be TPCP without affecting property* (8)-(9).

Augmenting a Kraus map $\mathcal{E}$ with Kraus operators $\{M_k\}_{k=1,\dots,m}$ to a TPCP map means adding a finite number p of Kraus operators $\{M_k\}_{k=m+1,\dots,m+p}$ so that $\sum_k M_k^\dagger M_k = I$.

Remark: Property (9) ensure us that among all quantum operations mapping ρ_{t+1} back to ρ_t, $\mathcal{R}^\dagger_{\mathcal{E},\rho_t}$ is the natural *time-reversal* of $\mathcal{E}^\dagger$ with respect to ρ_t. In fact, notice that if ρ_t is full rank, (9) implies that $\mathcal{T}_{\rho_{t+1}} \circ \mathcal{T}_{\rho_t}$ is the the identity map on quantum operations. While studying quantum error correction problems, the same $\mathcal{R}^\dagger_{\mathcal{E},\rho}(\cdot)$ has been suggested by Barnum and Knill as a near-optimal correction operator [2] in the full rank case.

4. Application to information dynamics

Consider a reference quantum Markov evolution on a finite time interval, generated by an initial density matrix ρ_0 and a sequence of TPCP maps $\{\mathcal{E}_t^\dagger\}_{t\in[0,T-1]}$. A sequence of Hermitian operators $\{Y_t\}_{t\in[0,T-1]}$ is said to be *space-time harmonic* with respect to the family of identity-preserving maps $\{\mathcal{E}_t\}_{t\in[0,T-1]}$ if $Y_t = \mathcal{E}_t(Y_{t+1})$. On the other hand, $\{Y_t\}_{t\in[0,T-1]}$ is said to be *space-time harmonic in reverse-time* with respect to the family $\{\mathcal{R}_{\mathcal{E}_T,\rho_t}\}$ if $Y_{t+1} = \mathcal{R}_{\mathcal{E}_t,\rho_t}(Y_t)$. The sequence is called *space time subharmonic* if $Y_t \le \mathcal{E}_t(Y_{t+1})$, where we are referring to the natural partial order between Hermitian matrices. We then have the following useful result:

Proposition 4.1. *Let Y_t be a space-time harmonic process with respect to $\{\mathcal{E}_t\}_{t\ge 0}$, with eigenvalues $\lambda_{t,i} \in \mathcal{I} \subset \mathbb{R}$ at all times, and $f : \mathcal{I} \to \mathbb{R}$ be operator convex*[b]*. Then $Z_t := f(Y_t)$ is space-time subharmonic.*

While the usual definition of quantum relative entropy is due to Umegaki [12], the natural definition of relative entropy in our setting is the Belavkin-Staszewski's (BS) relative entropy [4]:

$$\mathbb{D}_{BS}(\rho\|\sigma) = \mathrm{trace}\left(\sigma\left(\sigma^{-\frac{1}{2}}\rho\sigma^{-\frac{1}{2}}\right)\log\left(\sigma^{-\frac{1}{2}}\rho\sigma^{-\frac{1}{2}}\right)\right). \tag{10}$$

The BS relative entropy is consistent with the classical relative entropy, which is recovered by considering commuting matrices, and with the von Neumann entropy.

Consider two quantum Markov evolutions, corresponding to different initial conditions $\rho_0 \ne \sigma_0$, but with same family of trace-preserving quantum operations $\{\mathcal{E}_t^\dagger\}$. Define the observable $Y_t = \sigma_t^{-\frac{1}{2}}\rho_t\sigma_t^{-\frac{1}{2}}$. We thus have that: $\mathcal{R}_{\mathcal{E},\sigma_t}(Y_t) = \sum_k \sigma_{t+1}^{-\frac{1}{2}} M_k \sigma_t^{\frac{1}{2}} \sigma_t^{-\frac{1}{2}} \rho_t \sigma_t^{-\frac{1}{2}} \sigma_t^{\frac{1}{2}} M_k^\dagger \sigma_{t+1}^{-\frac{1}{2}} = Y_{t+1}$. This shows that Y_t evolves in the forward direction with the backward transition mechanism of σ_t, which makes it quantum *space-time harmonic in reverse time* with respect to the transition of σ_t. As a consequence of Proposition 4.1, we have the following result.

Corollary 4.1. *Consider two quantum Markov evolutions associated to the initial conditions $\rho_0 \ne \sigma_0$ and to the same family of TPCP maps $\{\mathcal{E}_t^\dagger\}$. Suppose that ρ_t, σ_t are invertible, for all t's. Let $Y_t = \sigma_t^{-\frac{1}{2}}\rho_t\sigma_t^{-\frac{1}{2}}$ and let $Z_t := g(Y_t)$, with $g(x) = x\log(x)$. Then Z_t is a reverse time, space-time*

[b]A function f is called *operator convex* on $\mathcal{I}$ if $f(\lambda A + (1-\lambda)B) \le \lambda f(A) + (1-\lambda)f(B)$, for any $\lambda \in [0,1]$, and matrices A, B with spectrum in $\mathcal{I}$ [3].

subharmonic process with respect to the quantum operations $\{\mathcal{R}_{\mathcal{E},\sigma_t}(\cdot)\}$, *i.e.*

$$Z_{t+1} = g\left(Y_{t+1}\right) \leq \mathcal{R}_{\mathcal{E},\sigma_t}\left(g\left(Y_t\right)\right) = \mathcal{R}_{\mathcal{E},\sigma_t}\left(Z_t\right). \tag{11}$$

This can be seen as an H-Theorem in operator form: In fact, the reverse time subharmonic property of $\{Z_t\}$ of Theorem 4.1 implies under expectation a more usual, Lindblad-Araki-Uhlmann-like [1,7,11] form of the H-theorem:

$$\mathbb{D}_{BS}(\rho_{t+1}||\sigma_{t+1}) \leq \mathbb{D}_{BS}(\rho_t||\sigma_t). \tag{12}$$

If $\bar{\sigma}$ is the unique stationary state of $\{\mathcal{E}_t^\dagger\}$, we get a quantum version of the second law.

Acknowledgments

Work partially supported by Department of Information Engineering QUINTET project and and by the University of Padova QFUTURE and CPDA080209/08 Grants.

References

1. H. Araki, Relative entropy for states of von Neumann algebras, *Publ. RIMS Kyoto Univ.*, **11** (1976), 809-833.
2. H. Barnum and E. Knill. Reversing quantum dynamics with near-optimal quantum and classical fidelity. *Journal of Mathematical Physics*, 43(5):2097–2106, 2002.
3. R. Bhatia. *Matrix Analysis.* Springer-Verlag, New York, 1997.
4. V. P. Belavkin and P. Staszewski. C*-algebraic generalization of relative entropy and entropy. *Annales de l´Institute Henri Poincaré*, 37(1):51–58, 1982.
5. F. Hiai and D. Petz. The proper formula for relative entropy and its asymptotics in quantum probability. *Communication in Mathematical Physics*, 143:99–114, 1991.
6. K. Kraus. *States, Effects, and Operations: Fundamental Notions of Quantum Theory.* Lecture notes in Physics. Springer-Verlag, Berlin, 1983.
7. G. Lindblad. Completely positive maps and entropy inequalities, *Communication in Mathematical Physics*, **40**, 147–151 1975.
8. E. Nelson. The adjoint markov process. *Duke Math. J.*, 25:671–690, 1958.
9. E.Nelson,*Dynamical Theories of Brownian Motion*, Princeton University Press, Princeton, 1967.
10. F. Ticozzi and M. Pavon. On Time-reversal and space-time harmonic processes for Markovian quantum channels. Submitted, 2008. On-line preprint: `http://arxiv.org/abs/0811.0929v2`
11. A. Uhlmann, Relative entropy and the Wigner-Yanase-Dyson-Lieb concavity in an interpolation theory, *Commun. Math. Phys.*, **54** (1977), 21-32.
12. H. Umegaki. Conditional expectations in an operator algebra iv (entropy and information). *Kodai Math. Sem. Rep.*, 14:59–85, 1962.

GLOBAL CONTROLLABILITY OF MULTIDIMENSIONAL RIGID BODY

A. V. SARYCHEV

Dipartimento di Matematica per le Decisioni, Università di Firenze
v. C.Lombroso 6/17, Firenze, 50134, Italy
E-mail: asarychev@unifi.it

We study global controllability of 'rotating' multidimensional rigid body (MRB) controlled by application of few torques. Study by methods of geometric control requires analysis of algebraic structure introduced by the quadratic term of Euler-Frahm equation. We discuss problems, which arise in the course of this analysis, and establish several global controllability criteria for damped and non damped cases.

Keywords: Multidimensional rigid body, geometric control, global controllability criteria, Navier-Stokes equation.

1. Introduction

In recent work [1–3] one studied controllability of Navier-Stokes (NS) equation, controlled by forcing applied to few modes on a 2D domain. Geometric control approach has been employed for establishing approximate controllability criteria for NS/Euler equation on 2D torus, sphere, hemisphere, rectangle and generic Riemannian surface with boundary.

In the present contribution we address controllability issues for a finite-dimensional "kin" of NS equation - Euler-Frahm equation for rotation of multidimensional rigid body (**MRB**) subject to few controlling torques and to possible damping. The equation evolves on $so(n)$. We formulate global controllability criteria which are structurally stable with respect to the choice of 'directions' of controlled torques.

According to geometric approach to studying controllability, one starts with a system controlled by low-dimensional input and proceeds with a sequence of *Lie extensions* ([5,6]) which add to the system new controlled vector fields. The latter are calculated via iterated Lie-Poisson brackets of the controlled vector fields and the drift (zero control vector field). The core of the method and the main difficulty is in finding proper Lie extensions and in tracing results of their implementation.

The Lie extension employed in [1–3] for studying controllability of NS equation, and similar one used equation below (see Subsection 4.1), involves double Lie bracket of drift vector field with a couple of *constant* controlled vector fields (they are identified with their values, or *directions* belonging to $so(n)$). At least one of the directions must be a *steady state* of **MRB**, i.e. an 'equilibrium points' of Euler-Frahm equation.

The double Lie bracket results in constant controlled vector field (extending direction); the correspondence between couple of original controlled directions and the extending one defines bilinear operator β on $so(n)$. More extending controlled directions are obtained by *iterated* application of β. For proving *global controllability of **MRB*** we must verify *saturating property* - coincidence of the set of extending directions with $so(n)$ after a number of iterations.

Tracing the iterations is by no means easy. For NS equation all cases, successfully analyzed in [1–3], are related to an explicit description of the basis of steady states and to specific representation of the operator β with respect to this basis. The results, so obtained, are heavily dependent on choice of original controlled directions and on geometry of the domain where the NS equation evolves.

Below we manage to establish several controllability criteria for damped and non damped **MRB** controlled by one, two or three torques. We pay special attention to deriving criteria which are structurally stable with respect to perturbation of (some of) the controlled directions.

2. Euler equation for generalized rigid body and Euler-Frahm equation for MRB

We follow [4] for definition of 'generalized rigid body'. Let $\mathcal{G}$ be a Lie group, $\mathfrak{g}$ its Lie algebra and let left-invariant Riemannian metric on $\mathcal{G}$ be defined by scalar product $\langle\cdot,\cdot\rangle$ on $\mathfrak{g}$.

Introduce $\mathcal{I} : \mathfrak{g} \mapsto \mathfrak{g}^*$ - a symmetric operator, which corresponds to the Riemannian metrics by formula: $\langle\xi,\eta\rangle = \mathcal{I}\xi|\eta$, where $\cdot|\cdot$ is the natural pairing between $\mathfrak{g}$ and $\mathfrak{g}^*$. The operator $\mathcal{I}$ is called *inertia operator* of generalized rigid body.

The trajectory of the motion of generalized rigid body is a curve $g(t) \in \mathcal{G}$. *Angular velocity*, corresponding to this motion is: $\Omega = L_{g^{-1}*}\dot{g} \in \mathfrak{g}$, where L_g is left translation by g. The image of angular velocity Ω under $\mathcal{I}$ is *angular momentum* $M \in \mathfrak{g}^*$. Energy of the body equals $\langle\Omega,\Omega\rangle = M|\Omega$.

Euler equation for the motion of generalized rigid body is $\dot{\Omega} = \mathcal{B}(\Omega,\Omega)$, where bilinear operator $\mathcal{B} : \mathfrak{g} \times \mathfrak{g} \mapsto \mathfrak{g}$ is defined by formula:

$$\langle [a,b], c\rangle = \langle \mathcal{B}(c,a), b\rangle, \tag{1}$$

$[\cdot,\cdot]$ staying for Lie-Poisson bracket in $\mathfrak{g}$.

MRB is particular case of generalized rigid body, where the Lie group $\mathcal{G} = SO(n)$, and *angular velocities* $\Omega \in \mathfrak{g} = so(n)$ are skew-symmetric matrices.

Identifying $so(n)$ with $so^*(n)$ by means of Killing form, we may think of momentum M as of skew-symmetric matrix. Then the inertia operator is a map $\mathcal{I}_C : \Omega \mapsto (\Omega C + C\Omega) = M \in so(n) \overset{\mathcal{K}}{\cong} so^*(n)$, where C is some positive semidefinite matrix.

Operator $\mathcal{I}_C$ is symmetric with respect to Killing form and is invertible (Sylvester theorem), whenever C is positive definite.

We compute $\mathcal{B}$ according to (1) ($[\cdot,\cdot]$ being matrix commutator):

$$\mathcal{B}(\Omega^1,\Omega^2) = \mathcal{I}_C^{-1}\left[\mathcal{I}_C\Omega^1,\Omega^2\right],\ \mathcal{B}(\Omega,\Omega) = \mathcal{I}_C^{-1}\left[\mathcal{I}_C\Omega,\Omega\right] = \mathcal{I}_C^{-1}[C,\Omega^2].$$

Euler-Frahm equation for the motion of **MRB** is:

$$\dot\Omega = \mathcal{I}_C^{-1}\left[\mathcal{I}_C\Omega,\Omega\right] = \mathcal{I}_C^{-1}[C,\Omega^2], \tag{2}$$

The motion, subject to damping, is described by the equation

$$\dot\Omega = \mathcal{I}_C^{-1}[C,\Omega^2] - \nu\Omega,\ \nu \geq 0.$$

3. Controllability of rotating MRB: problem setting and main results

Controlled rotation of **MRB** is described by equation

$$\dot\Omega = \mathcal{I}_C^{-1}[C,\Omega^2] - \nu\Omega + \sum_{i=1}^{r} G^i u_i(t),\ \nu \geq 0,\ G^i \in so(n). \tag{3}$$

We are interested in *global controllability* of (3), meaning that for any $\tilde\Omega, \hat\Omega \in so(n)$ system (3) can be steered from $\tilde\Omega$ to $\hat\Omega$ in some time $T \geq 0$. We are interested in achieving global controllability by small number of controls; we prove that r can be taken ≤ 3 for all $n \geq 3$.

Equation (3) is particular case of control-affine system with quadratic(+linear) *drift vector field* and *constant controlled vector fields.*

The following *genericity condition* is assumed to hold furtheron: symmetric matrix C *is positive definite and has distinct eigenvalues.*

Our first result claims *global controllability of* ***MRB*** *by means of two controlled torques.*

Theorem 3.1. *There exists a pair of directions $G^1, G^2 \in so(n)$ (depending on C), such that the system (3) with $r = 2$ is globally controllable.* □

The proof of this Theorem, sketched below, is based on direct computation of Lie extensions in specially selected basis, related to C. More difficult is formulating criteria, which are *structurally stable* with respect to perturbation of controlled directions.

We start with *non damped* **MRB**, controlled by one torque. In this case - given *recurrence* of Euler-Frahm dynamics (2) - *bracket generating property* suffices for guaranteeing global controllability. This property means that evaluations (at each point) of iterated Lie brackets of drift and controlled vector fields span $so(n)$. Given high dimension of $so(n)$, verification of the bracket generating property for *generic* controlled direction is nontrivial task. We do this analyzing linearization of quadratic Euler operator. The result is

Theorem 3.2. *For generic* $G \in so(n)$ *the system* $\dot{\Omega} = \mathcal{I}_C^{-1}[C, \Omega^2] + Gu(t)$, *is globally controllable, also if control is bounded:* $|u| \leq b,\ b > 0$. □

We now pass to the damped case. Our method requires one of the controlled directions to be *steady state* for **MRB**. Recall that steady state or steady direction of **MRB** is equilibrium point of (2) - a matrix $\hat{G}$ for which $[\mathcal{I}_C\hat{G}, \hat{G}] = [C, \hat{G}^2] = 0$. Matrix $\hat{G}$ is *principal axis* of **MRB**, if $\mathcal{I}_C\hat{G} = \mu\hat{G},\ \mu \in \mathbb{R}$. These two sets coincide for $n = 3$, while for $n \geq 4$ the set of steady directions is much richer.

The results obtained for the damped case differ for *odd* and *even* n.

Theorem 3.3. *Let* $r = 2$, n *be odd in (3). For generic stationary direction* G^1 *and generic* $G^2 \in so(n)$ *the system (3) is globally controllable.* □

An additional symmetry in the case of even n, obliges one to involve additional controlled direction for achieving global controllability.

Theorem 3.4. *Let* $r = 3$, n *be even in (3). For generic stationary direction* $G^1 \in so(n)$ *and generic pair* (G^2, G^3) *of directions the system (3) is globally controllable.* □

Generic element of a subset $W \subseteq so(n)$ means an element of open dense subset of W in induced topology.

4. Sketch of the proof of Theorem 3.1

4.1. *Key Lie extension*

Lie extensions mean finding vector fields X, which are *compatible* with control system, in the sense that closures of attainable sets of the control system are invariant for X. If one is able to prove global controllability of the system *extended* by some compatible vector fields, then controllability of the original system can be concluded by standard argument.

Key Lie extension, we employ, is described by the following

Proposition 4.1. *Let for control system*

$$\dot{x} = f(x) + \tilde{g}(x)u + \bar{g}(x)v, \tag{4}$$

evolving on a manifold Q, hold the relations

$$\{\tilde{g}, \bar{g}\} = 0, \ \{\tilde{g}, \{\tilde{g}, f\}\} = 0, \tag{5}$$

($\{\cdot,\cdot\}$ stays for Lie brackets of vector fields on Q). Then the system $\dot{x} = f(x) + \tilde{g}(x)u + \bar{g}(x)v + \{\bar{g}, \{\tilde{g}, f\}\}(x)w$ is Lie extension of (4). □

Remark 4.1. Vector fields $\pm\{\bar{g}, \{\tilde{g}, f\}\}$ are *extending controlled vector fields*; they are also compatible with (4). □

We will repeatedly employ Proposition 4.1 for extending control system (3). At each step the first of the relations (5) will be trivially satisfied since all original and extending controlled vector fields will be constant. For drift vector field $f(\Omega) = \mathcal{I}_C^{-1}[C, \Omega^2]$ in (3), and constant controlled vector field $\tilde{g} \equiv \tilde{G} \in so(n)$, the Lie bracket $\{\tilde{g}, \{\tilde{g}, f\}\} \equiv \mathcal{I}_C^{-1}[C, \tilde{G}^2]$ is constant vector field. The second relation (5) would hold if and only if $\tilde{G}$ is steady state. When repeating the extension it is crucial to guarantee at each step disponibility of steady state controlled direction.

For two constant controlled vector fields $\tilde{g} \equiv \tilde{G}, \bar{g} \equiv \bar{G}, \ \tilde{G}, \bar{G} \in so(n)$ the value of constant *extending controlled vector field* $\{\bar{g}, \{\tilde{g}, f\}\}$ is

$$\beta(\tilde{G}, \bar{G}) = \mathcal{I}^{-1}[C, \tilde{G}\bar{G} + \bar{G}\tilde{G}]; \tag{6}$$

(6) defines symmetric bilinear operator β on $so(n)$.

4.2. *Algebra of principal axes and controllability proof*

Diagonalize matrix C presenting it as $C = \mathrm{Ad}\, SD = SDS^{-1}$ with S orthogonal and $D = \mathbf{diag}\{I_1, \ldots, I_n\}, \ I_1 < I_2 < \cdots < I_n$.

Introduce matrices $\Theta^{rs} = \mathbf{1}_{rs} - \mathbf{1}_{sr} \in so(n), \ (1 \leq r < s \leq n)$, with $\mathbf{1}_{rs}$ being matrix with (the only nonvanishing) unit element at row r and column s. Matrices $\Omega^{rs} = \mathrm{Ad}\, S\Theta^{rs}$ turn out to be 'eigenvectors' of the operators (ad C) and $\mathcal{I}_C$. They form set of principal axes of the **MRB**.

'Multiplication table' for β with respect to the basis Ω^{rs} is

$$\beta(\Omega^{rs}, \Omega^{rs}) = 0, \ \beta(\Omega^{rs}, \Omega^{r\ell}) = (I_\ell - I_s)(I_s + I_\ell)^{-1}\Omega^{s\ell},$$

$\beta(\Omega^{rs}, \Omega^{k\ell}) = 0$, whenever r, s, k, ℓ are distinct.

Take

$$G^1 = \Omega^{12} \text{ - principal axis, } G^2 = \Omega^{23} + \Omega^{34} + \cdots + \Omega^{n-1,n}.$$

It suffices to prove that iterated applications of β to G^1, G^2 result in a basis of $so(n)$, because then the extended system would possess full-dimensional input and therefore would be globally controllable. The original system (3) would be globally controllable as well.

According to the multiplication table $G^3 = \beta(G^2, G^1) = \beta(\Omega^{12}, \Omega^{23})$ coincides up to a multiplier with principal axis Ω^{13}. Calculating subsequently extending controlled 'directions' $G^i = \beta(G^{i-1}, G^2)$, $i > 2$, we see that all G^i coincide up to a nonzero multiplier with $\Omega^{1,i}$, i.e. are principal axes. Also $\beta(\Omega^{1i}, \Omega^{1k})$ coincides up to a multiplier with Ω^{ik}; this means that iterating applications of β to G^1, G^2 generate basis of $so(n)$.

Remark 4.2. If one perturbs G^2 in generic way then computation of $G^3 = \beta(G^2, G^1)$ will not result in a stationary direction. Thus Proposition 4.1 can not be iterated. Hence the provided construction would not allow to conclude structural stability of global controllability property.

Remark 4.3. Publication [8] studies controllability of non damped **MRB** by using of a pair of controlled 'flywheels' - different type of "internal-force controls", described by bilinear control system on Lie group.

References

1. A.A.Agrachev and A.V.Sarychev, *J. Math. Fluid Mechan.* **7**, 108 (2005).
2. A.A.Agrachev and A.V.Sarychev, in: *Instability in Models Connected with Fluid Flows I* (C. Bardos, A. Fursikov Eds., Springer, New York, 1–35,2008).
3. S.S.Rodrigues, *J. Dyn. Control Systems* **12**, 517 (2006).
4. V.I.Arnold and B.M.Khesin, *Topological Methods in Hydrodynamics* (Springer, New York, 1998).
5. V.Jurdjevic, *Geometric Control Theory* (Cambridge Univ. Press, 1997).
6. A.V.Sarychev, *J. Mathem. Sciences* **135**, 3195 (2006).
7. Yu.N.Fedorov and V.V.Kozlov, in: *Dynamical systems in classical mechanics* (Amer. Math. Soc. Transl. **168**(2), Providence, 141–171, 1995).
8. M.V.Deryabin, *J. Mathem. Sciences* **161**, 181 (2009).

GENERIC CONTROLLABILITY OF THE BILINEAR SCHRÖDINGER EQUATION

U. BOSCAIN

CMAP École Polytehnique CNRS, Route de Saclay, 91128 Palaiseau, France
E-mail: boscain@cmap.polytechnique.fr

T. CHAMBRION and M. SIGALOTTI

INRIA Nancy - Grand Est, Équipe-projet CORIDA
and Institut Élie Cartan, UMR CNRS/INRIA/Nancy Université
BP 239, 54506 Vandœuvre-lès-Nancy, France
E-mail: Thomas.Chambrion(Mario.Sigalotti)@iecn.u-nancy.fr

P. MASON

Laboratoire des Signaux et Systèmes, CNRS, Supélec
3 rue Joliot-Curie, 91192 Gif-sur-Yvette, France
E-mail: Paolo.Mason@lss.supelec.fr

In [1] we proposed a set of sufficient conditions for the approximate controllability of a discrete-spectrum bilinear Schrödinger equation. These conditions are expressed in terms of the controlled potential and of the eigenpairs of the uncontrolled Schrödinger operator. The aim of this paper is to show that these conditions are generic with respect to the uncontrolled and the controlled potential.

Keywords: Schrödinger equation, controllability, generic conditions.

1. Introduction

In this paper we consider controlled Schrödinger equations of the type

$$i\frac{\partial\psi}{\partial t}(t,x) = (-\Delta + V(x) + u(t)W(x))\psi(t,x), \tag{1}$$

where $u(t) \in U$, $\psi : I \times \Omega \to \mathbf{C}$ for some $\Omega \subset \mathbf{R}^d$ open bounded, I is a subinterval of $\mathbf{R}$, $\psi|_{I\times\partial\Omega} = 0$. Here V, W are suitable real valued functions and U is a nonempty subset of $\mathbf{R}$.

As proved in [2], the control system (1) is never exactly controllable in $L^2(\Omega)$. Nevertheless, several positive controllability results have been proved in recent years. Among them, let us mention the exact controllability among regular enough wave functions for $d = 1$ and $V = 0$ [3,4] and the recently obtained L^2-approximate controllability [5]. The result we will consider for the discussion below is the L^2-approximate controllability obtained by the authors in [1].

The scope of this paper is to establish that the sufficient conditions for controllability proposed in [1] are robust and frequent enough. The mathematical framework for this analysis is provided by the standard notion of genericity. Due to the lack of space, in this paper we will just state our genericity results without providing the corresponding proofs (except in one case). Proofs will be presented in a future work.

Let us mention that the genericity question for the Schrödinger equation is already addressed in [5], where some partial results are given. In particular, genericity for the case $d = 1$ is essentially proven in [5, Lemma 3.12]. Further genericity results on the controllability of a linearized Schrödinger equation can be found in [6] and are further discussed in Sec. 6.

2. Notations and definition of solutions

We denote by $\mathbf{N}$ the set of positive integers, by A^* the adjoint of an operator A. We fix $d \in \mathbf{N}$ to denote the dimension of the space in which the Schrödinger equation is considered. We denote by Ξ the set of nonempty, open and bounded subsets of $\mathbf{R}^d$.

In the following we consider Equation (1) assuming that the potentials V, W are taken in $L^\infty(\Omega, \mathbf{R})$. Then, for every $u \in U$, $-\Delta + V + uW : H^2(\Omega, \mathbf{R}) \cap H_0^1(\Omega, \mathbf{R}) \to L^2(\Omega, \mathbf{C})$ is a skew-adjoint operator on $L^2(\Omega, \mathbf{C})$ with discrete spectrum. (See [7].) In particular, $-\Delta + V + uW$ generates a group of unitary transformations $e^{it(-\Delta+V+uW)} : L^2(\Omega) \to L^2(\Omega)$. Therefore, $e^{it(-\Delta+V+uW)}(\mathcal{S}) = \mathcal{S}$ where $\mathcal{S}$ denotes the unit sphere of $L^2(\Omega)$.

For every $u \in L^\infty([0,T], U)$ and every $\psi_0 \in L^2(\Omega)$ there exists a unique weak (and mild) solution $\psi(\cdot; \psi_0, u) \in \mathcal{C}([0,T], \mathcal{H})$. Moreover, if $\psi_0 \in D(A)$ and $u \in \mathcal{C}^1([0,T], U)$ then $\psi(\cdot; \psi_0, u)$ is differentiable and it is a strong solution of (1). (See [8] and references therein.)

Definition 2.1. We say that the quadruple (Ω, V, W, U) is approximately controllable if for every $\psi_0, \psi_1 \in \mathcal{S}$ and every $\epsilon > 0$ there exist $T > 0$ and $u \in L^\infty([0,T], U)$ such that $\|\psi_1 - \psi(T; \psi_0, u)\| < \epsilon$.

In order to state the approximate controllability result obtained in [1], we need to recall the following two definitions.

Definition 2.2. The elements of a sequence $(\mu_n)_{n \in \mathbf{N}} \subset \mathbf{R}$ are said to be $\mathbf{Q}$-linearly independent (equivalently, the sequence is said to be non-resonant) if for every $N \in \mathbf{N}$ and $(q_1, \ldots, q_N) \in \mathbf{Q}^N \setminus \{0\}$ one has $\sum_{n=1}^N q_n \mu_n \neq 0$.

Definition 2.3. A $n \times n$ matrix $C = (c_{jk})_{1\le j,k\le n}$ is said to be connected if for every pair of indices $j, k \in \{1, \dots, n\}$ there exists a finite sequence $r_1, \dots, r_l \in \{1, \dots, n\}$ such that $c_{jr_1} c_{r_1 r_2} \cdots c_{r_{l-1} r_l} c_{r_l k} \neq 0$.

In the following we denote by $\sigma(V, \Omega) = (\lambda_j(V, \Omega))_{j\in\mathbf{N}}$ the non-decreasing sequence of eigenvalues of $-\Delta + V$ (on $H^2(\Omega, \mathbf{R}) \cap H_0^1(\Omega, \mathbf{R})$), counted according to their multiplicity and by $(\phi_j(V, \Omega))_{j\in\mathbf{N}}$ the corresponding sequence of eigenfunctions (unique up to the sign if the corresponding eigenvalue is simple). In particular $(\phi_j(V, \Omega))_{j\in\mathbf{N}}$ forms an orthonormal basis of $L^2(\Omega, \mathbf{C})$.

The theorem below recalls the controllability results obtained by the authors in [1, Theorems 3.4, 5.2].

Theorem 2.1. *Let $\Omega \in \Xi$, V, W belong to $L^\infty(\Omega, \mathbf{R})$, and U contain the interval $[0, \delta)$ for some $\delta > 0$. Assume that the elements of $\big(\lambda_{k+1}(V, \Omega) - \lambda_k(V, \Omega)\big)_{k\in\mathbf{N}}$ are $\mathbf{Q}$-linearly independent and that for infinitely many $n \in \mathbf{N}$ the matrix*

$$B^{(n)}(\Omega, V, W) := \left(\int_\Omega W(x)\phi_j(V, \Omega)\phi_k(V, \Omega)\, dx \right)_{j,k=1}^{n}$$

is connected (i.e., $B^{(n)}(\Omega, V, W)$ is frequently connected). Then (Ω, V, W, U) is approximately controllable.

Remark 2.1. In [1] we prove that the conditions of Theorem 2.1 are also sufficient for the approximate controllability of (Ω, V, W, U) in the more general sense of density matrices. Moreover, in [1] the case Ω unbounded is also considered. The potentials V and W are allowed to be unbounded as well, and Theorem 2.1 still holds, though the notion of solution of (1) gets more delicate. In this presentation we restrict our attention to the bounded case, although the results presented below admit suitable counterparts in the unbounded setting.

We say that (Ω, V, W) is *fit for control* if $-\Delta + V$ is non-resonant and $B^{(n)}(\Omega, V, W)$ is frequently connected.

We say that the quadruple (Ω, V, W, U) is *effective* if $(\Omega, V + uW, W)$ is fit for control for some u such that $[u, u + \delta) \subset U$.

Theorem 2.1 states that being effective is a sufficient condition for controllability.

3. Genericity: topologies and definitions

Let us recall that every complete metric space X is a Baire space, that is, any intersection of countably many open and dense subsets of X is dense in X. The intersection of countably many open and dense subsets of a Baire space is called a *residual* subset of X. Given a Baire space X and a boolean function $P : X \to \{0,1\}$ we say that P is a *generic property* if there exists a residual subset Y of X such that every x in Y satisfies property P, that is, $P(x) = 1$.

In the following the role of X will be played by $L^\infty(\Omega) \times L^\infty(\Omega)$ or $L^\infty(\Omega)$.

4. The triple (Ω, V, W) is generically fit for control with respect to the pair (V, W)

Here below we prove that, given $\Omega \in \Xi$, for a generic pair $(V, W) \in L^\infty(\Omega)\times L^\infty(\Omega)$ the triple (Ω, V, W) is fit for control.

Let us start by recalling a known result on the generic simplicity of eigenvalues (see [9,10]).

Proposition 4.1 (Albert). *Let $\Omega \in \Xi$. For every $k \in \mathbf{N}$ the set*

$$\mathcal{R}_k = \{V \in L^\infty(\Omega) \mid \lambda_1(V,\Omega), \dots, \lambda_k(V,\Omega) \text{ simple}\} \tag{2}$$

is open and dense in $L^\infty(\Omega)$. Hence, the spectrum $\sigma(V,\Omega)$ is, generically with respect V, simple.

We generalize Proposition 4.1 as follows.

Proposition 4.2. *Let $\Omega \in \Xi$. For every $K \in \mathbf{N}$ and $q = (q_1, \dots, q_K) \in \mathbf{Q}^K \setminus \{0\}$, the set*

$$\mathcal{O}_q = \left\{ V \in L^\infty(\Omega) \mid \sum_{j=1}^{K} q_j \lambda_j(V,\Omega) \neq 0 \right\} \tag{3}$$

is open and dense in $L^\infty(\Omega)$. Hence, the spectrum $\sigma(V,\Omega)$ forms, generically with respect V, a non-resonant family.

Proposition 4.1 is clearly a special case of Proposition 4.2, since $\mathcal{R}_k = \cap_{j=1}^{k} \mathcal{O}_{e_{j+1}-e_j}$, where $e_1, \dots, e_{k+1}$ denotes the canonical basis of $\mathbf{R}^{k+1}$.

The following theorem extends the analysis from V to the pair (V, W), combining the generic non-resonance of the spectrum of $-\Delta + V$ with a genericity connectedness condition on the matrices $B^{(n)}(\Omega, V, W)$.

Theorem 4.1. *Let $\Omega \in \Xi$. Then, generically with respect to $(V, W) \in L^\infty(\Omega) \times L^\infty(\Omega)$ the triple (Ω, V, W) is fit for control and, in particular, (Ω, V, W, U) is approximately controllable for every $U \subset \mathbf{R}$ with nonempty interior.*

Proof. Recall that $\mathcal{R}_k$, defined in (2), is open and dense in $L^\infty(\Omega)$. If V belongs to $\mathcal{R}_k$, then the eigenfunctions $\phi_1(V, \Omega), \ldots, \phi_k(V, \Omega)$ are uniquely defined in $H_0^1(\Omega)$ up to sign. It makes sense, therefore, to define

$$\mathcal{U}_k = \{(V, W) \in \mathcal{R}_k \times L^\infty(\Omega) \mid \int_\Omega W\phi_{j_1}(V, \Omega)\phi_{j_2}(V, \Omega) \neq 0 \text{ for every } 1 \leq j_1, j_2 \leq k\}.$$

As it follows from the unique continuation theorem, for every $1 \leq j_1, j_2 \leq k$ the product $\phi_{j_1}(V, \Omega)\phi_{j_2}(V, \Omega)$ is a nonzero function on Ω. Therefore, $\mathcal{U}_k$ is dense in $L^\infty(\Omega) \times L^\infty(\Omega)$. Its openness follows, moreover, from the continuity of $V \mapsto \{\phi_j(V, \Omega), -\phi_j(V, \Omega)\}$ on $\mathcal{R}_k$ for $j = 1, \ldots, k$ (see, for instance, [11]).

The proof is concluded by noticing that (Ω, V, W) is fit for control if (V, W) belongs to

$$\left(\cap_{k \in \mathbf{N}} \mathcal{U}_k\right) \cap \left(\cap_{q \in \cup_{k \in \mathbf{N}} \mathbf{Q}^k \setminus \{0\}} \mathcal{O}_q \times L^\infty(\Omega)\right),$$

which is a countable intersection of open and dense subsets of $L^\infty(\Omega) \times L^\infty(\Omega)$. □

5. Generic controllability with respect to one single argument

The following result states that, for a fixed potential $W \in L^\infty(\Omega)$, the triple (Ω, V, W) is generically fit for control.

Proposition 5.1. *Let $\Omega \in \Xi$. Fix W non-constant and absolutely continuous on Ω. Then, generically with respect to $V \in L^\infty(\Omega)$, (Ω, V, W) is fit for control.*

We shall now show that, for a fixed potential V, generically with respect to $W \in L^\infty(\Omega)$, (Ω, V, W, U) is effective. Notice that (Ω, V, W) cannot be fit for control if the spectrum of V is resonant, independently of V. In this regard the result is necessarily weaker than Proposition 4.2, where the genericity of the fit for controlness was proved. The precise statement of our result is given by the following proposition.

Proposition 5.2. *Let $\Omega \in \Xi$, V an absolutely continuous function on Ω and $U \subset \mathbf{R}$ with nonempty interior. Then, generically with respect to W, (Ω, V, W, U) is effective.*

6. Conclusion

In this paper we presented some results showing that, once (Ω, V) or (Ω, W) is fixed, the bilinear Schrödinger equation on Ω having V as uncontrolled and W as controlled potential is generically approximately controllable with respect to the other element of the triple (Ω, V, W).

Acknowledgments

The authors are grateful to Yacine Chitour, Antoine Henrot and Yannick Privat for helpful discussions.

This work was supported by the BQR "Contrôle effectif des systèmes quantiques", Région Lorraine – Nancy Université.

References

1. T. Chambrion, P. Mason, M. Sigalotti and U. Boscain, *Ann. Inst. H. Poincaré Anal. Non Linéaire* **26**, 329 (2009).
2. G. Turinici, On the controllability of bilinear quantum systems, in *Mathematical models and methods for ab initio Quantum Chemistry*, eds. M. Defranceschi and C. Le Bris, Lecture Notes in Chemistry, Vol. 74 (Springer, 2000).
3. K. Beauchard, *J. Math. Pures Appl. (9)* **84**, 851 (2005).
4. K. Beauchard and J.-M. Coron, *J. Funct. Anal.* **232**, 328 (2006).
5. V. Nersesyan, *Comm. Math. Phys.* **290**, 371 (2009).
6. K. Beauchard, Y. Chitour, D. Kateb and R. Long, *Journal of Functional Analysis* (to appear).
7. K. Friedrichs, *Math. Ann.* **109**, 465 (1934).
8. J. M. Ball, J. E. Marsden and M. Slemrod, *SIAM J. Control Optim.* **20**, 575 (1982).
9. J. H. Albert, *Proc. Amer. Math. Soc.* **48**, 413 (1975).
10. K. Uhlenbeck, *Amer. J. Math.* **98**, 1059 (1976).
11. T. Kato, *Perturbation theory for linear operators*Die Grundlehren der mathematischen Wissenschaften, Band 132, Die Grundlehren der mathematischen Wissenschaften, Band 132 (Springer-Verlag New York, Inc., New York, 1966).

A NOTE ON THE FOUCAULT PENDULUM AND THE SUB-RIEMANNIAN FORMALISM

A. ANZALDO-MENESES* and F. MONROY-PÉREZ†

Departamento de Ciencias Básicas, UAM-Azcapotzalco,
México D.F., 02200, México
*E-mail: * answald@ymail.com*
† fmp@correo.azc.uam.mx

The well known Foucault pendulum is studied within the formalism of sub-Riemannian geometry on a step-2 nilpotent Lie group. For small oscillations, trajectories are explicitly calculated, they turn out to be hypotrochöids obtained by rolling without slipping a circle onto another circle.

Keywords: Sub-Riemannian geometry, non-holonomic distribution, Foucault pendulum.

1. Introduction

The Foucault pendulum is recognized as a feasible demonstration of the rotation movement of Earth, and has been extensively studied. We present in this paper the viewpoint of sub-Riemannian geometry for this classical problem. This viewpoint is based on the fact that the non-holonomic constraints are encoded by means of a Pfaffian system, that represents part of the kinematics. Furthermore, the distribution of vector fields that generates the kernel of the system, satisfy the full rank condition, and spans a step-2 nilpotent Lie algebra. A Riemannian metric, restricted to the distribution, can be defined and naturally associated to the kinetic energy. For details on the general framework of sub-Riemannian geometry, we refer the reader to the survey by A.M. Vershik *et al.* [1] and the recent book by R. Montgomery [2]. The approach for tackling problems in mechanics by analyzing the geometry of the non-holonomic constrains through an optimal control formulation is not new and can be seen for instance in R. Brockett and L. Dai [3].

In Sec. 2 we set the Foucault pendulum as a geodesic sub-Riemannian problem in a six dimensional step-2 nilpotent Lie group. We then derive the equations of motion in Sec. 3. A geometric analysis for trajectories in the case of small oscillation is carried out in Secs. 4, and 5 . At the end, in Sec. 6 we derive some conclusions and perspectives.

2. Sub-Riemannian setting for the Foucault pendulum

We take an inertial frame (X, Y, Z), in such a way that the rotation axis of Earth, coincides with the Z direction. For a pendulum of length ℓ and point mass $m, \vec{\omega}$ denotes the angular velocity and (x, y, z) the position of the mass measured from a fixed coordinate system with origin located at latitude α measured from equator. The x direction is taken on a meridian, the y direction on a parallel, and the z direction is perpendicular to the tangent plane at the intersection of both circles. For the gravitational force field $\vec{F}_G = -\vec{\nabla} V_G$, the trajectories are determined by the minimization of the functional

$$S_0 = \int \left(\frac{m}{2} \left| \frac{d\vec{r}}{dt} \right|^2 - V_G \right) dt.$$

A vector $\vec{r}$ in the non-inertial system on Earth's surface behaves as

$$\frac{d\vec{r}}{dt} = \dot{\vec{r}} + \vec{\omega} \times \vec{r},$$

where $\dot{\vec{r}} = (\dot{x}, \dot{y}, \dot{z})$ and $\vec{\omega} = (-\omega \cos(\alpha), 0, \omega \sin(\alpha))$. The kinetic energy is given as follows

$$\frac{m}{2} \left| \frac{d\vec{r}}{dt} \right|^2 = \frac{m}{2} (|\dot{\vec{r}}|^2 + |\vec{\omega} \times \vec{r}|^2 + 2\dot{\vec{r}} \cdot (\vec{\omega} \times \vec{r})).$$

The second term leads to a centrifugal force perpendicular to the rotation axis, and together with the central gravitational potential, yields a vertical force of magnitude mg, perpendicular to the tangent plane of to the planet surface, we set then: $-V_G + (1/2)|\vec{\omega} \times \vec{r}|^2 = -mgz$. The third term, yields $\dot{\vec{r}} \cdot (\vec{\omega} \times \vec{r}) = \vec{\omega} \cdot (\vec{r} \times \dot{\vec{r}}) = \omega_x (y\dot{z} - z\dot{y}) + \omega_y (z\dot{x} - x\dot{z}) + \omega_z (x\dot{y} - y\dot{x})$. And the holonomic constraint $x^2 + y^2 + z^2 - \ell^2 = 0$, has to be taken into account. The complete functional is then written as

$$\begin{aligned} S &= \int \Big((\frac{m}{2}(\dot{x}^2 + \dot{y}^2 + \dot{z}^2 - 2gz) + \lambda_0 r^2 + \lambda_1 (\dot{\xi} + y\dot{z} - z\dot{y}) \\ &+ \lambda_2 (\dot{\eta} + z\dot{x} - x\dot{z}) + \lambda_3 (\dot{\zeta} + x\dot{y} - y\dot{x}) \Big) dt. \end{aligned}$$

where the Lagrange parameter λ_0 has been introduced, set $\lambda_1 = -m\omega \cos \alpha$, $\lambda_2 = 0$ and $\lambda_3 = m\omega \sin \alpha$ and the differentials $\dot{\xi} dt$, $\dot{\eta} dt$ and $\dot{\zeta} dt$. The parameters λ_i can be seen as the Lagrange parameters associated with the non-holonomic constraints $w_1 = d\xi + ydz - zdy, w_2 = d\eta + zdx - xdz$, and $w_3 = d\zeta + xdy - ydx$.

Consider now the six-dimensional manifold M with local coordinates $q = (x, y, z, \xi, \eta, \zeta)$, together with the distribution of smooth vector fields $\Delta = \{X_1, X_2, X_3\}$, dual to the aforementioned Pfaffian system, that is $X_1 = \partial x + y\partial\zeta - z\partial\eta$, $X_2 = \partial y + z\partial\xi - x\partial\zeta$, and $X_3 = \partial z + x\partial\eta - y\partial\xi$. These vector fields generate a six dimensional step-2 nilpotent Lie algebra $\mathfrak{g}$ with non-zero brackets, $X_{12} = -2\partial\zeta$, $X_{13} = 2\partial\eta$, and $X_{23} = -2\partial\xi$. Since $T_q M$ is six dimensional for all $q \in M$, the distribution Δ is bracket generating.

A sub-Riemannian structure for this problem is given by the pair formed by the distribution Δ and the Euclidean metric given by the kinetic energy, horizontal curves satisfy $\dot{q} = \dot{x}X_1(q) + \dot{y}X_2(q) + \dot{z}X_3(q)$, and horizontal energy minimizers corresponds to extremals of the functional S. It is clear that this formulation of the problem coincides with that of an optimal control problem with quadratic cost and affine in controls plant [4], by considering the velocities as the control parameters.

3. Equations of motion

The Euler-Lagrange equations readily yield $\dot{\lambda}_1 = 0$, $\dot{\lambda}_2 = 0$, $\dot{\lambda}_3 = 0$. Therefore λ_1, λ_2 and λ_3 are constants and the extremals are solutions of the following system

$$m\ddot{x} = 2\lambda_3\dot{y} - 2\lambda_2\dot{z} + 2\lambda_0 x, \tag{1}$$

$$m\ddot{y} = -2\lambda_3\dot{x} + 2\lambda_1\dot{z} + 2\lambda_0 y, \tag{2}$$

$$m\ddot{z} = 2\lambda_2\dot{x} - 2\lambda_1\dot{y} + 2\lambda_0 z - mg, \tag{3}$$

which together with the relations

$$\dot{\xi} = z\dot{y} - y\dot{z}, \tag{4}$$

$$\dot{\eta} = x\dot{z} - z\dot{x}, \tag{5}$$

$$\dot{\zeta} = y\dot{x} - x\dot{y}, \tag{6}$$

yield the equations of motion in $\mathbb{R}^6$. The coordinates can be selected in such a way that the y component of the angular velocity is zero, i.e. $\lambda_2 = 0$. In base space (x, y, z) the equations are gauge invariant.

4. Small oscillations

We reduce our study to small oscillations, observe that ξ and η are integrable, now by taking $z = -\ell$ to first order, equation (3) yields $2\lambda_0 =$

$-mg/\ell$. By neglecting the terms containing $\dot{z}$ with respect to those having $\dot{x}$ and $\dot{y}$, we end up with the following system defined on $\mathbb{R}^3$

$$\begin{aligned} \ddot{x} &= 2\tilde{\omega}\dot{y} - \omega_0^2 x, && (7) \\ \ddot{y} &= -2\tilde{\omega}\dot{x} - \omega_0^2 y, && (8) \\ \dot{\zeta} &= -x\dot{y} + y\dot{x}, && (9) \end{aligned}$$

with the frequencies $\omega_0 = \sqrt{g/\ell}$ and $\tilde{\omega} = \lambda_3/m$; which is the Euler-Lagrange system for $L^0 = m/2(\dot{x}^2+\dot{y}^2) - mg/2\ell(x^2+y^2) + 2\lambda_3(\dot{\zeta}+x\dot{y}-y\dot{x})$.

4.1. *The canonical momenta and the Hamiltonian*

As customary take $p_x = L^0_x, p_y = L^0_y$ and $p_\zeta = L^0_\zeta$, it readily follows that $p_x = m\dot{x} - 2\lambda_3 y$, $p_y = m\dot{y} + 2\lambda_3 x$, and $p_\zeta = 2\lambda_3$, and consequently the adjoint system is written as $\dot{p}_x = -m\omega_0^2 x, \dot{p}_y = -m\omega_0^2 y$, and $\dot{p}_\zeta = 0$. For then $\dot{p}_\zeta$ is constant along extremals, and furthermore, the first two equations of motion imply that $\dot{x}^2+\dot{y}^2+\omega_0^2x^2+\omega_0^2y^2$ is also constant. In consequence, the Hamiltonian is a constant of motion given as follows

$$H = \frac{1}{2m}(p_x + p_\zeta y)^2 + \frac{1}{2m}(p_y - p_z x)^2 + \frac{m}{2}(x^2 + y^2)\omega_0^2.$$

5. Horizontal trajectories

We consider the complex variable $u = x + iy$, which implies $\dot{\zeta} = \mathrm{Im}\,(u\dot{u}^*)$ and

$$\ddot{u} = -i\frac{2\lambda_3}{m}\dot{u} - \frac{g}{\ell}u.$$

For $u(t) = e^{i\alpha_\pm t}$ we obtain the eigenvalues $\alpha_\pm = -\tilde{\omega} \pm \tilde{\omega}_0$, with frequencies $\tilde{\omega}_0 = \sqrt{\tilde{\omega}^2 + \omega_0^2}$, $\omega_0 = \sqrt{g/\ell}$ and $\tilde{\omega} = \lambda_3/m$. Here, $\tilde{\omega}$ is equal to the rotation angular speed ω times the sinus of the geographical latitude.

The general solution writes as $u = A_+e^{i\alpha_+ t} + A_- e^{i\alpha_- t}$, from where we obtain

$$\begin{aligned} u &= e^{-i\tilde{\omega}\,t}(A_+e^{i\tilde{\omega}_0\,t} + A_-e^{-i\tilde{\omega}_0\,t}), \\ \dot{u} &= i\,e^{-i\tilde{\omega}\,t}(\alpha_+A_+e^{i\tilde{\omega}_0\,t} + \alpha_-A_-e^{-i\tilde{\omega}_0\,t}), \\ \zeta &= -(\alpha_+|A_+|^2 + \alpha_-|A_-|^2)\,t - 2\mathrm{Re}\left(A_+A_-^*\,(e^{2i\,\tilde{\omega}_0\,t} - 1)\right). \end{aligned}$$

The first two relations yield a rotation given by the slow mode, with frequency $\tilde{\omega}$, of the fast mode motion, with frequency $\tilde{\omega}_0$. Therefore, the trajectory in base space performs a precession with frequency $\omega \sin\alpha$, whereas ζ increases by the same amount after $2\pi/\tilde{\omega}_0$ seconds. The conservation of energy is written as $2H/m = |\dot{u}|^2 + \omega_0^2|u|^2$. We can compute explicitly the products to get

$$
\begin{aligned}
|u|^2 &= |A_+|^2 + |A_-|^2 + 2\,\mathrm{Re}(A_+ A_-^*\, e^{2i\tilde{\omega}_0 t}),\\
|\dot{u}|^2 &= \alpha_+^2|A_+|^2 + \alpha_-^2|A_-|^2 - 2\omega_0^2\,\mathrm{Re}(A_+ A_-^*\, e^{2i\tilde{\omega}_0 t}),\\
\frac{2}{m}H &= \alpha_+^2|A_+|^2 + \alpha_-^2|A_-|^2 + \omega_0^2(|A_+|^2 + |A_-|^2).
\end{aligned}
$$

The initial conditions shall fix the remaining constants for u, and in the trajectories in base space can be written explicitly.

5.1. *Starting from an arbitrary point with a given velocity*

We take $x(0) = x_0$, $y(0) = 0$, and $\dot{x}(0) = v_0 \cos\beta$ and $\dot{y}(0) = v_0 \sin\beta$. In this case we have $u(0) = 1$ and $\dot{u}(0) = v_0 e^{i\beta}$, from where we get $A_+ + A_- = x_0$, $A_+\alpha_+ + A_-\alpha_- = -i\,v_0 e^{i\beta}$, and $2\tilde{\omega}_0 A_\pm = \mp\alpha_\mp x_0 \mp v_0 e^{i\beta + i\pi/2}$.
We obtain first,

$$
\begin{aligned}
4\tilde{\omega}_0^2|A_+|^2 &= \alpha_-^2 x_0^2 + v_0^2 + 2\alpha_- x_0\, v_0 \cos(\beta + \pi/2),\\
4\tilde{\omega}_0^2|A_-|^2 &= \alpha_+^2 x_0^2 + v_0^2 + 2\alpha_+ x_0\, v_0 \cos(\beta + \pi/2),
\end{aligned}
$$

and then $4\tilde{\omega}_0^2\mathrm{Re}(A_+A_-^*) = -\alpha_-\alpha_+ x_0^2 + 4\tilde{\omega} x_0\, v_0 \cos(\beta + \pi/2) - v_0^2$, and $4\tilde{\omega}_0^2\mathrm{Im}(A_+A_-^*) = -4\tilde{\omega}_0 x_0\, v_0 \sin(\beta + \pi/2)$. In these relations we have $\alpha_+\alpha_- = -\omega_0^2$, $\alpha_+ + \alpha_- = 2\tilde{\omega}_0$ and $\alpha_-^2 + \alpha_+^2 = 4\tilde{\omega}^2 + 2\omega_0^2$. In consequence, trajectories in total space are written as follows

$$
\begin{aligned}
x &= \frac{1}{2\tilde{\omega}_0}(-x_0[\alpha_- \cos\alpha_+ t - \alpha_+ \cos\alpha_- t] + v_0[\sin(\alpha_+ t + \beta) - \sin(\alpha_- t + \beta)],\\
y &= \frac{1}{2\tilde{\omega}_0}(-x_0[\alpha_- \sin\alpha_+ t - \alpha_+ \sin\alpha_- t] - v_0[\cos(\alpha_+ t + \beta) - \cos(\alpha_- t + \beta)],\\
\zeta &= -a_0 t - \frac{a_1}{2\omega_0^2}(\cos 2\tilde{\omega}_0 t - 1) + \frac{a_2}{2\omega_0^2}\sin 2\tilde{\omega}_0 t,
\end{aligned}
$$

where $a_0 = \alpha_+^2|A_+|^2 + \alpha_-^2|A_-|^2$, $a_1 = -2\omega_0^2\mathrm{Re}(A_+A_-^*)$ and $a_2 = +2\omega_0^2\mathrm{Im}(A_+A_-^*)$. The curves in base space correspond to hypotrochöids, family of curves generated by rolling without slipping, one circle over another circle. This family of curves is very well known, can be drawn by spirographs, and contains the particular cases of hypocycloids and rhodonée.

6. Conclusions and perspectives

We study the classical Foucault's pendulum under the framework of sub-Riemannian geometry. This approach allows to set a differential system for horizontal curves that can be explicitly integrated. This formalism leads also to some other physical models. The calculation of sub-Riemannian spheres and the associated wave fronts, as well as the general non-symmetric Foucault's pendulum are part of our current research and shall be reported elsewhere.

The problem under discussion, although classic, and in some sense standard textbook material, provides under the sub-Riemannian approach, an interesting source of new theoretical and applied problems. For instance, it shall be interesting to establish a connection between the different curves obtained and the other physical models, such as two-level systems, quantum computing. Recently has been reported some interesting applications of the Foucault-like problems satellite formation lying, satellite constellation and space terminal rendezvous [5].

References

1. A.M. Vershik and V. Ya. Gershkovich, *Nonholonomic dynamical systems, geometry of distributions and variational problems.* In: Encyclopaedia of Mathematical Sciences, Vol. 16; Dynamical systems VII, eds. V.I. Arnold and S.P. Novikov, (Springer-Verlag, 1991).
2. R. Montgomery, *A tour of subriemannian geometries, their geodesics and applications*, Mathematical Surveys and Monographs, Vol. 91, (American Mathematical Society, 2002).
3. R.W. Brockett and L. Dai, *Non-holonomic kinematics and the role of elliptic functions in constructive controllability*, in Non-holonomic motion planning, A. Li and J. Canny, Eds. (Kluwer Academic publishers, Boston, pp. 1–21, 1993).
4. V. Jurdjevic, *Geometric control theory*, Cambridge studies in advanced mathematics 51, (Cambridge University Press , 1997).
5. D. Condurache and V. Martinusi, *Foucault Pendulum-like problems: A tensorial approach*, (International Journal of Non-linear Mechanics, pp. 43, 743–760, 2008).

TIME-OPTIMAL CONTROL OF A DISSIPATIVE SPIN 1/2 PARTICLE

D. SUGNY, M. LAPERT and E. ASSÉMAT

Laboratoire Interdisciplinaire Carnot de Bourgogne (ICB), UMR 5209
CNRS-Université de Bourgogne, 9 Av. A. Savary
BP 47 870, F-21078 DIJON Cedex, France

We present an analysis of the time-optimal control of a dissipative spin 1/2 particle whose dynamics is governed by the Bloch equation. We use the Pontryagin Maximum Principle to determine the optimal control for a realistic example in Nuclear Magnetic Resonance. A gain of 60% is obtained over the standard solution.

Keywords: Time-optimal control, Pontryagin maximum principle, nuclear magnetic resonance.

1. Introduction

Optimal control theory can be viewed as a generalization of the classical calculus of variations for problems with dynamical constraints. Optimal control theory was born in its modern version with the Pontryagin Maximum Principle (PMP) in the late 1950's [1]. Its development was originally inspired by problems of space dynamics, but it is now a key tool to study a large spectrum of applications as robotics, economics, and quantum mechanics. Solving an optimal control problem means finding a particular control law (i.e. a particular pulse sequence), the optimal control, such that the corresponding trajectory satisfies given boundary conditions and minimizes a cost criterion. The cost functionals of physical interests are the energy of the field and the minimization of the control duration. The strategy for solving an optimal control problem consists in finding extremal trajectories which are solutions of a generalized Hamiltonian system subject to the maximization condition of the PMP. In a second step, one selects among the extremals the ones which effectively minimize the cost criterion. Although its implementation looks straightforward, the practical use of the PMP is far from being trivial and each control has to be analyzed using geometric and numerical methods. The first applications of optimal control theory in the control of quantum dynamics began in the mid 80's. Continuous advances have been done both theoretically and experimentally until today [2,3]. The optimal equations can be solved by purely numerical techniques or by using geometric tools [4,5]. Surprisingly, the geometric aspects have been largely ignored in the chemical-physics literature and only few results exist in quantum mechanics mainly for closed quantum systems (see [6,7] to cite a few). The aim of this paper is to present another aspect of the geometric control of quantum systems, i.e. the control by magnetic fields of a spin 1/2 particle in a dissipative environment [8–10].

2. Methodology

To simplify the discussion, we assume that the frequency of the control field is resonant with the frequency of the transition of the spin 1/2 [6]. In this case, the state of the system can be completely represented by a two-dimensional state vector $X \in \mathbb{R}^2$ and a single control u is sufficient [8]. The corresponding controlled system is defined by differential equations of the form $\dot{X} = F_0(X) + uF_1(X)$ and the control parameter satisfies $|u| \leq 1$, i.e. the interval $[-1, 1]$ is the set of admissible controls. In this open loop control problem, the objective is to determine a function $u(t)$ such that the system goes in minimum time from the initial point X_0 to a target state X_1 which is assumed to belong to the accessibility set of X_0. To solve this optimal control problem [1,4,5], we use the PMP which can be stated as follows. We introduce the pseudo-Hamiltonian $\mathcal{H}(X, P, u) = P \cdot (F_0 + uF_1)$ where the adjoint state $P \in \mathbb{R}^{2*}$ for any time t. An optimal trajectory is solution of the equations

$$\begin{aligned} \dot{X}(t) &= \tfrac{\partial \mathcal{H}}{\partial P} \\ \dot{P}(t) &= -\tfrac{\partial \mathcal{H}}{\partial X} \end{aligned}$$

where u is obtained from the maximization condition $\mathcal{H}(X, P, u) = H(X, P) = \max_{v \in [-1,1]}[\mathcal{H}(X, P, v)]$ with the condition $H(X, P) \geq 0$ [5]. Since the PMP is only a necessary condition of optimality, the set of optimal trajectories is a subset of the set of all the Hamiltonian trajectories of $\mathcal{H}$. For controlled systems on the plane $\mathbb{R}^2$, the solutions of the PMP take a very simple form. Introducing the switching function $\phi(X, P) = P \cdot F_1$ [5], one immediately sees that the maximization condition leads to bang-bang controls $u = \pm 1$ when $\phi(X, P) \neq 0$. If ϕ vanishes in an isolated point then the control field switches from one bound to the other. The singular situation is encountered when the switching function vanishes on a given time interval. In this case, the control cannot be directly determined by the maximization condition from the state variables X and P. The control parameter is instead computed by requiring that $\phi(X, P) = 0$ on the singular arc, which leads to the relations $\phi(X,P) = \dot{\phi}(X,P) = \ddot{\phi}(X,P) = \cdots = 0$. This condition allows to identify the manifolds, here lines of the plane $\mathbb{R}^2$ on which the singular trajectory lies. Note also that in this case, the control field is not constantly equal to $+1$ or -1, but can belong to the interval $[-1,1]$. As detailed in the example below, additional study has to be done to determine the optimality of singular extremals. The final optimal control law will be determined by gluing together singular and non-singular arcs.

One of the most promising fields of applications of quantum control is the control of spin systems in Nuclear Magnetic Resonance (NMR) [11]. In a first step, we consider a spin 1/2 particle in a dissipative environment whose dynamics is governed by the Bloch equation:

$$\begin{pmatrix} \dot{M}_x \\ \dot{M}_y \\ \dot{M}_z \end{pmatrix} = \begin{pmatrix} -M_x/T_2 \\ -M_y/T_2 \\ +(M_0 - M_z)/T_1 \end{pmatrix} + \begin{pmatrix} \omega_y M_z \\ -\omega_x M_z \\ \omega_x M_y - \omega_y M_x \end{pmatrix}$$

where $\vec{M}$ is the magnetization vector and $\vec{M}_0 = M_0 \vec{e}_z$ is the equilibrium point of the dynamics. We assume that the control field $\vec{\omega} = (\omega_x, \omega_y, 0)$ satisfies the constraint $|\vec{\omega}| \leq \omega_{max}$. We introduce the normalized coordinates $\vec{x} = (x,y,z) = \vec{M}/M_0$, which implies that at thermal equilibrium the z component of the scaled vector $\vec{x}$ is by definition $+1$. The normalized control field which satisfies $|u| \leq 2\pi$ is defined as $u = (u_x, u_y, 0) = 2\pi\vec{\omega}/\omega_{max}$, while the normalized time τ is given by $\tau = (\omega_{max}/2\pi)t$. Dividing the previous system by $\omega_{max} M_0/(2\pi)$, one deduces that the dynamics of the normalized coordinates is ruled by the following system of differential equations:

$$\begin{pmatrix} \dot{x} \\ \dot{y} \\ \dot{z} \end{pmatrix} = \begin{pmatrix} -\Gamma x \\ -\Gamma y \\ \gamma - \gamma z \end{pmatrix} + \begin{pmatrix} u_y z \\ -u_x z \\ u_x y - u_y x \end{pmatrix}$$

where $\Gamma = 2\pi/(\omega_{max} T_2)$ and $\gamma = 2\pi/(\omega_{max} T_1)$.

We consider the control problem of bringing the system from the equilibrium point $\vec{M}_0$ to the zero-magnetization point which is the center of the Bloch ball. Since the initial point belongs to the z- axis, it can be shown that the controlled system is equivalent to a single input system where, e.g., $u_y = 0$ [8,9]. Roughly speaking, this means that the meridian planes play all the same role for the optimal trajectory. Taking $u_y = 0$, we are thus considering a single input problem in a plane of the form:

$$\begin{pmatrix} \dot{y} \\ \dot{z} \end{pmatrix} = \begin{pmatrix} -\Gamma y \\ \gamma - \gamma z \end{pmatrix} + u \begin{pmatrix} -z \\ y \end{pmatrix}$$

where the subscript x has been omitted for the control parameter. We can then apply for this system the theoretical description of the previous paragraph where $F_0 = (-\Gamma y, \gamma - \gamma z)$ and $F_1 = (-z, y)$.

3. Time-optimal control

We introduce the switching function $\phi = -p_y z + p_z y$ [5]. Using the fact that $\frac{d\phi}{dt} = P \cdot V$ where $V = (-\gamma + \gamma z - \Gamma z, -\Gamma y + \gamma y)$ and the relations $P \cdot F_1(X) = P \cdot V = 0$ on a singular arc, one deduces that the vectors F_1 and V must be parallel on this set since P is non zero. This means that the singular trajectories belong to the set $S = \{X \in \mathbb{R}^2 | \det(F_1, V) = 0\}$ which corresponds to the union of the vertical line $y = 0$ and of the horizontal one with z given by

$$z_0 = -\frac{\gamma}{2(\Gamma - \gamma)} = -\frac{T_2}{2(T_1 - T_2)}$$

if $\Gamma \neq \gamma_+$ (or equivalently if $T_1 \neq T_2$). The corresponding singular control u_s, which is determined from the condition $\ddot{\phi}(X, P) = 0$, is given by

$$u_s(y, z) = \frac{-y\gamma(\Gamma - 2\gamma) - 2yz_0(\gamma^2 - \Gamma^2)}{2(\Gamma - \gamma)(y^2 - z_0^2) - \gamma z_0}. \tag{1}$$

One deduces that the singular control vanishes on the vertical singular line and that it is admissible, i.e. $|u_s| \leq 2\pi$, on the horizontal one if $|y| \geq |\gamma(\gamma - 2\Gamma)|/[2\pi(2\Gamma - 2\gamma)]$. For smaller values of y, the system cannot follow the horizontal singular arc and a switching curve appears from the point where the admissibility is lost [5]. A switching curve is a line in the plane (y, z) where the optimal control changes sign when crossing it.

The optimality of the singular trajectories can be determined geometrically by introducing the clock form α which is a 1-form such that $\alpha(F_0) = 1$ and $\alpha(F_1) = 0$. The form α is defined on points where F_0 and F_1 are not collinear. Here this set is the union of two parabolas of equation $-\Gamma y^2 + \gamma z(1 - z) = 0$. Let γ_1 and γ_2 be two extremals starting and ending at the same points and τ_1 and τ_2 the corresponding times needed to follow the two trajectories. The clock form allows to determine the time taken to travel a path since, for instance, $\int_{\gamma_1} \alpha = \int_0^{\tau_1} \alpha(\dot{X}) d\tau = \int_0^{\tau_1} \alpha(F_0) d\tau = \tau_1$. To compare τ_1 and τ_2, we consider the loop $\gamma_1 \bigcup \gamma_2^{-1}$ where γ_2^{-1} is γ_2 run backward. Introducing the surface D delimited by γ_1 and γ_2, a simple computation leads to $\int_{\gamma_1 \bigcup \gamma_2^{-1}} \alpha = \int_D d\alpha$. Since $d\alpha$ is equal to zero only on the singular set and remains of constant sign outside [4], one obtains that $\tau_1 - \tau_2 = \int_D d\alpha$. In particular, it can be shown that the horizontal singular line is locally optimal and that the vertical one is optimal if $z > z_0$.

We consider the control problems defined by the relaxation parameters γ^{-1} and Γ^{-1} (expressed in the normalized time unit defined above) of 23.9 and 1.94, respectively and $M_0 \approx 2.15 \times 10^{-5}$. We compare the optimal control law with an intuitive one used in NMR. The intuitive solution is composed of a bang pulse to reach the opposite point of the initial state along the z- axis followed by a zero control where we let the dissipation act up to the center of the Bloch ball. The optimal and the IR solutions are plotted in Fig. 1. Geometric tools allow to show that the optimal control is the concatenation of a bang pulse, followed successively by a singular control along the horizontal singular line, another bang pulse and a zero singular control along the vertical singular line. Figure 1 displays also the switching curve which has been determined numerically by considering a series of trajectories with $u = +2\pi$ originating from the horizontal singular set where $\phi = 0$. The points of the switching curve correspond to the first point of each trajectory where the switching function vanishes. To determine the optimal control law, we have also checked that the second bang pulse of the optimal sequence does not cross the switching curve up to the vertical singular axis. In this example, a gain of 58% is obtained for the optimal solution over the intuitive one.

4. Conclusion

We hope that this example of geometric control of a spin 1/2 particle in a dissipative environment will motivate systematic investigations of controls in quantum mechanics. The next step of this study could be the analysis of the optimal control of coupled spins with bounded fields generalizing thus the different works in this domain with unbounded controls [7].

Acknowledgments
We acknowledge B. Bonnard and S. J. Glaser for many helpful discussions.

References

1. L. S. Pontryagin, *The mathematical theroy of optimal processes* (John Wiley and Sons, New-York, London, 1962).
2. S. Rice and M. Zhao, *Optimal control of quantum dynamics* (Wiley, New-York, 2000).
3. M. Shapiro and P. Brumer, *Principles of quantum control of molecular processes* (Wiley, New-York, 2003).
4. B. Bonnard and M. Chyba, *Singular trajectories and their role in control theory* (Mathématiques and Applications, Springer-Verlag, Berlin, 2003).

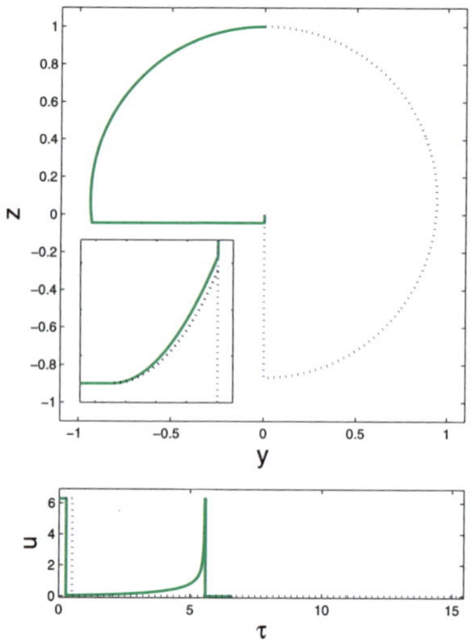

Fig. 1. Plot of the optimal trajectories (solid green curve) and of the intuitive one (dashed blue curve) in the plane (y, z) for $T_1 = 740$ ms, $T_2 = 60$ ms and $\omega_{max}/(2\pi) =$ 32.3 Hz. The corresponding control laws are represented in the lower panel. In the upper panel, the small insert represents a zoom of the optimal trajectory near the origin. The dotted line is the switching curve originating from the horizontal singular line. The vertical dashed line corresponds to the intuitive solution. The solid green curve is the optimal trajectory near the origin.

5. U. Boscain and B. Piccoli, *Optimal syntheses for control systems on 2D manifolds* (Mathématiques and Applications, Springer-Verlag, Berlin, 2004).
6. U. Boscain and P. Mason, *J. Math. Phys.* **47**, p. 062101 (2006).
7. N. Khaneja, R. Brockett and S. J. Glaser, *Phys. Rev. A* **63**, p. 032308 (2001).
8. B. Bonnard and D. Sugny, *SIAM J. on Control and Optimization* **48**, p. 1289 (2009).
9. B. Bonnard, M. Chyba and D. Sugny, *IEEE Transactions on Automatic control* **54**, p. 2598 (2009).
10. D. Sugny, C. Kontz and H. R. Jauslin, *Phys. Rev. A* **76**, p. 023419 (2007).
11. M. H. Levitt, *Spin dynamics: Basis of Nuclear Magnetic Resonance* (John Wiley and Sons, New-York, London, Sydney, 2008).

WHAT CAN WE HOPE ABOUT OUTPUT TRACKING OF BILINEAR QUANTUM SYSTEMS?

U. BOSCAIN

CMAP École Polytehnique CNRS, Route de Saclay, 91128 Palaiseau, France
E-mail: boscain@cmap.polytechnique.fr

T. CHAMBRION and M. SIGALOTTI

INRIA Nancy - Grand Est, Équipe-projet CORIDA
and Institut Élie Cartan, UMR CNRS/INRIA/Nancy Université
BP 239, 54506 Vandœuvre-lès-Nancy, France
E-mail: Thomas.Chambrion(Mario.Sigalotti)@iecn.u-nancy.fr

P. MASON

Laboratoire des Signaux et Systèmes, CNRS, Supélec
3 rue Joliot-Curie, 91192 Gif-sur-Yvette, France
E-mail: Paolo.Mason@lss.supelec.fr

We consider a non-resonant bilinear Schrödinger equation with discrete spectrum driven by a scalar control. We prove that this system can approximately track any given trajectory, up to the phase of the coordinates, with arbitrary small controls. The result is valid both for bounded and unbounded Schrödinger operators. The method used relies on finite-dimensional control techniques applied to Lie groups.

Keywords: Quantum systems, output tracking.

1. Introduction

1.1. *Physical context*

We will be interested in this paper in a non-relativistic and non-stochastic Schrödinger equation on a domain (i.e., an open connected subset) Ω of $\mathbf{R}^d$ that is either bounded or equal to the whole $\mathbf{R}^d$ ($d \in \mathbf{N}$). Let ψ denotes the wave function and the real-valued function V be the *potential* of the Schrödinger operator. The wave function verifies $\int_\Omega \psi^2 = 1$. We assume moreover that V is extended as $+\infty$ on $\mathbf{R}^d \setminus \Omega$, so that, in the case Ω bounded, ψ satisfies the boundary condition $\psi|_{\partial\Omega} = 0$. The controlled Schrödinger equation with one scalar control is the evolution equation

$$i\frac{d\psi}{dt} = -\Delta\psi(x,t) + V(x)\psi(x,t) + u(t)W(x)\psi(x,t), \tag{1}$$

where the real-valued function W is the *controlled potential.*

The control function $u : [0,T] \to \mathbf{R}$ is chosen in order to steer the quantum particle from its initial state to a prescribed target. A classical result asserts that exact controllability in L^2 is hopeless (see [1] and [2]).

The approximate controllability of one particular system of the type (1) has already be proved by Beauchard using Coron's return method (see [3] and references therein). Approximate controllability results for general systems under generic hypotheses were proved later with completely different methods independently in [4] and [5].

1.2. *Mathematical framework*

We reformulate below the Problem (1) in an abstract mathematical framework (see [4, Section 3] for details).

Let U be a subset of $\mathbf{R}$. Let H be an Hilbert space, $A : D(A) \subset H \to H$ be a densely defined (not necessarily bounded) essentially skew-adjoint operator and $B : D(B) \subset H \to H$ be a densely defined (not necessarily bounded) linear operator.

We assume that (A, B, U) satisfies the following three conditions: (H1) A and B are skew-adjoint, (H2) there exists an orthonormal basis $(\phi_k)_{k=1}^{\infty}$ of H made of eigenvectors of A, and all these eigenvectors are associated to simple eigenvalues (H3) $\phi_k \in D(B)$ for every $k \in \mathbf{N}$. A crucial consequence of these hypotheses is that for every $u \in U$, an extension of $A+uB$ generates a group of unitary transformations $e^{t(A+uB)} : H \to H$. For $(k,l) \in \mathbf{N}^2$, we define also the numbers $a(k,l) = \langle A\phi_k, \phi_l\rangle$ and $b(k,l) = \langle B\phi_k, \phi_l\rangle$.

We consider the conservative diagonal single input control systems

$$\frac{d\psi}{dt}(t) = A(\psi(t)) + u(t)B(\psi(t)). \tag{2}$$

A point ψ^0 of H and a piecewise constant function $u : [0,T] \to U$, $u = \sum_{l=1}^{L} \chi_{[t_l,t_{l+1})}u_l$ being given, we say that the solution of (2) with initial condition $\psi^0 \in H$ and corresponding to the control function $u : [0,T] \to U$ is the curve $\psi : [0,T] \to H$ defined by $\psi(t) = e^{(t-\sum_{l=1}^{l_t-1} t_l)(A+u_l B)} \circ \cdots \circ e^{t_1(A+u_1B)}(\psi^0)$ where $\sum_{l=1}^{l_t-1} t_l \leq t < \sum_{l=1}^{l_t} t_l$ and $u(\tau) = u_j$ if $\sum_{l=1}^{j-1} t_l \leq \tau < \sum_{l=1}^{j} t_l$. A piecewise constant function $u : [0,T] \to \mathbf{R}$ being given, the propagator of the control system (2) will be denoted by Φ.

1.3. *Main result*

Theorem 1.1. *Assume that the system (2) satisfies (i) the spectrum of A is a $\mathbf{Q}$-linearly independent family (ii) for every integer k, $b(k, k+1) \neq 0$ and (iii) zero is in the closure of the interior of U. Let $c : [0,T] \to L(H,H)$ be a continuous curve taking value in the set of the unitary operators of H and such that $c(0) = Id_H$. Let N be an integer. Then, for every $\epsilon > 0$, there exist $T_u > T$, a continuous non-decreasing bijection $s : [0,T] \to [0, T_u]$ and a piecewise constant control $u : [0, T_u] \to U$ such that the corresponding propagator $\Phi : [0, T_u] \times H \to H$ of system (2) satisfies (i) for every t in $[0,T]$, $\big||\langle \Phi(s(t), \phi_l), \phi_k\rangle| - |\langle c(t)(\phi_l), \phi_k\rangle|\big| < \epsilon$ for every k in $\mathbf{N}$, $1 \leq l \leq N$, and (ii) $\|\Phi(T_u, \phi_l) - c(T)(\phi_l)\| < \epsilon$ for every $1 \leq l \leq N$.*

1.4. *Content of the paper*

In Sec. 2, we explain how to choose a Galerkyn approximation of the original infinite dimensional control problem (2) in some space $SU(m)$. In Sec. 3, we use the Lie group structure of $SU(m)$ to compute the dimensions of some Lie subalgebras of $\mathfrak{su}(m)$ and to prove that the Galerkyn approximations obtained in Sec. 2 have some good tracking properties. A sketch of the proof of Theorem 1.1 is given in Sec. 4.

2. Choice of Galerkyn approximations

2.1. *Control and time-reparametrization*

We may assume without loss of generality that U has the special form $U = (0, \delta]$. Remark that, if $u \neq 0$, $e^{t(A+uB)} = e^{tu((1/u)A+B)}$. Associate with any piecewise constant $u = \sum_{l=1}^{k-1} \chi_{[t_l, t_{l+1}[} u_l \in PC([0,T_u], U)$ the function $v = \sum_{l=1}^{k-1} \chi_{[\tau_l, \tau_{l+1}[} 1/u_l \in PC([0, T_v], 1/U)$, with $T_v = \sum_l u_l(t_{l+1} - t_l)$ and τ_l defined by $\tau_1 = t_1$ and $\tau_{l+1} = \tau_l + u_l(t_{l+1} - t_l)$. Up to the time and control reparametrization given above, it is enough to prove Theorem 1.1 for the system

$$\frac{d\psi}{dt}(t) = v(t)A(\psi(t)) + B(\psi(t)) \tag{3}$$

where the set of admissible controls is the set $PC\left(\mathbf{R}^+, (1/\delta, +\infty)\right)$.

2.2. *Galerkyn approximation*

For a fixed piecewise constant control $v : \mathbf{R}^+ \to 1/U$ and a fixed ψ^0 in H, we consider the solution ψ of the system (3) of conservative diagonal single-input control systems with initial condition ψ^0.

For $k \in \mathbf{N}$, we define the function $x^k = \langle \psi, \phi_k \rangle : \mathbf{R} \to \mathbf{C}$. With our definition of solution, x^k is absolutely continuous and for almost all t in $\mathbf{R}^+$

$$\frac{d}{dt} x^k = v(t) a(k,k) + \sum_{l \in \mathbf{N}} b(k,l) x^l.$$

Prop 2.1. For every continuous curve $s : [0, T_s] \to H$ taking value in the unit sphere of H (that is, $\|s(t)\| = 1$ for all t in $[0, T_s]$), define the family $f_l = |\langle s, \phi_l \rangle|^2$, $l \in \mathbf{N}$. Then, for any strictly positive ϵ, there exists an integer $N(\epsilon)$ such that for all t in $[0, T_s]$, $\sum_{l=1}^{N(\epsilon)} f_l(t) > 1 - \epsilon$.

We define $\pi^m : H \to \mathbf{C}^m$ by $\pi^m(v) = \sum_{k=1}^{m} \langle v, \phi^k \rangle e_k^m$ for every v in H, where e_k is the kth element of the canonical basis of $\mathbf{C}^m$.

Prop 2.2. Fix a reference curve $c : [0, T] \to L(H, H)$ as in the hypotheses of Theorem 1.1, $\epsilon > 0$ and N a positive integer. Then there exists a continuous curve $M : [0, T] \to SU(m)$ such that $\|\pi^m(c(t)\phi^k) - M(t)\pi^m \phi^k\| < \epsilon$ for every t in $[0, T]$ and every k in $\{1..N\}$.

For $r \geq 1$, define the $r \times r$ matrices $A^r = [a(k,l)]_{1 \leq k,l \leq r}$ and $B^r = [b(k,l)]_{1 \leq k,l \leq r}$. The Galerkyn approximation of order m of the system (3) is

$$\frac{dx}{dt}(t) = v(t) A^m(x(t)) + B^m(x(t)) \tag{4}$$

The system (4) defines a control system on the differentiable manifold $\mathbf{C}^m$. Since the system (4) is linear, it is possible to lift it to the group of matrices of the resolvent.

We now proceed to a technical change of variable (variation of the constant) and define for every piecewise constant control function v in $PC(\mathbf{R}, 1/U)$ and every positive t, $y(t) = e^{-A^m \int_0^t v(s) ds} x(t)$.

Recalling that for all $m \times m$ matrices a, b, $e^{-a} b e^{a} = e^{\mathrm{ad}_a} b$, one checks that y verifies

$$\frac{dy}{dt}(t) = \exp\left(\mathrm{ad}_{\int_0^t v(s) ds A^m}\right) B^m y(t) \tag{5}$$

The system (5) defines a control system on the differentiable manifold $SU(m)$, and for every positive t and every $1 \leq k, l \leq m$, $|\langle x(t)\phi_k, \phi_l \rangle| = |\langle y(t)\phi_k, \phi_l \rangle|$.

3. Tracking properties of the Galerkyn approximations

First, we have to recall some classical definitions and results for invariant Lie groups control systems, see [6] and [7].

Let G be a semi-simple compact Lie groups, with Lie algebra $\mathfrak{g} = T_{Id}G$. Consider a smooth right invariant control system on G of the form

$$\begin{cases} \frac{d}{dt}g(t) &= dR_{g(t)}f(u(t)) \\ g(0) &= g_0 \end{cases} \tag{Σ}$$

where U is a subset of $\mathbf{R}$, $u : \mathbf{R} \to U$ is a L^∞ control function, $f : U \to \mathfrak{g}$ is a smooth application, g_0 is a given initial condition and $dR_a b$ denotes the value of the differential of the right translation by a taken at point b. We define the set $\mathcal{V} = \overline{\text{conv}\{f(u), u \in U\}}$ as the topological closure of the convex hull of all admissible velocities at point Id. The topological closure of the convex hull of all admissible velocities at point g is $dR_g(\mathcal{V})$.

Prop 3.1. Let P be a Lie-subgroup of G with Lie algebra $\mathfrak{p}$. If $\mathcal{V}$ contains some bounded symmetric set S such that $\mathfrak{p} \subset \text{Lie}(S)$, then for any continuous curve $c : [0,T] \to P$, for any $\epsilon > 0$, there exist $T_u > 0$, a L^∞ control function $u : [0, T_u] \to U$, and an increasing continuous bijection $\phi : [0, T_u] \to [0, T]$ such that the trajectory $g : [0, T_u] \to G$ of (Σ) with control u and initial condition $c(0)$ satisfies (i) $d_G(c(\phi(t)), g(t)) < \epsilon$ for every t in $[0, T_u]$ and (ii) $\phi(T_u) = T$.

To obtain trackabillity properties for the system (5), it is enough to check that the finite dimensional systems (5) satisfies the conditions on S given in Proposition 3.1 with $\mathfrak{p} = \mathfrak{su}_m$. We define S as the set of matrices $S = \pm\{b(k,l)E_{k,l} + b(l,k)E_{l,k}, 1 \leq k,l \leq m\}$. S is a symmetric and bounded subset of $\mathcal{V}$ (see [7, Appendix A]). The fact that $\text{Lie}(S) = \mathfrak{p}$ follows by the connectedness hypothesis (H2) (see [4, Proposition 4.1] for a detailed computation).

4. Infinite dimensional tracking

4.1. *Tracking in the phase variables*

For the proof of Theorem 1.1, we follow the method introduced in [4]. From the application $c : \mathbf{R} \to L(H, H)$ and the tolerance ϵ given in the hypotheses of Theorem 1.1, we use the results presented in Sec. 2 to find an integer m, the finite dimensional control system (5) and the trajectory $t \mapsto M(t)$ to be tracked in $SU(m)$. Proposition 3.1 gives the existence of some time $T_v > 0$ and some control function v in $PC([0, T_v], 1/U)$ such that the corresponding trajectory of (5) tracks the trajectory $t \mapsto M(t)$ with an error less than ϵ on each coordinate.

Since for every $1 \leq k \leq m$, the sequence $(b(k,l))_{l\geq 1}$ is in ℓ^2, there exists some N_1 in $\mathbf{N}$ such that $\sum_{l=N_1+1}^{\infty} |b(k,l)|^2 < \frac{\epsilon}{NT_v}$ for every $1 \leq k \leq m$. The next result asserts that any trajectory of the system (5) can actually be tracked (up to ϵ), with the N_1-Galerkyn approximation of system (3).

Prop 4.1. There exists a sequence $(v_k)_k$ in $PC\left(\mathbf{R}^+, (\frac{1}{\delta}, +\infty)\right)$ such that the sequence of matrix valued curves $t \mapsto e^{\mathrm{ad}_{\int vA^{(m)}}} B^{(m)}$ converges in the integral sense to $t \mapsto \left(\begin{array}{c|c} M(t) & 0_{m,N_1-m} \\ \hline 0_{N_1-m,m} & G(t) \end{array}\right)$, where $t \mapsto G(t)$ is some continuous curve in $U(N_1 - m)$.

Proof. The proof is a direct application of [4, Claim 4.3]. □

A direct application of [4, Claim 4.4] proves that, for k large enough, the control function $v = v_k$ given by Proposition 4.1 satisfies the conclusion (i) of Theorem 1.1.

4.2. *Final phase adjustment*

After time reparametrization, we get a control function $u \in PC([0, T_u], U)$ from v. Up to prolongation with the constant zero function, the control function $u : [0, T_u] \to U$ obtained in last paragraph can always be assumed to satisfy $T_u > T$ (the prolongation obviously still satisfies conclusion (i) of Theorem 1.1).

To achieve the proof of Theorem 1.1, one has to change u in such a way that it satisfies the conclusion (ii) of Theorem 1.1. One gets the result with a straightforward application of [4, Proposition 4.5].

References

1. J. M. Ball, J. E. Marsden and M. Slemrod, *SIAM J. Control Optim.* **20**, 575 (1982).
2. G. Turinici, **74**, 75 (2000).
3. K. Beauchard, *J. Math. Pures Appl. (9)* **84**, 851 (2005).
4. T. Chambrion, P. Mason, M. Sigalotti and U. Boscain, *Ann. Inst. H. Poincaré Anal. Non Linéaire* **26**, 329 (2009).
5. V. Nersesyan, *Comm. Math. Phys.* **290**, 371 (2009).
6. Y. L. Sachkov, *J. Math. Sci. (New York)* **100**, 2355 (2000), Dynamical systems, 8.
7. A. Agrachev and T. Chambrion, *ESAIM Control Optim. Calc. Var.* **12**, 409 (2006).

PART I

Control problems for dynamical systems under uncertainty and conflict

REFINED ASYMPTOTICS FOR SINGULARLY PERTURBED REACHABLE SETS

E. V. GONCHAROVA

Institute for System Dynamics and Control Theory
Siberian Branch of Russian Academy of Sciences
134, Lermontov Street, Irkutsk, 664033, Russia
E-mail: goncha@icc.ru

A. I. OVSEEVICH

Institute for Problems in Mechanics
Russian Academy of Sciences
101, Vernadsky Avenue, Moscow, 119526, Russia
E-mail: ovseev@ipmnet.ru

We study the limit behavior of the reachable sets to singularly perturbed linear dynamic systems with time dependent coefficients under geometric constraints on control. The system data are assumed to be Lipschitz continuous functions of time. The fast component of the phase vector is governed by a strictly stable linear system. It is shown that the reachable sets converge as the small parameter ε of singular perturbation tends to zero, and the rate of convergence is $O(\varepsilon \log 1/\varepsilon)$. Under an extra assumption on strict convexity of the set of admissible controls we find the coefficient of $\varepsilon \log 1/\varepsilon$ in the asymptotic expansion for the support function of the reachable set.

Keywords: Singularly perturbed linear dynamic control systems, reachable sets, asymptotics.

1. Problem statement

The paper concerns the study of the asymptotic behavior of the reachable sets to singularly perturbed linear control systems. The motions described by such systems evolve in the two different time scales: "slow" and "fast" times, and, in a natural way, split into the two dynamics. The system coefficients vary slowly with respect to the "fast" time scale.

The purpose of this work is to find an asymptotic estimate for the reachable set to singularly perturbed linear systems, when the small parameter of singular perturbations tends to zero.

Consider the following singularly perturbed linear dynamic system under geometric constraint on control

$$
\begin{aligned}
\dot{x} &= Ax + By + Fu, \\
\varepsilon \dot{y} &= Cx + Dy + Gu, \ u \in U,
\end{aligned} \tag{1}
$$

with a given initial state $x(0) = 0$, $y(0) = 0$, and $t \in [0, T]$, where $T > 0$ is fixed. Here, $\varepsilon > 0$ is a small parameter, the phase vector $z = (x, y)$ consists of "slow" $x \in \mathbb{R}^n$, and "fast" $y \in \mathbb{R}^m$ components. An admissible control is by definition a measurable function $u(\cdot)$ such that $u(t) \in U$ for almost all $t \in [0, T]$, where $U \subset \mathbb{R}^k$ is a nonvoid convex compact.

We assume the matrix functions $A, \ldots, G$ and the convex compact U are Lipschitz continuous with respect to t. The latter means that the support function $h = H_U$ of the set U is Lipschitz continuous in t. The matrix D is assumed to be asymptotically stable, i.e. $\operatorname{Re}\operatorname{Spec} D < 0$, for any t.

Denote by $\mathcal{D}_\varepsilon(T)$ the reachable set to system (1) at time T, i.e. the set of ends at time T of all admissible trajectories of system (1). We will study the limit behavior of the set $\mathcal{D}_\varepsilon(T)$ as $\varepsilon \to 0$.

Under the same assumptions, this issue was addressed in [1,2], where it was shown that the sets $\mathcal{D}_\varepsilon(T)$ have a limit $\mathcal{D}_0(T)$ with respect to the Hausdorff metric as $\varepsilon \to 0$, and the rate of convergence of the reachable sets is $O(\varepsilon^\alpha)$, where $0 < \alpha < 1$ is arbitrary. In this paper we succeeded to get an essential refinement of the results [1,2]. We proved that the said rate of convergence of the reachable sets is, in fact, $O(\varepsilon \log 1/\varepsilon)$. In suitable coordinates the "slow" and "fast" state components split, and the limit set $\mathcal{D}_0(T)$ is the direct product of the limit reachable sets of the "slow" and "fast" subsystems. Each factor in the product is a (topological) manifold with boundary, while the product is a manifold with corners. Thus, in the first approximation, slow and fast controlled motions are independent. Under an extra assumption that the support function h of the set U is C^1-smooth outside the origin, we found the exact coefficient of $\varepsilon \log 1/\varepsilon$ in the asymptotic expansion for the support function of the prelimit reachable set. Thus, the estimate $O(\varepsilon \log 1/\varepsilon)$ for the rate of convergence is sharp. The correction term in the asymptotics turns out to be negative. Geometrically, this looks like "rounding off" the corners of the limit reachable set $\mathcal{D}_0(T)$.

The assumption on Lipschitz continuity of the system coefficients is essential. There are examples where the system parameters are Hölder continuous with exponent $0 < \alpha < 1$, while the rate of convergence of the reachable sets is $\Omega(\varepsilon^\alpha)$.

2. Splitting dynamic system

Following [1,3] we can simplify the original problem by using gauge transformations. If $z = \begin{pmatrix} x \\ y \end{pmatrix}$, $\mathfrak{A} = \begin{pmatrix} A & B \\ C/\varepsilon & D/\varepsilon \end{pmatrix}$, and $\mathfrak{B} = \begin{pmatrix} F \\ G/\varepsilon \end{pmatrix}$ one can perform a substitution $z = Xw$, where X is an invertible 2×2 block matrix, and we get a new control system $\dot{w} = \widetilde{\mathfrak{A}}w + \widetilde{\mathfrak{B}}u$. Here, $\widetilde{\mathfrak{A}} = X^{-1}\mathfrak{A}X - X^{-1}\dot{X}$, and $\widetilde{\mathfrak{B}} = X^{-1}\mathfrak{B}$. If X is Lipschitz continuous with respect time t, then such a transformation does not have an essential influence on the behavior of the reachable sets, but allows us to simplify the system matrix so that the Lipschitz continuity and stability of the corresponding system coefficients are preserved. We aim at reducing the system matrix to a block-diagonal form $\widetilde{\mathfrak{A}} = \begin{pmatrix} \widetilde{A} & 0 \\ 0 & \widetilde{D}/\varepsilon \end{pmatrix}$, to separate slow and fast variables. Suffice it to do an approximate reduction, so that the blocks $\widetilde{B}$ and $\widetilde{C}$ are $O(\varepsilon)$, because this gives us approximate reachable sets $\mathcal{D}_\varepsilon(T)$ with the same order $O(\varepsilon)$ of precision. By using the lower-triangular transformation $X = \begin{pmatrix} 1 & 0 \\ -D^{-1}C & 1 \end{pmatrix}$, we obtain a new block $\widetilde{C}/\varepsilon$, where $\widetilde{C} = \varepsilon\frac{d}{dt}(D^{-1}C) + \varepsilon D^{-1}C(A - BD^{-1}C) = O(\varepsilon)$. Note that the Lipschitz continuity implies that the derivative $\frac{d}{dt}(D^{-1}C) = O(1)$ is bounded. Similarly, by applying the upper-triangular transformation $X = \begin{pmatrix} 1 & \varepsilon BD^{-1} \\ 0 & 1 \end{pmatrix}$, one can ensure that $\widetilde{B} = O(\varepsilon)$. Then, we arrive at the split case:

$$\begin{aligned} \dot{x} &= \widetilde{A}x + \widetilde{F}u, \\ \varepsilon\dot{y} &= \widetilde{D}y + \widetilde{G}u, \ u \in U, \end{aligned} \tag{2}$$

where all matrices and the convex compact U are Lipschitz continuous with respect to time, and $\widetilde{D}$ is a stable matrix at each time instant. The reachable sets to systems (1) and (2) coincide up to an error of order $O(\varepsilon)$. This follows immediately from the stability of $\widetilde{D}$ and the classical Tikhonov–Levinson theorem (see, e.g., [3–5]). The asymptotics we are looking for is, in fact, rather crude. Its remainders are of order $o(\varepsilon \log 1/\varepsilon)$ so that an error of order $O(\varepsilon)$ is negligible.

The direct computations show that the matrix coefficients of systems (1), and (2) are related by $\widetilde{A} = A - BD^{-1}C$, $\widetilde{D} = D$, $\widetilde{F} = F - BD^{-1}G$, $\widetilde{G} = G$. In the next section, we state our main results in terms of the original system (1), while, in the proofs, the system splitting is heavily used.

3. Asymptotics of support functions to reachable sets

Denote by $H_\varepsilon(\xi,\eta)$ the support function of the reachable set $\mathcal{D}_\varepsilon(T)$ to system (1), where $\varepsilon > 0$, and ξ,η are dual to the phase variables x,y. Define the function

$$H_0(\xi,\eta) \;=\; \int_0^T h_t(\widetilde{F}(t)^*\Phi(T,t)^*\widetilde{\xi})\,dt + \int_0^\infty h_T(G(T)^* e^{D(T)^*t}\eta)\,dt, \tag{3}$$

where the function Φ is the fundamental matrix for the linear system

$$\dot{x} = (A - BD^{-1}C)x, \tag{4}$$

$\widetilde{\xi} = \xi - C^*D^{*-1}\eta$, $\widetilde{F} = F - BD^{-1}G$, and $h = h_t$ is the support function of the set $U = U_t$ of controls. The function H_0 is, in fact, the support function for the limit reachable set $\mathcal{D}_0(T) = \lim_{\varepsilon\to 0}\mathcal{D}_\varepsilon(T)$.

Theorem 3.1. *Let $H_\varepsilon(\xi,\eta)$ be the support functions of the reachable sets $\mathcal{D}_\varepsilon(T)$ to system (1), then $H_\varepsilon(\xi,\eta)$ converge $H_0(\xi,\eta)$ uniformly on compacts as $\varepsilon \to 0$. Moreover, we have the asymptotic equivalence:*

$$H_\varepsilon(\xi,\eta) = H_0(\xi,\eta) + O(\varepsilon\log 1/\varepsilon(|\xi| + |\eta|)) \text{ as } \varepsilon \to 0.$$

In other words, $d_H(\mathcal{D}_\varepsilon,\mathcal{D}_0) = O(\varepsilon\log 1/\varepsilon)$, where d_H is the Hausdorff metric.

The idea underlying the proof goes back at least to [6], and, basically, says that the reachable set to a linear control system can be decomposed in accordance with the decomposition of the spectrum of the system matrix into stable, unstable, and neutral components. Each composing part is formed by using controls supported on non-overlapping time intervals. In our case, there are just two spectral components: the neutral one corresponding to "slow" variables, and the stable one corresponding to "fast" variables. We divide the time interval $[0,T]$ into the two subintervals $[0,T-\delta]$ and $[T-\delta,T]$, where δ is a small positive parameter. The controls supported on the "long" interval are responsible for the "slow" part of the reachable set, while the controls supported on the "short" interval form the "fast" part of it. The proper choice of δ is crucial for the accuracy of approximation. By choosing a small $\delta > 0$ such that $\delta \to 0$ and $\delta/\varepsilon \to \infty$ as $\varepsilon \to 0$, we get an approximation to the set $\mathcal{D}_0(T)$. The main difference of the present paper with [1] stems from the final choice $\delta \sim \varepsilon\log 1/\varepsilon$ instead of $\delta \sim \varepsilon^\alpha$.

Our main asymptotic result consists in finding the remainder in the previous theorem in a more precise form $\mathbf{c}(\xi,\eta)\varepsilon\log 1/\varepsilon + o(\varepsilon\log 1/\varepsilon)$. We can do this under an extra assumption that the support function h_T of the set U_T of controls is C^1-smooth outside the origin. This means that the set U_T is strictly convex. Denote the argument of the function $\bar{h} = h_T$ by ζ, and consider the average

$$\mathrm{Av}_{\,\tau}(\frac{\partial \bar{h}}{\partial v})(\eta) = \frac{1}{\tau}\int_0^{\tau} \frac{\partial \bar{h}}{\partial v}(G(T)^* e^{D(T)^* t}\eta)\, dt.$$

The limit $\mathrm{Av}(f) = \lim\limits_{\tau\to\infty} \mathrm{Av}_\tau(f)$ does exist for any homogeneous of zero degree function f which is continuous for any $v \neq 0$. Indeed, the function $\phi(t) = G(T)^* e^{D(T)^* t}\eta$ has the form of a vector-valued quasipolynomial, and, therefore,

$$\phi(t) = e^{(\mathrm{Re}\,\lambda)t} t^N \sum e^{i\omega t} a_\omega + o(e^{(\mathrm{Re}\,\lambda)t} t^N),$$

where λ is an eigenvalue of the matrix $D(T)$, $N+1$ is the maximal size of the corresponding Jordan block, sum is taken over all real ω such that $\mathrm{Re}\,\lambda + i\omega$ is an eigenvalue of $D(T)$, and a_ω is a time-independent vector. Since the function f is homogeneous of zero degree, we have

$$\mathrm{Av}_{\,\tau}(f) \;=\; \frac{1}{\tau}\int_0^{\tau} f(\phi(t))\, dt = \frac{1}{\tau}\int_0^{\tau} f\left(\phi_0(t)\right)\, dt + o(1),$$

where $\phi_0(t) = \sum e^{i\omega t} a_\omega$ is a trigonometric polynomial. "Generically", the trigonometric polynomial $\phi_0(t)$ is comprised either of one or two harmonics. The limit

$$\mathrm{Av}\,(f) = \lim_{\tau\to\infty} \frac{1}{\tau}\int_0^{\tau} f\left(\phi_0(t)\right)\, dt,$$

does exist, since the integrand in (3) is an almost periodic function which can be averaged.

Theorem 3.2. *Assume that the support function $\bar{h} = h_T$ of the control set U_T is C^1-smooth outside the origin. Then, the support function $H_\varepsilon(\xi,\eta)$ of the reachable set $\mathcal{D}_\varepsilon(T)$ to system (1) has the following asymptotic representation:*

$$H_\varepsilon(\xi,\eta) \;=\; H_0(\xi,\eta) + \mathbf{c}(\xi,\eta)\,\varepsilon \log 1/\varepsilon + o(\varepsilon \log 1/\varepsilon) \quad \text{as } \varepsilon \to 0, \quad (5)$$

where

$$\mathbf{c}(\xi,\eta) = \frac{1}{\Lambda}\left[Av\left\langle \widetilde{F}(T)^*\widetilde{\xi}, \frac{\partial \bar{h}}{\partial v}\right\rangle(\eta) - \bar{h}(\widetilde{F}(T)^*\widetilde{\xi})\right],$$

$\Lambda = \Lambda(\eta)$ is the absolute value of the first Lyapunov exponent of the function $t \mapsto e^{D(T)^* t}\eta$, $\widetilde{\xi} = \xi - C(T)^* D(T)^{*-1}\eta$, and $\widetilde{F}(T) = F(T) - B(T)D(T)^{-1}G(T)$.

Note that the coefficient $\mathbf{c}(\xi,\eta)$ of $\varepsilon\log\frac{1}{\varepsilon}$ in the asymptotic expansion (5) is nonpositive. Indeed, $\langle\zeta,\dfrac{\partial\bar{h}}{\partial v}(v)\rangle \le \bar{h}(\zeta)$ for any ζ and $v\neq 0$, and the averaging operation preserves the inequality.

4. Conclusion

In the generic case, when the set U_T of admissible controls at the time instant T is not a singleton, the coefficient $\mathbf{c}(\xi,\eta)$ is not identically (for all ξ,η) equal to zero. Otherwise, this would imply the equality $\bar{h}(\zeta)=\langle\zeta,\varphi\rangle$, $\varphi = \mathrm{Av}\,(\dfrac{\partial\bar{h}}{\partial\zeta}(G^*e^{D^*t}\eta))$, for all ζ and η. The latter necessarily means that $\bar{h}(\zeta)$ is the support function for the singleton $U_T=\{\varphi\}$. The fact that $c(\xi,\eta)\not\equiv 0$ means that the estimate given by Theorem 1 is sharp, i.e., for some ξ, η, we have $|H_\varepsilon(\xi,\eta)-H_0(\xi,\eta)|\ge C\varepsilon\log 1/\varepsilon$, where $C>0$ does not depend on ε.

The above results can be illustrated by the following simple example of a singularly perturbed linear system:

$$\dot{x}=u,\ \ \varepsilon\dot{y}=-y+u,$$

where x,y,u are scalars, and $|u|\le 1$. This example is also presented in [1]. An easy calculation reveals that in this case the difference of the support functions of the prelimit and limit reachable sets equals

$$\Delta H \ = \ H_\varepsilon(\xi,\eta)-H_0(\xi,\eta) = -2t_\varepsilon|\xi| - |\eta|(2e^{-t_\varepsilon/\varepsilon}-e^{-T/\varepsilon})$$

provided that $\xi\eta<0$. Here, $t_\varepsilon=\varepsilon\log\frac{1}{\varepsilon}\frac{|\eta|}{|\xi|}$. Thus, for fixed ξ,η in this range, the difference ΔH has the form $-2|\xi|\varepsilon\log\frac{1}{\varepsilon}+C\varepsilon+r$, where C is a constant, and the remainder r is exponentially small as $\varepsilon\to 0$. This proves again that the estimate in Theorem 1 is sharp.

Acknowledgments

The work is supported by RFBR (grants 08-01-00156, 08-08-00292).

References

1. A. L. Dontchev and J. I. Slavov, *Systems & Control Letters* **11**, 385 (1988).
2. A. L. Dontchev and V. M. Veliov, *SIAM J. Control and Optim.* **21**, 566 (1983).
3. P. V. Kokotovich, *SIAM Review* **6**, 501 (1984).
4. L. Flatto and N. Levinson, *J. Rat. Mech. Anal.* **4**, 943 (1955).
5. A. B. Vasilieva and V. F. Butuzov, *Asymptotic Expansions of Solutions of Singularly Perturbed Differential Equations* (Nauka, Moscow, 1973).
6. A. L. Dontchev and V. M. Veliov, *C. R. Acad. Bulg. Sci.* **36**, 1021 (1983).

MULTIESTIMATES FOR LINEAR-GAUSSIAN CONTINUOUS SYSTEMS UNDER COMMUNICATION CONSTRAINTS

B. I. ANAN'EV

Institute of Mathematics and Mechanics of UB of RAS
Yekaterinburg, Russia
E-mail: abi@imm.uran.ru

Estimation problems for linear systems are considered under mixed disturbances. The structure of defined multiestimates is found out: they are the sum of a random vector and a determinate set, which depend on parameters. The parameters define the conditional probability of inclusion of the multiestimate in a covering set. The form of the covering set is discussed. The presence of communication constraints is taken into account.

Keywords: Multiestimates, communication channel, estimation, control.

1. Introduction

There are many control problems under incomplete information in which state estimation algorithms are used. A motion correction problem is one of them [1]. For linear multistage systems, in [2] estimates were offered in the form of sets. In [3], the concept of random information set for multistage systems with the mixed uncertainty is introduced. In [4], a generalization of the random information sets is offered for multistage stochastic inclusions. An inclusion of the multiestimates in a covering set with the communication constraints is considered in [5]. Here we consider a linear system in Ito's form: $dx = (A(t)x + v(t))dt + d\xi(t)$, $t \in [0,T]$, $dy = (C(t)x + w(t))dt + d\eta(t)$, $y_0 = 0$, where $x(t) \in R^n$, $y(t) \in R^m$. The x_0 has Gaussian distribution $\mathrm{Law}\,(x_0) = N(\bar{x}_0, \gamma_0)$. Uncertain parameters are restricted by the convex and compact constraints: $\bar{x}_0 \in \bar{X}_0$, $v(t) \in V$, $w(t) \in W$. The processes $\xi(t)$, $\eta(t)$ are supposed to be Wiener ones with $\mathrm{cov}(d\xi, d\xi) = Q(t)dt$, $\mathrm{cov}(d\eta, d\eta) = R(t)dt$. The processes $\xi(t)$, $\eta(t)$ are independent and do not depend on the initial vector x_0. Let $\ker C = \{x : Cx = 0\}$, $\mathrm{im}\, C = \{y : y = Cx, \exists x\}$, then the equality $\{x : Cx \in W\} = C^-(W \cap \mathrm{im}\ C) \oplus \ker C$ takes place. We introduce determinate functions $x^1 = \mathrm{E}x$, $y^1 = \mathrm{E}y$, where E is the expectation, and random processes $x^0 = x - x^1$, $y^0 = y - y^1$. Using the function

$y^{1t}(\cdot) = \{y^1(s) : s \in [0,t]\}$ we construct the determinate information set [1], which is denoted by $\mathcal{X}^1(t,y)$. To this end for given $\Delta > 0$ and for any $t \in [0,T]$, consider the finite sequence $y^{1t}(\cdot\Delta) = \{y^1(\Delta), \ldots, y^1(j_t^\Delta \Delta), y^1(t)\}$ consisting of $j_t^\Delta + 1$ elements, where $j_t^\Delta = \lfloor t/\Delta \rfloor$. Here $\lfloor\cdot\rfloor$ denotes the integer part of a real number. Denote by $X(t,s)$ the fundamental matrix of the equation of motion. Let's put $(j_t^\Delta + 1)\Delta = t$ by definition and introduce the designations: $C_i^\Delta = \int_{(i-1)\Delta}^{i\Delta} C(s)X(s,(i-1)\Delta)ds$, $X_i^\Delta = X(i\Delta,(i-1)\Delta)$, $\mathcal{W}_i^\Delta = \{\int_{(i-1)\Delta}^{i\Delta} w(s)ds : w(s) \in W\} \subset R^n$, $\mathcal{V}_i^\Delta = \left\{\left[\int_{(i-1)\Delta}^{i\Delta} X(i\Delta,s)v(s)ds; \int_{(i-1)\Delta}^{i\Delta}\int_s^{i\Delta} C(t)X(t,s)dtv(s)ds\right] : v(s) \in V\right\} \subset R^{n+m}$, $\tilde{C}_i^\Delta = [C_i^\Delta, F]$, $\tilde{X}_i^\Delta = [X_i^\Delta, D]$, where $F = [O_{m\times n}, I_m]$, $D = [I_n, O_{n\times m}]$, I_n is a unity matrix, $O_{m\times n}$ is a zero matrix. Consider the multistage system: $x_i^1 = X_i^\Delta x_{i-1}^1 + Dv$, $v \in \mathcal{V}_i^\Delta$, $y^1(i\Delta) = y^1((i-1)\Delta) + C_i^\Delta x_{i-1}^1 + Fv + w$, $w \in \mathcal{W}_i^\Delta$, $x_0^1 \in \bar{X}_0$, $y^1(0) = 0$, $i = 1, \ldots, j_t^\Delta + 1$, and introduce the recurrent sets defined by the equations: $\bar{\mathcal{X}}_i^\Delta = \left(\mathcal{X}_{i-1}^\Delta \times \mathcal{V}_i^\Delta\right) \bigcap \left(\tilde{C}_i^{\Delta-}((y_i^1 - \mathcal{W}_i^\Delta)\bigcap \operatorname{im}\tilde{C}_i^\Delta) \oplus \ker \tilde{C}_i^\Delta\right)$, $y_i^1 = y^1(i\Delta) - y^1((i-1)\Delta)$, $\mathcal{X}_i^\Delta = \tilde{X}_i^\Delta \bar{\mathcal{X}}_i^\Delta$, $\mathcal{X}_0^\Delta = \bar{X}_0$. Then the following convergence takes place in Hausdorf metric uniformly with respect to $t \in [0,T]$: $\lim_{\Delta\to 0} \mathcal{X}_{j_t^\Delta+1}^\Delta = \mathcal{X}^1(t,y)$. We call the sets $\mathcal{X}(t,y) = \mathcal{X}^1(t,y) + x^0(t)$, $t \in [0,T]$, the random information sets or the multiestimates. The sets $\mathcal{X}^1(t,y)$ and $\mathcal{X}(t,y)$ coincide in the absence of random components. In this work, we construct a covering set $\mathcal{Z}(t)$, depending on the measured signal $y^t(\cdot)$ and providing inclusion $\mathcal{X}(t,y) \subset \mathcal{Z}(t)$ with given conditional probability irrespective of realization of uncertain parameters $\{\bar{x}_0, v(\cdot), w(\cdot)\}$. The problem can be complicated due to the presence of communication constraints.

2. Constructing of covering set

Let the symbol $\mathcal{Y}^{1t}$ mean the set of all determinate signals on the segment $[0,t]$. Fixing $y^{1t} \in \mathcal{Y}^{1t}$ and marking out the random element $y^{0t} = y(\cdot) - y^{1t}$, we can find the conditional distribution $\mathrm{Law}\,(x^0(t) \mid y^{0t}, y^{1t}) = N(m^0(t), \gamma(t))$ for the vector $x^0(t)$. The parameters $m^0(t)$, $\gamma(t)$ satisfy the equations of Kalman's filter [6]. Let $k(t)$ be a vector satisfying the Kalman's equation, where the signal y^0 is replaced by y, and $\alpha(t)$ be a vector satisfying to the similar equation, where the signal y^0 is replaced by $-y^1$. Let's form the set $\mathcal{Z}(t) = k(t) + \bigcup_{y^{1t}\in\mathcal{Y}^{1t}}(\alpha(t) + \mathcal{X}^1(t,y))$ and use it as a covering one for the multiestimate $\mathcal{X}(t,y)$. Now our goal is to count up the conditional probability $P(\mathcal{X}(t,y) \subset \mathcal{Z}(t) \mid y^{0t}, y^{1t})$. We have the equality of events $\{\mathcal{X}(t,y) \subset Z(t)\} = \{x^0(t) - m^0(t) \in \mathcal{D}(t)\}$, where $\mathcal{D}(t) = \left\{x : x + \alpha(t) + \mathcal{X}^1(t,y) \subset \bigcup_{y^{1t}\in\mathcal{Y}^{1t}} \left(\alpha(t) + \mathcal{X}^1(t,y)\right)\right\}$. The set $\mathcal{D}(t)$ depends on the concrete realization y^{1t} and contains zero at any such realization. Thus, the required conditional probability is equal to $P(\mathcal{X}(t,y) \subset \mathcal{Z}(t) \mid y^{0t}, y^{1t}) = \int_{\mathcal{D}(t)} f(x,\gamma(t))dx$, where $f(x,\gamma)$ is the density of a Gaussian distribution $N(0,\gamma)$. The value y^{1t} is unknown. Therefore, it is possible to guarantee the inclusion $\mathcal{X}(t,y) \subset \mathcal{Z}(t)$ with conditional probability not smaller than the minimum of value of the integral over all $y^{1t} \in \mathcal{Y}^{1t}$. In order to provide the high probability of covering of the multiestimates, consider the concentration ellipsoid $\mathcal{E}_t(l) = \{x \in \text{im}\,\gamma(t) : x'\gamma(t)^- x \le l^2\}$ such that $P(x^0(t) - m^0(t) \in \mathcal{E}_t(l)) = \int_{\|x\|\le l} f(x, I_k)dx \ge 1 - \varepsilon$, $k = \text{rank}\,\gamma(t)$. Then we have the equality of the events $\{\mathcal{X}(t,y) \subset \mathcal{Z}(t) + \mathcal{E}_t(l)\} = \{x^0(t) - m^0(t) \in \tilde{\mathcal{D}}(t)\}$, where $\tilde{\mathcal{D}}(t) = \left\{x : x + \alpha(t) + \mathcal{X}^1(t,y) \subset \bigcup_{y^{1t}\in\mathcal{Y}^{1t}} \left(\alpha(t) + \mathcal{X}^1(t,y)\right) + \mathcal{E}_t(l)\right\}$. Whereas $\mathcal{E}_t(l) \subset \tilde{\mathcal{D}}(t)$, the inequality $P(\mathcal{X}(t,y) \subset \mathcal{Z}(t) + \mathcal{E}_t(l) \mid y^{0t}, y^{1t}) \ge 1 - \varepsilon$ is valid for any signal y^{1t}.

We can reduce the exhaustive search of uncertain signals y^{1t} under construction of covering set for a multiestimate and simultaneously construct the evolutionary covering set. Note that Law $(y^0(t)) = N(0, J(t))$. For any $t \in [0,T]$ let us choose the concentration ellipsoid $\mathbf{Y}_t(l) = \{x \in \text{im}\,J(t) : x'J(t)^- x \le l^2\}$ such that $P(y^0(t) \in \mathbf{Y}_t(l)) = \int_{\|x\|\le l} f(x, I_k)dx > 0.99$, $k = \text{rank}\,J(t)$. Then it is possible to assume practically that $y^1(t) \in y(t) - \mathbf{Y}_t(l)$, i.e. the confidence set $y(t) - \mathbf{Y}_t(l)$ covers the signal $y^1(t)$. Therefore we can consider the narrower set $\bar{\mathcal{Y}}^{1t} = \{y^{1t} \in \mathcal{Y}^{1t} : y^1(s) \in y(s) - \mathbf{Y}_s(l),\ s \in [0,t]\}$.

In case of small matrices Q, R and γ_0 this inclusion essentially reduces the search of the determined signals under construction of covering set. Consider an evolutionary construction of such covering set, which in the absence of random parameters, coincides with $\mathcal{X}^1(t,y)$. Let us pass to the approximating discrete scheme. Purely random system in discrete scheme can be written as $x_i^0 = X_i^\Delta x_{i-1}^0 + \xi_i^\Delta$, $\text{cov}\,(\xi_i^\Delta, \xi_i^\Delta) = Q_i^\Delta$; $y_i^0 = C_i^\Delta x_{i-1}^0 + \eta_i^\Delta$, $\text{cov}\,(\eta_i^\Delta, \eta_i^\Delta) = R_i^\Delta$, $\text{cov}\,(\xi_i^\Delta, \eta_i^\Delta) = S_i^\Delta$. Note that Law $(y_i^0) = N(0, J_i)$. Choosing the concentration ellipsoid $\mathbf{Y}_i^\Delta(l) = \{x \in \text{im}\,J_i : x'J_i^- x \le l^2\}$ such that $\int_{\|x\|\le l} f(x, I_k)dx = P(y_i^0 \in \mathbf{Y}_i^\Delta(l)) > 0.99$, $k = \text{rank}\,J_i$, we can assume that the inclusion $y_i^{1\Delta} \in y_i - \mathbf{Y}_i^\Delta(l)$ is practically valid.

Introduce the sets $\tilde{\mathcal{X}}_i^\Delta = \tilde{X}_i^\Delta(\tilde{\mathcal{X}}_{i-1}^\Delta \times \mathcal{V}_i^\Delta)$, $\tilde{\mathcal{X}}_0^\Delta = \bar{X}_0$, $\mathcal{Y}_i^\Delta = \tilde{C}_i^\Delta(\tilde{\mathcal{X}}_{i-1}^\Delta \times \mathcal{V}_i^\Delta) + \mathcal{W}_i^\Delta$, $\mathbf{Y}_i^\Delta(l,y) = \mathcal{Y}_i^\Delta \bigcap \left(y_i - \mathbf{Y}_i^\Delta(l)\right)$. The set $\mathbf{Y}_i^\Delta(l,y)$ represents the family of all signals y_i^1 that can be realized accordance with the confidence set $y_i - \mathbf{Y}_i^\Delta(l)$. In the absence of random parameters we obtain the singleton $\mathbf{Y}_i^{\Delta}(l,y) = \{y_i^1\}$. Consider Law $(x_i^0 \mid y^{0i}, y^{1i}) = N(m_i^0, \gamma_i)$. Here $y^{1i} = \{y_1^1, \ldots, y_i^1\}$ and y^{0i} is defined similarly. Parameters m_i^0, γ_i satisfy Kalman's filter. Let k_i be a vector satisfying the Kalman equation, where the signal y^0 is replaced by y, and α_i be a vector satisfying to the similar equation, where the signal y^0 is replaced by $-y^{1\Delta}$. Then we can form the set $\mathcal{Z}_i = k_i + \bigcup_{y_\cdot^1 \in \mathbf{Y}_\cdot^{\Delta}(l,y)} (\alpha_i + \mathcal{X}_i^{\Delta})$ and use it as a covering one for the discrete multiestimate $\mathcal{X}_i^{\Delta}(y) = x_i^0 + \mathcal{X}_i^{\Delta}$. Note that $\mathcal{Z}_i = \mathcal{X}_i^{\Delta}$ if the random parameters are absent. Unfortunately, covering sets $\mathcal{Z}_i$ are not recurrent as a rule. Now we a little expand $\mathcal{Z}_i$ and make their recurrent ones. Let $\mathcal{A}_i$, $\tilde{\mathcal{X}}_i^{\Delta}$, $\hat{\mathcal{X}}_i^{\Delta}$ be the sets defined by the relations: $\mathcal{A}_i = (X_i^{\Delta} - K_{i-1} C_i^{\Delta}) \mathcal{A}_{i-1} - K_{i-1} \mathbf{Y}_i^{\Delta}(l,y)$, $\mathcal{A}_0 = \{0\}$, $\tilde{\mathcal{X}}_i^{\Delta} = \left(\hat{\mathcal{X}}_{i-1}^{\Delta} \times \mathcal{V}_i^{\Delta} \right) \bigcap \left\{ z \in R^{2n+m} : \tilde{C}_i^{\Delta} z \in \mathbf{Y}_i^{\Delta}(l,y) - \mathcal{W}_i^{\Delta} \right\}$, $\hat{\mathcal{X}}_i^{\Delta} = \tilde{X}_i^{\Delta} \tilde{\mathcal{X}}_i^{\Delta}$, $\hat{\mathcal{X}}_0^{\Delta} = \bar{X}_0$, $i = 1, \ldots, j_t^{\Delta} + 1$. Then with the account of above restrictions we receive $\bigcup_{y_\cdot^1 \in \mathbf{Y}_\cdot^{\Delta}(l,y)} (\alpha_i + \mathcal{X}_i^{\Delta}) \subset \mathcal{A}_i + \hat{\mathcal{X}}_i^{\Delta}$, $i \geq 1$. The set $\hat{\mathcal{Z}}_i = k_i + \mathcal{A}_i + \hat{\mathcal{X}}_i^{\Delta}$ covers the multiestimate $\mathcal{X}_i^{\Delta}(y)$ with conditional probability $P(\mathcal{X}_i^{\Delta}(y) \subset \hat{\mathcal{Z}}_i \mid y^{0i}, y^{1i}) = \int_{\mathcal{B}_i} f(x, \gamma_i) dx$, where the function f is the same as above, and the set $\mathcal{B}_i$ is of the form $\mathcal{B}_i = \left\{ x : x + \alpha_i + \mathcal{X}_i^{\Delta} \subset \mathcal{A}_i + \hat{\mathcal{X}}_i^{\Delta} \right\}$. The minimum of the conditional probability over all $y_\cdot^1 \in \mathbf{Y}_\cdot^{\Delta}(l,y)$ is the guaranteed result of inclusion's probability. Note that $\hat{\mathcal{Z}}_i = \mathcal{X}_i^{\Delta}$ if the randomness is absent. The same as above it is possible to add the ellipsoid $\mathcal{E}_i(l)$ to set $\hat{\mathcal{Z}}_i$ in order to obtain the guaranteed result. Note that for any given signal y^{1t} $\lim_{\Delta \to 0} k_{j_t^{\Delta}+1} = k(t)$ a.s. Thus, we obtain that $d(\mathcal{Z}_{j_t^{\Delta}+1}, \mathcal{Z}(t)$ converges almost surely to zero when $\Delta \to 0$, where $d(\cdot,\cdot)$ is the Hausdorf metric. The same convergence takes place for $\mathcal{X}_{j_t^{\Delta}+1}^{\Delta}(y)$ and $\mathcal{X}(t,y)$.

3. Accounting of communication constraints

Note that equations and characteristics of noises are known in the center of processing information (CPI). Hence, for the construction of covering set for a multiestimate, it is necessary to transfer a signal observed on object in CPI as precisely as possible. For vector $y \in R^m$ consider the sup-norm $\|y\|_\infty = \max_i |y_i|$, $\|\cdot\|$ is the Euclidean norm. Norms for matrices are designated similarly. Let $B_a = \{y : \|y\|_\infty \leq a\}$ be a sphere of radius a. Given natural number q, we divide the sphere B_a into q^m subspheres of type $I_{j_1}^1 \times \cdots \times I_{j_m}^m$, where indexes $j_i \in \{1, \ldots, q\}$ and $I_j^i = \{y_i : -a + 2(j-1)a/q \leq y_i < -a + 2ja/q\}$, $i = 1, \ldots, m$, $j = 1, \ldots, q-1$; $I_q^i = \{y_i : a - 2a/q \leq y_i \leq a\}$. The vector $y \in B_a$ is coded by the sequence $\eta(y) = (j_1, \ldots, j_m)$, if $y \in I_{j_1}^1 \times \cdots \times I_{j_m}^m$. On the contrary, each set $(j_1, \ldots, j_m)$ of natural numbers is assigned to the vector $\gamma(j_1,\dots,j_m)$ with coordinates $\gamma_i = -a + (2j_i - 1)a/q$, $i = 1,\dots,m$. This vector is the geometric center of the set $I^1_{j_1} \times \cdots \times I^m_{j_m}$. Let $a = \max\{\|y\|_\infty : y \in \mathbf{Y}^\Delta_i(l) + \mathcal{Y}^\Delta_i,\ i = 1,\dots,n_1\}$. In the instant i, the coding-decoding is given by the formulas $(j_1,\dots,j_m) = \eta(y_i)$, if $y_i \in B_a$; $\overline{y}_i = \gamma(j_1,\dots,j_m)$. By construction we receive $\|y_i - \overline{y}_i\|_\infty \le a/q$, $y_i \in \overline{y}_i + B_{a/q}$. Therefore, $\mathbf{Y}^\Delta_i(l,y) \subset (\overline{y}_i + B_{a/q} - \mathbf{Y}^\Delta_i(l)) \cap \mathcal{Y}^\Delta_i = \overline{\mathbf{Y}}_i(l,y)$. For constructing of covering sets in CPI, the sets from this inclusion are used in corresponding relations instead of previous ones. Now let us choose the value q so that values of the vector k_i and the vector $\overline{k}_i$ defined from the equation $\overline{k}_i = X^\Delta_i \overline{k}_{i-1} + K_{i-1}(\overline{y}_i - C^\Delta_i \overline{k}_{i-1})$, $\overline{k}_0 = 0$, differed slightly. Suppose also that matrices A, C, Q, R are constant. Introduce the matrices $\tilde{A} = X^\Delta - S^\Delta R^{\Delta -} C^\Delta$, $B = (Q^\Delta - S^\Delta R^{\Delta -} S^{\Delta'})^{1/2}$. Let the following conditions be fulfilled: (1) the pair $(C^\Delta, \tilde{A})$ is observable; (2) the pair (A^Δ, B) is controllable; (3) $R^\Delta > 0$. Then it is known [6] that $\lim\limits_{i\to\infty} \gamma_i = \gamma^0$. In addition, the matrix $K = (S^\Delta + X^\Delta \gamma^0 C^{\Delta'})(R^\Delta + C^\Delta \gamma^0 C^{\Delta'})^{-1}$ is stable, i.e. $\|K\| < 1$. Suppose that the number a defined above is bounded for all n_1, and let $\overline{n}_1$ be a number such that $\|K_i\| \le \beta < 1$ for all $i > \overline{n}_1$.

Theorem 1. *Let the conditions (1)–(3) be fulfilled. For any $\varepsilon > 0$ and $i > \overline{n}_1$ we choose the number q so that $\delta\delta_1 a/(1-\beta)/q < \varepsilon$, where $\|K_i\| \le \delta$, $\forall i$, $\delta_1 = \max\limits_{1\le i\le \overline{n}_1} (\|A\| + \delta\|C\|)^i$. Then for all $i > \overline{n}_1$ we have $\|\overline{k}_i - k_i\| < \varepsilon$.*

If we replace the vector k_t on $\overline{k}_t$ in the relation, the received set will differ from previous one in Hausdorf's metric also on value ε. Note that received inequality establishes a connection between the accuracy of approximation of covering sets (it is defined by parameter ε) and the constraint on the capacity of the data transfer channel (is defined by parameter q).

One-dimensional Gaussian channels were considered in [6]. Here we generalize the results for multidimensional case and adapt these. This is not quite trivial. Consider the system in R^{2n} exited by 'white noise': $z_i = \mathbf{A}_i z_{i-1} + \chi_i + \mathbf{K}_{i-1} y^1_i$, $z_0 = [0; x^0_0]$, where $\mathbf{A}_i = [X^\Delta_i - K_{i-1} C^\Delta_i, K_{i-1} C^\Delta_i; 0_n, X^\Delta_i]$, $z_i = [k_i; x^0_i]$, $\mathrm{cov}(\chi_i, \chi_i) = \mathcal{Q}_i = [K_{i-1} R^\Delta_i K'_{i-1}, K_{i-1} S^{\Delta'}_i;\ S^\Delta_i K'_{i-1}, Q^\Delta_i]$, $\mathbf{K}_i = [K_i; 0_{n\times m}]$, $\chi_i = [K_{i-1}\eta^\Delta_i, \xi^\Delta_i]$. We will use **Encoder Class** of the form $\theta_i = A^0_{i-1}(\theta) + B_i(\mathbf{C}_{i-1} z_{i-1} + \zeta_i)$, where $\mathbf{C}_i = (\lambda_i(\theta)\tilde{\beta}^-_i)^{1/2}[I_n, 0_n]$; A^0 and λ are nonanticipating values, $\det B_i \ne 0$; ζ_i is standard Gaussian vector independent of χ_i. The matrix $\tilde{\beta}_i$ is the left upper cell of matrix β_i that satisfies the Riccati equation $\beta_i = \mathbf{A}_i \beta_{i-1}(I_n - \mathbf{C}'_{i-1}(I_n + \mathbf{C}_{i-1}\beta_{i-1}\mathbf{C}'_{i-1})^{-1}\mathbf{C}_{i-1}\beta_{i-1})\mathbf{A}'_i + \mathcal{Q}_i$. The parameters of encoders are subject to the energy constraints

$\mathrm{E}\|B_{i+1}^{-1}A_i^0(\theta) + (\lambda_i(\theta)\tilde{\beta}_i^-)^{1/2}k_i\|^2 \leq \mathbb{P}$, $i \geq 1$. Generalizing the result from [6] we obtain the following. If the determinate vector y_i^1 is known, the mean-square optimal **encoder** is received when $\lambda_i = \mathbb{P}/\mathrm{rank}\,\tilde{\beta}_i$ and $B_{i+1}^{-1}A_i^0(\theta) = -\mathbf{C}_i\hat{z}_i$, where $\hat{z}_i = E(z_i \mid \theta^{0i}, \theta^{1i})$. The value $\hat{z}_i$ satisfies the corresponding Kalman filter, and its first component $\hat{k}_i$ is optimal **decoder**.

As the value $y_i^1 \in \mathcal{Y}_i^\Delta$ is unknown, we use the minimax scheme as follows. First, we define recurrently the value $\delta_i = \mathbb{P}/\big(\min_{y_i^{1*}} \max_{y^{1i}} p_i'\tilde{\beta}_i^- p_i + \mathrm{rank}\,\tilde{\beta}_i\big)$, $y_i^{1*} \in \mathcal{Y}_i^\Delta$, $y^{1i} \in \mathcal{Y}_1^\Delta \times \cdots \times \mathcal{Y}_i^\Delta$, $i \geq 1$. Here $\tilde{\beta}_i$ is as above the left upper cell of matrix β_i, where now $\lambda_i = \delta_i$, and p_i is the first component of the vector $\overline{p}_i$ satisfying the equation $\overline{p}_i = \mathbf{K}_{i-1}(y_i^{1*} - y_i^1) + \mathbf{A}_i\big(\overline{p}_{i-1} - \beta_{i-1}\mathbf{C}_{i-1}'(I_n + \mathbf{C}_{i-1}\beta_{i-1}\mathbf{C}_{i-1}')^{-1}\mathbf{C}_{i-1}\overline{p}_{i-1})\big)$, $\overline{p}_0 = 0$. If we consider the recurrently defined vector $y_i^{1*} = \mathrm{argmin}\,\max_{y^{1i}} p_i'\tilde{\beta}_i^- p_i$, we get:

Theorem 2. *Let the values β_i, δ_i, and y_i^{1*} be defined as above. Then the* ***encoder*** *of the form $\theta_i = B_i\big(\mathbf{C}_{i-1}(z_{i-1} - \tilde{z}_{i-1}) + \zeta_i\big)$, $\tilde{z}_i = \mathbf{K}_{i-1}y_i^{1*} + \mathbf{A}_i\big(\tilde{z}_{i-1} + \beta_{i-1}\mathbf{C}_{i-1}'(I_n + \mathbf{C}_{i-1}\beta_{i-1}\mathbf{C}_{i-1}')^{-1}B_i^{-1}\theta_i\big)$, ensures energy constraints for any determinate vector $y_i^1 \in \mathcal{Y}_i^\Delta$. The corresponding* ***decoder*** *$\hat{k}_i$ is the first component of the vector $\hat{z}_i$, for which $\hat{z}_i = \mathbf{A}_i\big(\hat{z}_{i-1} + \beta_{i-1}\mathbf{C}_{i-1}'(I_n + \mathbf{C}_{i-1}\beta_{i-1}\mathbf{C}_{i-1}')^{-1}(B_i^{-1}\theta_i + \mathbf{C}_{i-1}\overline{p}_{i-1})\big) + \mathbf{K}_{i-1}y_i^1$. Here $\hat{z}_i + \overline{p}_i = \tilde{z}_i$, and the error of restoring of k_i is equal to $E\|\hat{k}_i - k_i\|^2 = \mathrm{tr}\tilde{\beta}_i$.*

For equations with unknown vector y_i^1 we can apply the same method as above for covering of the multiestimate in CPI. We have to use the inclusion $y_i^1 \in \mathcal{Y}_i^\Delta$ as the vector y_i known in the object is unknown in CPI.

Acknowledgments

This work was supported by RFBR, Grants No. 07-01-00341, 10-01-00672, and the Program of Presidium RAN 'Mathematical Control Theory'.

References

1. A.B. Kurzhanski, *Control and Estimation under Uncertainty* (Moskow, Nauka, 1977, in Russian).
2. I.Ya. Katz and A.B. Kurzhanski, *Soviet Math. Doklady* **221**, (1975, in Russian).
3. B.I. Anan'ev, *Proc. of the Steklov Inst. of Math., Suppl. 2* (MAIK, 2000).
4. B.I. Anan'ev, *Automation and Remote Control* **68**, 11 (2007).
5. B.I. Anan'ev, *Proc. of the 3-rd Intern. Conference on Physics and Control (PhysCon 2007)*, pp. 1-6, `http://lib.physcon.ru/?item=1361`.
6. R.Sh. Liptser and A.N. Shiryayev, *Statistics of Random Processes. V.1 General Theory. V.2 Applications* (New York, Springer, 2000).

ESTIMATION PROBLEM FOR LINEAR IMPULSIVE CONTROL SYSTEMS UNDER UNCERTAINTY

O. G. MATVIYCHUK

Department of Optimal Control, Institute of Mathematics and Mechanics
Ural Branch of the Russian Academy of Sciences
Ekaterinburg, 620219, Russia
E-mail: vog@imm.uran.ru

The paper deals with the state estimation problem for impulsive control system described by linear differential equations with impulsive terms (or measures). The problem is studied under uncertainty conditions with set-membership description of uncertain variables, which are taken to be unknown but bounded with given bounds (e.g., the model may contain unpredictable errors without their statistical description). Based on the techniques of approximation of the generalized trajectory tubes by the solutions of usual differential systems without measure terms and using the techniques of ellipsoidal calculus, we present a new state estimation algorithm for the studied impulsive control problem.

Keywords: Impulsive control, differential inclusions, reachable sets, estimation.

1. Introduction

The topics of this paper come from the theory of dynamical systems with unknown, but bounded uncertainties (the case of the so-called "set-membership" description of uncertainties) [1–5]. In this paper the impulsive control and estimation problem for a dynamic systems with unknown but bounded initial states is studied. Such problems arise from mathematical models of dynamical and physical systems for which we have an incomplete description of time dependence of their generalized coordinates.

We discuss the approaches to solution concepts for such uncertain dynamical systems based on ideas of well known discontinuous time substitution [6] and techniques of differential inclusions theory [7]. The approaches are based on techniques of approximation of the discontinuous generalized trajectory tubes by the solutions of usual differential systems without measure terms.

Furthermore, in this paper we use well known results of the theory [2,8] of ellipsoidal estimating of states of dynamical control systems with classical (measurable) controls and we develop these results to find the upper set-valued bounds for reachable sets of linear impulsive control problem.

In this paper the approach to the construction of ellipsoidal estimates of the reachable set of the linear impulsive control systems is suggested. To solve this problem we study the related differential inclusion of a classical type (without measure or impulsive controls) and we find ellipsoidal estimates of its trajectory tubes. Projections of these tubes coincide with the upper ellipsoidal bounds for reachable sets of the linear impulsive control system.

2. Problem formulation

Consider a dynamical linear control system described by a differential equation with an impulsive control $u(\cdot)$

$$dx = A(t)xdt + b(t)du, \ x \in \mathbb{R}^n, \ x(t_0 - 0) = x_0, \ t \in [t_0, T]. \qquad (1)$$

Here we assume that $A(t)$ is a continuous $n \times n$ - matrix function, $b(t)$ is continuous n-vector function ($t \in [t_0, T]$)

The initial state x_0 is unknown but bounded with a given bound

$$x_0 \in \mathcal{X}_0 = \mathcal{E}(a, R) = \{x_0 \in \mathbb{R}^n | (x_0 - a)' R^{-1} (x_0 - a) \leq 1\}, \qquad (2)$$

where R is a symmetric positive definite $n \times n$ matrix, $a \in \mathbb{R}^n$ is a center of the ellipsoid $\mathcal{X}_0$.

The impulsive control $u(t)$ ($u(\cdot) : [t_0, T] \to \mathbb{R}$) is continuous from the right and has bounded variation

$$\operatorname*{Var}_{t \in [t_0, T]} u(t) = \sup_{\{t_i\}} \sum_{i=1}^{k} |u(t_i) - u(t_{i-1})| \leq \mu, \qquad (3)$$

where $u = (u_1, \ldots, u_n)$; $t_i : t_0 < \ldots < t_k = T$ and μ is a given positive number. We assume also that $u(t)$ is increasing on $[t_0, T]$. Denote by symbol $\mathcal{U}$ the class of impulsive functions $u(t)$ that satisfied (3).

The solution $x(t) = x(t; t_0, u, x_0)$ of the control system (1) under constraints (2)–(3) has form

$$x(t; t_0, u, x_0) = X(t)x_0 + \int_{t_0}^{t} X(t) X^{-1}(\tau) b(\tau) du(\tau),$$

where $X(t)$ is the Cauchy matrix solution $\dot{X} = A(t)X$ $(X(0) = I)$.

Denote

$$\mathcal{X}(t; t_0, \mathcal{X}_0) = \bigcup_{u(\cdot) \in \mathcal{U}} \bigcup_{x_0 \in \mathcal{X}_0} x(t; t_0, u, x_0).$$

Actually, the set $\mathcal{X}(t) = \mathcal{X}(t; t_0, \mathcal{X}_0)$ is the reachable set of the impulsive differential system (1) from the initial set $\mathcal{X}_0$ at the instant t under restriction (3) for all possible admissible controls $u(\cdot)$.

The main problem of the paper is to find the ellipsoidal estimate $\mathcal{E}(a^+(T), Q^+(T))$ for the reachable set $\mathcal{X}(T)$ basing on the special structure of the data $\mathcal{X}_0$ and restriction (3) on the impulsive control.

3. Main results

Basing on results of ellipsoidal calculus [2,8] developed for linear uncertain systems and discrete-time versions of the funnel equations [9] we present the modified state estimation approaches that allow to solve the problem.

Let us introduce a new time variable and a new state coordinate [6]:

$$\eta(t) = t + \int_{t_0}^{t} du(t), \qquad \tau(\eta) = \inf\{t \mid \eta(t) \geq \eta\}.$$

Consider the auxiliary differential inclusion [3]

$$\frac{d}{d\eta}\begin{pmatrix} z \\ \tau \end{pmatrix} \in G(\tau, z), \quad z(t_0) = x_0 \in \mathcal{X}_0, \quad \tau(t_0) = t_0, \quad t_0 \leq \eta \leq T + \mu. \tag{4}$$

Here $G(\tau, z) = \bigcup\limits_{0 \leq \nu \leq 1} \left\{(1-\nu)\begin{pmatrix} A(\tau)z \\ 1 \end{pmatrix} + \nu \begin{pmatrix} b(\tau) \\ 0 \end{pmatrix}\right\}.$

Denote $w = \{z, \tau\}$ the extended state vector of the system (4) and denote $w(\eta, t_0, w_0)$ the solution of the differential inclusion (4). Consider the trajectory tube of this differential inclusion:

$$\mathcal{W}(\eta) = \bigcup_{w_0 \in \mathcal{X}_0 \times \{t_0\}} w(\eta, t_0, w_0), \quad t_0 \leq \eta \leq T + \mu. \tag{5}$$

From the properties of trajectory tubes of ordinary differential inclusion and the properties of the system (4) we conclude that the following theorem is valid.

Theorem 3.1 (Filippova, 2005). *The reachable set $\mathcal{X}(T)$ is the projection of $\mathcal{W}(T + \mu)$ at the subspace of variables z:* $\quad \mathcal{X}(T) = \pi_z \mathcal{W}(T + \mu)$.

Applying Theorem 3.1 to construct the ellipsoidal estimates of $\mathcal{X}(T)$ we need to construct first the ellipsoidal estimates of $\mathcal{W}(T + \mu)$.

Consider the particular case of the funnel equation related to (4) [3,9]:

$$\begin{gathered} \lim_{\sigma \to +0} \sigma^{-1} h\Big(\mathcal{W}(\eta + \sigma), \bigcup \left(w + \sigma G(w_{n+1}, w_1, \ldots, w_n)\right| \\ w = (w_1, \ldots, w_{n+1}) \in \mathcal{W}(\eta)\Big)\Big) = 0, \ \mathcal{W}(t_0) = \mathcal{W}_0, \ \eta \in [t_0, T + \mu], \end{gathered} \tag{6}$$

here $h(A, B)$ is the *Hausdorff distance* between compact sets A, $B \subseteq \mathbb{R}^n$.

From properties of solutions of the evolution equation (6) we conclude that the following theorem is valid.

Theorem 3.2. *For all $\sigma > 0$ the following inclusion holds*

$$\mathcal{W}(t_0 + \sigma) \subseteq \bigcup_{0\le\nu\le 1} \begin{pmatrix} \mathcal{E}(a^+(t_0,\sigma,\nu), Q^+(t_0,\sigma,\nu)) \\ t_0+\sigma(1-\nu) \end{pmatrix} + o(\sigma)B_*(0,1), \tag{7}$$

where $\lim\limits_{\sigma\to+0} \sigma^{-1}o(\sigma) = 0,\ \ B_*(0,1) = \{x\in\mathbb{R}^{n+1} |\ \|x\|\le 1\}$,

$$\begin{aligned} a^+(t_0,\sigma,\nu) &= (I + \sigma(1-\nu)A(t_0))a + \sigma\nu b(t_0), \\ Q^+(t_0,\sigma,\nu) &= (I+\sigma(1-\nu)A(t_0))R(I+\sigma(1-\nu)A(t_0))'. \end{aligned} \tag{8}$$

Proof. The proof of this theorem is carried out under the scheme of proof of the Theorem 3 [4]. □

In the paper [5] we have constructed upper estimates of the union of ellipsoids under condition that ellipsoids are nondegenerate. Here the set

$$\mathcal{W}(t_0,\sigma) = \bigcup_{0\le\nu\le 1} \mathcal{W}(t_0,\sigma,\nu) = \bigcup_{0\le\nu\le 1} \begin{pmatrix} \mathcal{E}\left(a^+(t_0,\sigma,\nu), Q^+(t_0,\sigma,\nu)\right) \\ t_0+\sigma(1-\nu) \end{pmatrix} \tag{9}$$

introduce in (7) is the union of degenerate ellipsoids in the extended space $\mathbb{R}^{n+1}$ for each parameter ν.

So we fix an arbitrary $\epsilon>0$ and put the degenerated ellipsoid $\mathcal{W}(t_0,\sigma,\nu)$ into nondegenerated ellipsoid $\mathcal{E}_\epsilon(w(t_0,\sigma,\nu), O_\epsilon(t_0,\sigma,\nu))$:

$$\begin{gathered} \mathcal{W}(t_0,\sigma,\nu) \subseteq \mathcal{E}_\epsilon(w(t_0,\sigma,\nu), O_\epsilon(t_0,\sigma,\nu)), \\ w(t_0,\sigma,\nu) = \begin{pmatrix} a^+(t_0,\sigma,\nu) \\ t_0 + \sigma(1-\nu) \end{pmatrix},\quad O_\epsilon(t_0,\sigma,\nu) = \begin{pmatrix} Q^+(t_0,\sigma,\nu) & 0 \\ 0 & \epsilon^2 \end{pmatrix}. \end{gathered} \tag{10}$$

Therefore for any $\epsilon>0$ the following inclusion is true

$$\mathcal{W}(t_0,\sigma) \subset \mathcal{W}_\epsilon(t_0,\sigma) = \bigcup_{0\le\nu\le 1} \mathcal{E}_\epsilon(w(t_0,\sigma,\nu), O_\epsilon(t_0,\sigma,\nu)). \tag{11}$$

In order to construct the external estimate of $\mathcal{W}_\epsilon(t_0,\sigma)$, we need to apply techniques of external ellipsoidal approximations of a convex hull of the union of ellipsoids with different centers and equal matrices (follows lemma) and ones with equal centers and different matrices [5].

Lemma 3.1. *For given a_1, a_2, Q ($a_i \neq 0$, $Q = Q' > 0$) the following inclusions hold ($p \in \left(0,\ 0.5((a_2 - a_1)'Q^{-1}(a_2 - a_1))^{\frac{1}{2}}\right]$)*

$$\mathcal{E}(a_1, Q) \cup \mathcal{E}(a_2, Q) \subset \mathcal{E}(a_1, Q) + \mathcal{E}((a_2 - a_1)/2, (a_2 - a_1)(a_2 - a_1)'/4) \subset$$
$$\subset \mathcal{E}((a_2 + a_1)/2, (1+p)Q + (1+p^{-1})(a_2 - a_1)(a_2 - a_1)'/4).$$

Algorithm of ellipsoidal estimation of $\mathcal{W}_\epsilon(t_0, \sigma)$ is given below.

Algorithm 3.1. *We fix arbitrary $\epsilon > 0$ and $\sigma > 0$. Subdivide the time segment $[0, 1]$ into subsegments $[\nu_j, \nu_{j+1}]$ where $\nu_j = jh$ ($j = 0, \ldots, m$), $h = 1/m$, $\nu_0 = 0$, $\nu_m = 1$.*

For given $\mathcal{X}_0 = \mathcal{E}(a, R)$ we find parameters $a^+(t_0, \sigma, \nu_j)$, $Q^+(t_0, \sigma, \nu_i)$ of ellipsoids $\mathcal{E}(a^+(t_0, \sigma, \nu_j), Q^+(t_0, \sigma, \nu_j))$ defined in (8) for $j = 0, \ldots, m$. For given $\{\mathcal{E}(a^+(t_0, \sigma, \nu_j), Q^+(t_0, \sigma, \nu_j)) |\ j = 0, \ldots, m\}$ we find $m + 1$ ellipsoids

$$\mathcal{E}_\epsilon(w(t_0, \sigma, \nu_j), O_\epsilon(t_0, \sigma, \nu_j)), \quad j = 0, \ldots, m \tag{12}$$

using (10) in the extended space $\mathbb{R}^{n+1}$.

(1) Consider the ellipsoids defined in (12) for $j = 0$ and $j = 1$, namely, $\mathcal{E}_\epsilon(w(t_0, \sigma, \nu_0), O_\epsilon(t_0, \sigma, \nu_0))$; $\mathcal{E}_\epsilon(w(t_0, \sigma, \nu_1), O_\epsilon(t_0, \sigma, \nu_1))$. Basing on the results of the Lemma 3.1 and paper [5] we find the ellipsoid $E_\epsilon^{+1}(w^+(t_0, \sigma), O^+(t_0, \sigma))$ such that $\mathcal{E}_\epsilon(w(t_0, \sigma, \nu_0), O_\epsilon(t_0, \sigma, \nu_0)) \cup \mathcal{E}_\epsilon(w(t_0, \sigma, \nu_1), O_\epsilon(t_0, \sigma, \nu_1)) \subset E_\epsilon^{+1}(w^+(t_0, \sigma), O^+(t_0, \sigma))$.

(2) We take two ellipsoids: $E_\epsilon^{+1}(w^+(t_0, \sigma), O^+(t_0, \sigma))$; $\mathcal{E}_\epsilon(w(t_0, \sigma, \nu_2), O_\epsilon(t_0, \sigma, \nu_2))$, ($j = 2$). As at the iteration (1) we find the ellipsoid $E_\epsilon^{+2}(w^+(t_0, \sigma), O^+(t_0, \sigma))$ such that $E_\epsilon^{+1}(w^+(t_0, \sigma), O^+(t_0, \sigma)) \cup \mathcal{E}_\epsilon(w(t_0, \sigma, \nu_2), O_\epsilon(t_0, \sigma, \nu_2)) \subset E_\epsilon^{+2}(w^+(t_0, \sigma), O^+(t_0, \sigma))$.

(3) Next steps continue iterations (1)–(2).

At the end of the process we will get the external estimate $\bigcup_j \mathcal{E}_\epsilon(w(t_0, \sigma, \nu_j), O_\epsilon(t_0, \sigma, \nu_j)) \subset E_\epsilon^+(w^+(t_0, \sigma), O^+(t_0, \sigma))$.
Therefore we will have the estimate of the reachable set $\mathcal{W}_\epsilon(t_0, \sigma) \subset E_\epsilon^+(w^+(t_0, \sigma), O^+(t_0, \sigma))$.

Basing on the previous results we may formulate the following scheme that gives the external ellipsoidal estimate of $\mathcal{X}(T)$ of the system (1).

Algorithm 3.2. *We fix arbitrary $\epsilon > 0$ and subdivide the time segment $[t_0, T + \mu]$ into subsegments $[t_i, t_{i+1}]$ where $t_i = t_0 + i\sigma$ ($i = 1, \ldots, k$), $\sigma = (T + \mu - t_0)/k$, $t_k = T + \mu$.*

(1) *Consider the time segment* $[t_0, t_1]$. *We take the initial set* $\mathcal{X}_0 = \mathcal{E}(a, R)$ *in the Algorithm 3.1 and find the ellipsoid* $E_\epsilon^{+1}(w^+(\sigma), O^+(\sigma))$ *such that* $\mathcal{W}(t_0, \sigma) \subseteq E_\epsilon^{+1}(w^+(\sigma), O^+(\sigma))$, *where sets* $\mathcal{W}(t_0, \sigma)$ *are defined in (9).*

(2) *Consider next time interval* $[t_1, t_2]$. *The ellipsoid* $E_\epsilon^{+1}(w^+(\sigma), O^+(\sigma))$ *is considered as the start ellipsoid at the moment* t_1 *for the Algorithm 3.1. Apply Algorithm 3.1 again. The resulted set* $E_\epsilon^{+2}(w^+(\sigma), O^+(\sigma))$ *will be the start ellipsoid for the next moment* t_2 *in the Algorithm 3.1.*

(3) *Repeat iteration (2) for each moment* $t_i : t_i = t_0 + i\sigma$ $(i = 2, \ldots, k)$. *At the end of the process the ellipsoid* $E_\epsilon^+(w^+, O^+)$ *will be obtained so that* $\mathcal{W}(T + \mu) \subseteq E_\epsilon^+(w^+, O^+)$.

(4) *By Theorem 3.1 find the projection of the ellipsoid* $E_\epsilon(w^+, O^+)$ *at the subspace of variables* $\{z_1, \ldots, z_n\}$ $\mathcal{E}(a^+(T), Q^+(T)) = \pi_z E_\epsilon(w^+, O^+)$.

Therefore we will have the external estimate $\mathcal{E}(a^+(T), Q^+(T))$ of the reachable set $\mathcal{X}(T)$ of system (1) from initial set $\mathcal{X}_0$ under restriction (3).

Acknowledgments

The research was supported by the Russian Foundation for Basic Researches (RFBR) under the Project No. 09-01-00223, Collaboration Program in Fundamental Research of UB RAS and SB RAS "Qualitative and Numerical Analysis of Evolution Equations and Control Systems" and Research Grant of UB RAS.

References

1. A. B. Kurzhanski, *Control and Observation under Conditions of Uncertainty* (Nauka, Moscow, 1977).
2. A. B. Kurzhanski and I. Valyi, *Ellipsoidal Calculus for Estimation and Control* (Birkhauser, Boston, 1997).
3. T. F. Filippova, Impulsive Control Problem for Uncertain Dynamic Systems In *Proc. of Second International Conference "Physics and Control"* (Saint Petersburg, Russia, 2005).
4. T. F. Filippova and E. V. Berezina,
5. O. G. Vzdornova and T. F. Filippova,
6. R. Rishel, *SIAM J. Control* **3**, 191 (1965).
7. A. F. Filippov, *Differential Equations with Discontinuous Right-hand Side* (Nauka, Moscow, 1985).
8. F. L. Chernousko, *State Estimation for Dynamic Systems* (Nauka, Moskow, 1988).
9. A. I. Panasyuk, *J. Optimiz. Theory Appl.*, **64**, 349 (1990).

ON NUMERICAL METHODS OF SOLVING SOME OPTIMAL PATH PROBLEMS ON THE PLANE

V. N. USHAKOV*, A. R. MATVIYCHUK and A. G. MALEV

Department of Dynamic systems, Institute of Mathematics and Mechanics of UBr RAS
S.Kovalevskaya street, Ekaterinburg, 620219, Russia
*E-mail: *ushak@imm.uran.ru*
www.imm.uran.ru

Three numerical methods of solution of some time optimal control problems for a system under phase constraints are described in the paper. Two suggested methods are based on transition to the discrete time model, constructing attainability sets and application of the guide construction. The third method is based on the Deikstra algorithm.

Keywords: Optimal control problem, differential inclusions, attainability sets.

1. Introduction

The paper deals with the time optimal control problem connected with studying the dynamic system under phase constraints. The paper continues investigations in [1-6].

Three numerical methods of solution of some time optimal control problems for a system under phase constraints are described in the paper. They are polygons method, grid method and method, based on the Deikstra algorithm.

2. Problem formulation

Consider the controlled moving object Υ^* in the m-dimensional Euclidean space. Denote by the center O of moving object Υ^* some chosen point inside Υ^*. Orientation of the Υ^* is fixed. Behavior of the center O is described by the equation

$$\dot{x} = f(t, x, u), \quad u \in P, \quad t \in [t_0, \vartheta], \quad t_0 < \vartheta < \infty. \tag{1}$$

Here x is the m-dimensional phase vector of the system, u is the control and P is a compact set in the Euclidean space R^r. It is assumed that traditional conditions providing the existing, uniqueness and extendability of solutions of the system (1) at full length of the interval $[t_0, \vartheta]$ are satisfied.

Along with the system (1), a compacts Φ and X_f from $[t_0, \vartheta] \times R^m$ and compact X_0 from $\Phi(t_0)$ are given. Here the set Φ is a phase constraint for the system (1), the set X_f is a goal set, and the X_0 plays a role of the start set. Let's consider that sections $\Phi(t)$ and $X_f(t)$, $t \in [t_0, \vartheta]$, are changed continuously with a time (see Fig. 1).

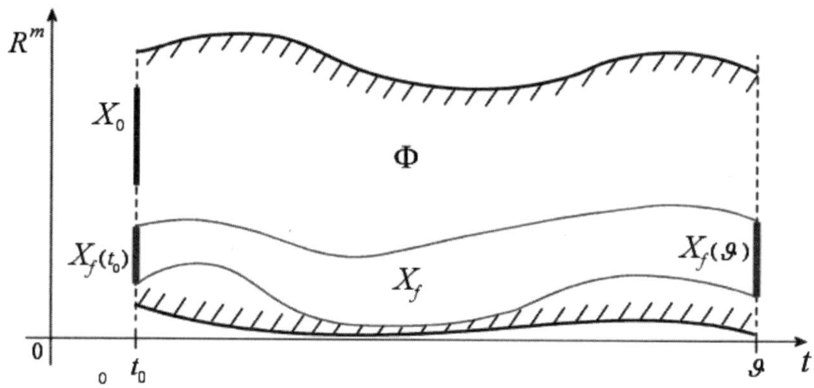

Fig. 1.

By an admissible control $u(t)$, $t \in [t_0, \vartheta]$, we mean any Lebesgue measurable function such, that $u(t) \in P$, $t \in [t_0, \vartheta]$.

Problem 2.1. *Construct an admissible control $u^*(t)$, $t \in [t_0, \vartheta]$, that steers the phase vector $x[t]$ (trajectory of the center O) of the system (1) from the X_0 into the X_f at minimal time so as $\Upsilon^*(t) \subset \Phi(t)$, $t \in [t_0, \vartheta]$.*

Remark 2.1. Exactly solve formulated problem for a general case is not possible. By this reason we solve this problem approximately. Namely we lead a movement of the center O of the Υ^* from the X_0 to a some chosen neighbourhood of the set X_f. At the same time we construct a control so as the moving object Υ^* is hold in the given neighbourhood of phase constraint Φ.

3. Scheme of solution

Let's consider the problem of the moving center O instead of the moving object Υ^*. We may do it because of fixed orientation of the Υ^*. To perform it the phase constraint is outlined by the Υ^*. In this case a phase constraint Φ for the Υ^* is substituted by a phase constraint Φ^* for the center O.

Define the differential inclusion (DI) $F(t,x)$ as following

$$\dot{x} \in F(t,x), \quad t \in [t_0, \vartheta], \tag{2}$$

where $F(t,x) = \text{co}\{f(t,x,u) : \ u \in P\}$.

Divide the interval $[t_0, \vartheta]$; i.e., specify the partition $\Gamma = \{t_0, t_1, \ldots, t_N = \vartheta\}$ of the interval $[t_0, \vartheta]$ such, that the diameter $\Delta = \max\{(t_{i+1} - t_i) : \ 0 \leq i \leq N-1\}$, of the partition Γ is sufficiently small. Further we will consider system (1) only at time moments of partition Γ

Associate a sequence $\{\tilde{X}(t_i)\}$ of sets $\tilde{X}(t_i) \subset R^m$ (attainability sets) with this partition. This sequence is defined recursively as following

$$\begin{gathered} \tilde{X}(t_0) = X_0, \\ \tilde{X}(t_{i+1}) = \Phi^*(t_{i+1}) \cap \tilde{Z}(t_{i+1}; t_i, \tilde{X}(t_i)), \\ i = 0,\ 1,\ \ldots,\ N_f - 1. \end{gathered}$$

Here, $\tilde{Z}(t^*; t_*, x_*) = x_* + (t^* - t_*)F(t_*, x_*)$, $t_0 \leq t_* < t^* \leq \vartheta$, $x_* \in R^m$; $\tilde{Z}(t^*; t_*, X_*) = \bigcup\limits_{x_* \in X_*} \tilde{Z}(t^*; t_*, x_*)$.

It is assumed that there are instants $t_i \in \Gamma$ in the discrete scheme such, that $\tilde{X}(t_i) \cap X_f(t_i) \neq \emptyset$, and t_{N_f} is the first one. Choose any point $y[t_{N_f}]$ in the $\tilde{X}(t_{N_f}) \cap X_f$ and assign some number $\varepsilon^* > 0$. Formulate the Problem 3.1 whose solution is the approximate solution of the Problem 2.1.

Problem 3.1. *It is required to construct an admissible control $u^*(t)$, $t \in [t_0, t_{N_f}]$, that leads a phase vector $x[t]$ of the system (1) from the X_0 into ε^*-neighbourhood of the point $y[t_{N_f}]$ at the instant t_{N_f} so that $x[t] \in \Phi^*(t)_{\varepsilon^*}$, $t \in [t_0, t_{N_f}]$.*

To solve the Problem 3.1 we also consider the DI with a small parameter $\varepsilon > 0$:

$$\dot{x} \in F(t,x) + \varepsilon Q,\ t \in [t_0, t_{N_f}], \tag{3}$$

where $Q = \{w \in R^m : \ \|w\| \leq 1\}$.

Using greater possibilities of the DI (3) we construct the Euler polygon for the DI (3), which goes through sets $\tilde{X}(t_i)$ at instants $t_i \in \Gamma^f$ and ends at the instant t_{N_f} in the point $y[t_{N_f}]$. Here $\Gamma^f = \{t_0,\ t_1,\ \ldots,\ t_{N_f}\}$. This Euler polygon will play a role of a "guide" in construction of the control $u^*(t)$, $t \in [t_0, t_{N_f}]$, solving the Problem 3.1 for the system (1).

We construct the Euler polygon $y[t]$, $t \in [t_0, t_{N_f}]$, starting from the point $y[t_{N_f}]$ and proceeding to the initial instant t_0 (see Fig. 2). Here

$$y[t] = y[t_i] + (t - t_i)f^*(t_i) + (t - t_i)\varepsilon w^*(t_i),\ t \in [t_i, t_{i+1}],$$

$f^*(t_i) \in F(t_i, y[t_i])$, $w^*(t_i) \in Q$, $i = 0, 1, \ldots, N_f - 1$. Nodes of this polygon-guide are points $y[t_0] \in \tilde{X}(t_0)$, $y[t_1] \in \tilde{X}(t_1)$, $\ldots$, $y[t_{N_f}] \in \tilde{X}(t_{N_f})$.

The following equality is hold:

$$y[t_{N_f}] = y[t_0] + \sum_{i=0}^{N_f-1} \Delta_i f^*(t_i) + \sum_{i=0}^{N_f-1} \Delta_i \varepsilon w^*(t_i),$$

where $f^*(t_i) \in F(t_i, y[t_i])$, $w^*(t_i) \in Q$, $i = 0, 1, \ldots, N_f - 1$.

We construct the control $u^*(t)$, $t \in [t_0, t_{N_f}]$, sequentially at steps $[t_i, t_{i+1})$, $i = 0, 1, \ldots, N_f - 1$, of the partition Γ^f in the form of a piecewise-constant control $u^*(t) \equiv u^*(t_i)$, $[t_i, t_{i+1})$, $i = 0, 1, \ldots, N_f - 1$.

Suppose that $u^*(t_0)$, $u^*(t_1)$, $\ldots$, $u^*(t_{i-1})$, corresponding to intervals $[t_0, t_1)$, $[t_1, t_2)$, $\ldots$, $[t_{i-1}, t_i)$, are constructed and the motion $x[t]$, $t \in [t_0, t_i]$, of the system (1) under the action of the control $u^*(t)$, $t \in [t_0, t_i)$ is realized. The vector function $x[t]$ satisfies the equation $\dot{x}[t] = f(t, x[t], u^*(t))$, $x[t_0] = y[t_0]$, almost everywhere on $[t_0, t_i]$. We choose the vector $u^*(t_i) \in P$, corresponding to the semiopen interval $[t_i, t_{i+1})$, from the condition

$$\begin{aligned} &\|(x(t_i) + \Delta_i f(t_i, x[t_i], u^*(t_i)) - y[t_{i+1}])\| = \\ &= \min_{u \in P} \|(x(t_i) + \Delta_i f(t_i, x[t_i], u) - y[t_{i+1}])\|. \end{aligned} \tag{4}$$

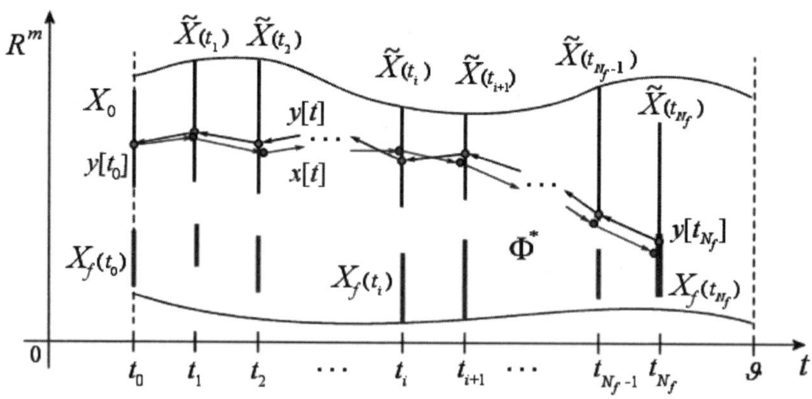

Fig. 2.

It is possible to achieve that in the case of sufficiently small diameter Δ the following is hold: $x[t] \in \Phi^*(t)_{\varepsilon^*}$, $t \in [t_0, t_{N_f}]$, $x[t_{N_f}] \in O_{\varepsilon^*}(y[t_{N_f}])$.

4. Numerical methods

Two of tree suggested methods use presented scheme. Third is based on Deikstra algorithm. Consider a peculiarity of these methods.

4.1. *Polygons method*

At this method all sets (the moving polygon Υ^*, start and final sets, attainability sets, the phase constraint) are presented as polygons. Polygons may be non-convex. Each polygon is specified by a set of closed broken lines. One of these broken lines is an external border, others form internal border of polygon (in the common case arbitrary polygon may have number of holes). All operations of constructing attainability sets are based on operations with polygons (union, subtraction and intersection). Because of all polygons are formed by number of closed broken lines it allows to save a lot of memory on personal computer (PC) and in many cases to safe a time of computations in comparison with grid methods. On a contrary, the polygons method has comparatively complicated logic of computations, require a very high calculation accuracy on PC and at the current realization can be applied only for the case on the plane (2-dimensional case).

4.2. *Grid method*

Grid method use not only discrete time model, but also use discrete space model. That is the m-dimensional space is broken with the regular grid and all sets are presented as sets of cells of this grid. The advantage of this method is the simple logic of calculations of attainability sets. This future allows to perform calculations on a m-dimensional space. On other side grid method is very time and memory consuming method (especially in the case of high precision of calculations). It leads us to the development of auxiliary methods which decrease a calculation time and decrease an amount of necessary PC memory. One of such algorithms is a border detecting algorithm that allows to exclude internal cells of sets from the calculation process.

4.3. *Method, based on Deikstra algorithm*

For a case of the stationary system (1) where the phase constraint Φ is fixed in a time we apply some kind of the grid method — method based on the Deikstra algorithm. Here we also replace the problem of moving object Υ^* by the problem of the moving center O, but we don't apply a three stage method here.

At this method m-dimensional space is broken with the regular grid and all sets are presented as sets of cells of this grid. Instead of previous approaches we consider all cells as vertexes of some weighed graph. The weigh of each rib of graph is the time that needed for moving along this rib. Ribs of graph and their weights are calculated during the calculation process and depend on the system (1) and the form of the set P. Advantages of this method are comparatively small calculation time and possibility to consider a changeable orientation of the moving object. In the case of changeable orientation we have additional dimensions. The shortcoming of this method is that we can apply this method only for the case of the fixed phase constraints Φ.

Acknowledgments

The research was supported by the Russian Foundation for Basic Researches projects no. 08-01-00587, the program "State Support of Leading Scientific Schools" (project no. N.Sh.-2640.2008.1), the programm of Presidium of RAS no. 29 "Mathematical control theory". and collaboration program in fundamental research of UB RAS and SB RAS "Qualitative and numerical analysis of evolution equations and control systems".

References

1. N. N. Krasovskii, *The Theory of Motion Control* (Nauka, Moscow, 1968).
2. N. N. Krasovskii and A. I. Subbotin, *Approximation in the Differential Games* (Nauka, Moscow, 1974).
3. A. R. Matviychuk and V. N. Ushakov, *J. The Theory and Control Systems* **1**, 5 (2006).
4. A. A. Neznakhin and V. N. Ushakov, *J. of Computational Mathematics and Math. Physics* **41**, 895 (2001).
5. P. Saint-Pierre and M. Quincampoix, *J. Math. System Estim. Control* **5**, 115 (1995).
6. G. A. Smirnov, *Moving Theory of Wheeled Cars* (Mechanical Engineering, Moscow, 1990)

QUANTILE OPTIMIZATION PROBLEM WITH INCOMPLETE INFORMATION

G. A. TIMOFEEVA

Department of Management of Transportation Process
Ural State University of Transport
Yekaterinburg, 620034, Russia
E-mail: Gtimofeeva@math.usurt.ru

A stochastic optimization problem with incomplete information is considered. Optimal solutions are selected using the minimax quantile criterion. This problem is related to a confidence estimation problem for a random vector with incompletely known distribution. Generalized confidence regions are used as confidence estimates for a statistically uncertain vector. The quantile stochastic optimization problem under incomplete information is solved by an optimal choice of the generalized confidence region.

Keywords: Stochastic optimization, incomplete information, quantile criterion.

1. Introduction

Stochastic optimization problems with incomplete information about distributions of random perturbations were studied in [1,2]. In this paper we deal with a quantile optimization problem under uncertainty. Stochastic optimization problems with the quantile criterion were studied in [3,4].

The quantile optimization problem is related to the confidence estimation problem which were considered for statistically uncertain systems in [5–7]. The confidence estimation problem is usually reformulated as the problem with the quantile criterion. On the other hand the quantile optimization problem may be reduced to a generalized minimax problem by an appropriate choice of the confidence set for random perturbations.

In the paper two connected problems are studied:

- to define and to construct generalized confidence regions for a random vector with incompletely known distribution;
- to solve the quantile optimization problem depended on both random and uncertain nonrandom parameters.

2. Statistically uncertain vectors and generalized confidence sets

Let us definite a notion of a statistically uncertain random vector [8] and consider its confidence estimate.

Definition 2.1. A map $\tilde{\xi}(\omega, Z) : \Omega \times Z \to \mathbb{R}^n$ is called *a statistically uncertain random vector* if:

(1) the function $\xi(\omega, z)$ is a random vector for any fixed $z \in Z$, i.e. the set $\{\omega : \xi(\omega, z) \in B\} \in \mathcal{A}$ is measurable for any Borel set $B \in \mathcal{B}^{(n)}$, $z \in Z$;
(2) the probability $P_z(B) = P\{\xi(\omega, z) \in B\}$ is a continuous function with respect to z for any fixed Borel set $B \in \mathcal{B}^{(n)}$;
(3) the set Z is a compact set consisted of more than one point.

Let $\tilde{\xi}(\omega, Z)$ be a statistically uncertain continuous random vector, $\{X_z^\alpha \mid z \in Z\}$ be a family of confidence sets with the level α, i.e. for any X_z^α the relation $P\{\xi(\omega, z) \in X_z^\alpha\} = \alpha$ holds. Denote by $\hat{X}_\alpha$ the union of the confidence sets

$$\hat{X}_\alpha = \bigcup_{z \in Z} X_z^\alpha. \tag{1}$$

Usually the union $\hat{X}_\alpha$ of confidence regions is taken as a confidence region in statistically uncertain case. It is shown in [8] that this estimator may be improved in the most cases by means of generalized confidence regions.

Definition 2.2. A measurable set $\tilde{X}_\alpha \subset \mathbb{R}^{(n)}$ is called *a generalized confidence set* with level α for a statistically uncertain random vector $\tilde{\xi}(\omega, Z)$, if $\mathcal{P}\{\tilde{\xi}(\omega, Z) \in \tilde{X}_\alpha\} \triangleq \min_{z \in Z} P\{\xi(\omega, z) \in \tilde{X}_\alpha\} = \alpha$.

Generalized confidence sets (as well as standard confidence sets) are not uniquely defined: there are many generalized confidence regions corresponding to a fixed probability α. The next statement [8] follows from the definitions and properties of the probability function.

Theorem 2.1. *If the union $\hat{X}_\alpha$ of the confidence sets is a measurable set then it is a generalized confidence set with a level $\alpha_1 \geq \alpha$ for the statistically uncertain continuous random vector $\tilde{\xi}(\omega, Z)$. The equality $\alpha_1 = \alpha$ holds if and only if there exists a parameter $z^* \in Z$ such that $P\{\xi(\omega, z^*) \in \hat{X}_\alpha\} = \alpha$.*

For the same statistically uncertain vector the union of confidence sets may be a generalized confidence set with the same probability level or with a greater level. It depends on the forms of the confidence sets. The following simple example illustrates the properties of generalized confidence regions.

Example 2.1. Let $\tilde{\xi}(\omega, Z) = \{z + \eta(\omega) \mid z \in Z\}$, where Z is the interval $Z = [-a, a]$, $\eta(\omega)$ is a normal distributed random value with given statistical moments: $E\xi = 0$, $E\xi^2 = \sigma^2$.

The set $X_\alpha(0) = [-t_{0.5\alpha}\sigma, t_{0.5\alpha}\sigma]$ is a confidence set for $\eta(\omega)$ with probability α, where t_β is the quantile of the standard normal distribution.

The sets $X_\alpha(z) = z + X_\alpha(0)$, $z \in Z$, are confidence sets with the level α for $\xi(\omega, z) = z + \eta(\omega)$, but $\hat{X}_\alpha = \bigcup_{z \in Z} X_\alpha(z) = Z + X_\alpha(0)$ is not a generalized confidence set with the same level for $\tilde{\xi}(\omega, Z)$ since

$$\min_{z \in [-a,a]} P\{z + \eta(\omega) \in \hat{X}_\alpha\} > \alpha.$$

The generalized confidence set has a form

$$\tilde{X}_\alpha = [-\sigma \mathrm{gal}(\alpha, \sigma^{-1}a) - a, a + \sigma \mathrm{gal}(\alpha, \sigma^{-1}a)],$$

where $g = \mathrm{gal}(\alpha, v)$ is the root of the equation

$$\Phi(g + 2v) + \Phi(g) = \alpha, \quad \Phi(z) = \frac{1}{\sqrt{2\pi}} \int_0^z e^{-\frac{x^2}{2}} dx \tag{2}$$

Since for all $v > 0$, $\alpha \in (0.5, 1)$ the inequalities $t_{\alpha - 0.5} < \mathrm{gal}(\alpha, v) < t_{\alpha/2}$ hold, then $\tilde{X}_\alpha \subset \hat{X}_\alpha = Z + X_\alpha(0)$.

The generalized confidence sets for a Gaussian n-vector with an incompletely known mean value have the same properties [8].

3. Statistically uncertain quantile optimization problem

Let us consider the stochastic optimization problem with the quantile criterion and incomplete information:

$$\max_{z \in Z} q_\alpha(u, z) \to \min, \quad u \in U, \tag{3}$$

$$q_\alpha(u, z) = \min\{q : P\{F(u, \xi(\omega, z)) \le q\} \ge \alpha\}. \tag{4}$$

If $\xi(\omega, z) = \varphi(\eta(\omega), z)$ for all $z \in Z$, where the function $\varphi(y, z)$ is measurable on y and continuous on z, $\eta(\omega)$ is a random vector with a given continuous distribution, then $F(u, \varphi(y, z)) = F_1(u, z, y)$ and the quantile has the form:

$$q_\alpha(u, z) = \min\{q : P\{F_1(u, z, \eta(\omega)) \le q\} \ge \alpha\}. \tag{5}$$

Theorem 3.1. *Let the following conditions hold:*

(1) $\eta(\omega)$ is a continuous random vector;

(2) $Z \subset \mathbb{R}^m$ *is a compact set;*
(3) the function $F_1(u,z,y)$ *is continuous one on* $U \times Z \times \mathbb{R}^n$*;*
(4) $P\{|F_1(u,z,\eta(\omega)) - q| \le \varepsilon\} > 0$ *for all* $u \in U$, $z \in Z$, $q \in (q_-(u,z), q_+(u,z))$, *where* $q_-(u,z) = \inf_y F_1(u,z,y)$, $q_+(u,z) = \sup_y F_1(u,z,y)$.

Then the optimal quantile in problem (3),(5)

$$\tilde{q}_\alpha^* = \inf_{u \in U} \max_{z \in Z} q_\alpha(u,z) \tag{6}$$

is equal to

$$\tilde{q}_\alpha^* = \inf_{u \in U} \max_{z \in Z} \min_{E_\alpha \in \mathcal{E}_\alpha} \max_{y \in E_\alpha} F(u,y), \tag{7}$$

where $\mathcal{E}_\alpha$ *is the family of confidence sets*

$$\mathcal{E}_\alpha = \{E_\alpha \in \mathcal{B}^{(n)} : P\{\eta(\omega) \in E_\alpha\} \ge \alpha\}. \tag{8}$$

Proof. From properties of quantile functions [4] it follows that the equality

$$q_\alpha(u,z) = \min_{E_\alpha \in \mathcal{E}_\alpha} \max_{y \in E_\alpha} F_1(u,z,y)$$

holds for any fixed $u \in U$, $z \in Z$. In the conditions of Theorem 3.1 the quantile function $q_\alpha(u,z)$ is continuous on the set $Z \times U$. Since the set Z is a compact one, an optimal vector $z^*(u)$ such that $q(u,z^*(u)) = \min_{z \in Z} q(u,z)$ exists. Thus quantile optimization problem is reduced to a generalized minimax deterministic problem (7)–(8). □

The obtained problem (7) seems more difficult than the initial quantile optimization problem (3),(5). But one can take an appropriate family of confidence sets $\{E_\alpha^1(z) \mid z \in Z\}$, then solve the problem

$$q_\alpha^1 = \inf_{u \in U} \max_{z \in Z} \max_{y \in E_\alpha^1(z)} F_1(u,z,y),$$

and consider its solution u^1 as a suboptimal solution of the quantile optimization problem (3), and obtain an estimate of the optimal quantile $q_\alpha^1 \ge \tilde{q}_\alpha^*$. The problem is how to choose the family of the confidence sets. If we take the same confidence set for all $z \in Z$, then $q_\alpha^1 \ge \hat{q}_\alpha$, where

$$\hat{q}_\alpha = \inf_{u \in U} \min_{E_\alpha \in \mathcal{E}_\alpha} \max_{z \in Z} \max_{y \in E_\alpha} F_1(u,z,y). \tag{9}$$

Criterion (9) was considered [4] for the statistically uncertain quantile optimization problem. Obviously an inequality $\tilde{q}_\alpha^* \le \hat{q}_\alpha$ is carried out.

Let us note that the inequality is as a rule a strict one and formulate sufficient conditions for the equality.

Theorem 3.2. *Let the conditions of Theorem 3.1 hold and for any* $u \in U$ *there exists* $z^* = z^*(u) \in Z$ *such that* $\max_{z \in Z} F_1(u, z, y) = F_1(u, z^*(u), y)$ *for all* $y \in \mathbb{R}^n$, *then the minima of the criteria coincide:* $\tilde{q}_\alpha^* = \hat{q}_\alpha$.

Proof. It follows from condition of the Theorem that

$$\hat{q}_\alpha = \inf_{u \in U} \min_{E_\alpha \in \mathcal{E}_\alpha} \max_{y \in E_\alpha} \max_{z \in Z} F_1(u, z, y) = \inf_{u \in U} \min_{E_\alpha \in \mathcal{E}_\alpha} \max_{y \in E_\alpha} F_1(u, z^*(u), y) \leq$$

$$\leq \inf_{u \in U} \max_{z \in Z} \min_{E_\alpha \in \mathcal{E}_\alpha} \max_{y \in E_\alpha} F_1(u, z, y) = \tilde{q}_\alpha^*.$$

Since the inequality $\tilde{q}_\alpha^* \leq \hat{q}_\alpha$ is carried out in any case, we get $\tilde{q}_\alpha^* = \hat{q}_\alpha$. □

Example 3.1. Let us consider the problem of minimization of the function $F_1(u, z, \eta(\omega)) = |z + u + \eta(\omega)|$, where $z \in Z = [z_0 - a, z_0 + a]$ is the an incompletely known parameter, $\eta(\omega)$ is a random Gaussian perturbation with the known parameters $E\eta = 0$, $E\eta^2 = \sigma^2$.

We choose a control $u \in \mathbb{R}^1$ according to the minimax quantile criterion (6). From Theorem 3.1 it follows that

$$\tilde{q}_\alpha^* = \min_u \max_{z \in Z} \min_{E_\alpha \in \mathcal{E}_\alpha} \max_{y \in E_\alpha} |z + u + y|.$$

From properties of the probability function [4] it follows that the optimal value

$$q_\alpha(u, z) = \min_{E \in \mathcal{E}_\alpha} \max_{y \in E} |z + u + y|$$

is reached on a symmetrical confidence set for any fixed u, z. According to Example 2.1 we get $q_\alpha(u, z) = \sigma \text{gal}(\alpha, |u + z| \cdot \sigma^{-1}) + |u + z|$ and

$$\tilde{q}_\alpha^* = \min_u \max_{z \in Z} (|u + z| + \sigma \text{gal}(\alpha, \sigma^{-1}|u + z|) = a + \sigma \text{gal}(\alpha, \sigma^{-1} a).$$

If we consider the random and uncertain perturbations together (9), then

$$\hat{q}_\alpha = \min_u \min_{E_\alpha \in \mathcal{E}_\alpha} \max_{z \in Z} \max_{y \in E_\alpha} |u + z + y| = a + t_{0.5\alpha} \cdot \sigma.$$

In the considered problem $\hat{q}_\alpha < \tilde{q}_\alpha^*$ for all $a > 0$, $\alpha \in (0.5, 1)$. The similar effect is observed in the most stochastic quantile optimization problems with uncertainty.

The optimization over confidence sets may be substituted by the optimization over generalized confidence sets.

Theorem 3.3. *Let $\tilde{\xi}(\omega, Z)$ be a statistically uncertain continuous random vector, a function $F(u,y)$ be continuous. Then*

$$\tilde{q}_\alpha^* = \inf_{u \in U} \min_{\tilde{E}_\alpha \in \tilde{\mathcal{E}}_\alpha} \max_{y \in \tilde{E}_\alpha} F(u,y), \tag{10}$$

where $\tilde{\mathcal{E}}_\alpha \subset \mathcal{B}^{(n)}$ is the family of the generalized confidence sets $\tilde{E}_\alpha$ with level not less than α: $\tilde{\mathcal{E}}_\alpha = \{\tilde{E}_\alpha \in \mathcal{B}^{(n)} : \mathcal{P}\{\tilde{\xi}(\omega, Z) \in \tilde{E}_\alpha\} \geq \alpha\}$.

Proof. Any generalized confidence set $\tilde{E}_\alpha$ can be presented as a union of confidence sets $E_\alpha^*(z)$ with the same level. There is $z^* \in Z$ such that $E_\alpha^*(z^*) = \tilde{E}_\alpha$ (see [8]). Therefore for any fixed $u \in U$ we have

$$\min_{\tilde{E}_\alpha \in \tilde{\mathcal{E}}_\alpha} \max_{y \in \tilde{E}_\alpha} F(u,y) = \max_{z \in Z} \min_{E_\alpha \in \mathcal{E}_\alpha} \max_{y \in E_\alpha} F(u,y). \qquad \square$$

The exact minimax solution of the minimax problem (10) requires a significant calculations. But we can find a suboptimal solution and estimate the optimal quantile.

4. Conclusion

The quantile optimization for problem with incomplete information about random parameters distributions is considered. The problem is reduced to the problem of the optimal choice of generalized confidence region for statistically uncertain vector.

References

1. J. Dupacova, *Stochastic Programming with Incomplete Information*, IIASA Working Paper, WP-86-008 (1986).
2. Yu. Ermoliev, A. A.Gaivoronski and C. Neveda, *SIAM J. Control and Optimization* **23**, 5, 134 (1986).
3. A. Precopa, *Stochastic Programming* (Dordrecht etc.: Kluwer Acad. Publ., 1995).
4. A. I. Kibzun and Yu. S. Kan, *Stochastic Programming Problem with Probability and Quantile Functions* (Chichester etc.: John Wiley & Sons, 1996).
5. I. Ya. Kats and G. A. Timofeeva, The most probable states in statistically uncertain dynamic systems, in *Proc. 10th IFAC Workshop Control Appl. of Optimiz.*, (London: Pergamon, 1995), pp. 29–32.
6. A. B. Kurzanski and M. Tanaka, *On a Unified Framework for Deterministic and Stochastic Treatment of Identification Problem* (Laxenburg, IIASA, 1989).
7. A. Matasov, *Estimators for Uncertain Dynamic Systems* (Dordrecht etc.: Kluwer Acad. Publ., 1999).
8. G. A. Timofeeva, *Automation and Remote Control* **63**, 6, 906 (2002).

ESTIMATES OF TRAJECTORY TUBES IN CONTROL PROBLEMS UNDER UNCERTAINTY

T. F. FILIPPOVA

Department of Optimal Control, Institute of Mathematics and Mechanics
Ural Branch of Russian Academy of Sciences
16 S. Kovalevskaya Street, Ekaterinburg, 620219, Russia
E-mail: ftf@imm.uran.ru

The paper deals with the problems of control and state estimation for nonlinear dynamical control system described by differential equations with unknown but bounded initial states. The nonlinear function in the right-hand part of a differential system is assumed to be of quadratic type with respect to state variable. Based on the well-known results of ellipsoidal calculus developed for linear uncertain systems, we present the modified state estimation approaches which use the special structure of the dynamical system.

Keywords: Reachable sets, trajectory tubes, ellipsoidal estimates.

1. Introduction

The topics of this paper come from the control theory for systems with unknown but bounded uncertainties related to the case of set-membership description of uncertainties which are taken to be unknown but bounded with given bounds (e.g., the model may contain unpredictable errors without their statistical description) [1–6]. The motivations for these studies come from applied areas ranged from engineering problems in physics to economics as well as to ecological modelling.

We will start by introducing the following basic notations. Let R^n be the n–dimensional Euclidean space and $x'y$ be the usual inner product of $x, y \in R^n$ with the prime as a transpose, with $\| x \| = (x'x)^{1/2}$. Denote also comp R^n to be the variety of all compact subsets $A \subseteq R^n$.

Consider the control system described by the ordinary differential equation

$$\dot{x} = f(t, x, u(t)), \quad t \in [t_0, T] \tag{1}$$

with function $f : T \times R^n \times R^n \to R^m$ measurable in t and continuous in other variables. Here x stands for the state vector, t stands for time and control $u(\cdot)$ is a measurable function satisfying the constraints

$$u(\cdot) \in U = \{u(\cdot) : \ u(t) \in U_0, \quad t \in [t_0, T]\}, \tag{2}$$

where $U_0 \in \text{comp}R^m$. Let us assume that the initial state $x(t_0)$ to the system (1) is unknown but bounded

$$x(t_0) = x_0, \quad x_0 \in X_0 \in \text{comp}R^n. \tag{3}$$

Let an absolutely continuous function $x(t) = x(t, u(\cdot), t_0, x_0)$ be a solution to (1) with an initial state x_0 satisfying (3) and with a control function $u(t)$ satisfying (2). The differential system (1)–(3) is studied here in the framework of the theory of uncertain dynamical systems (differential inclusions [7]) through the techniques of trajectory tubes [5]

$$X(\cdot) = \bigcup \{\ x(\cdot) = x(\cdot, u(\cdot), t_0, x_0) \mid x_0 \in X_0,\ u(\cdot) \in U\} \tag{4}$$

which combine all solutions $x(\cdot, u(\cdot), t_0, x_0)$ to (1)–(3). Note that time cross-sections $X(t)$ of the tube $X(\cdot)$ coincide with the reachable sets at instants t to control system (1)–(3).

Let us mention here the well-known result [7] from the theory of differential inclusion that the trajectory tube $X(\cdot)$ coincides with the set of all solutions $\{x(\cdot) = x(\cdot, t_0, x_0)\}$ to the following differential inclusions

$$\dot{x} \in F(t, x) = \bigcup \{\ f(t, x, u) \mid u \in U_0\},\ t_0 \le t \le T, \tag{5}$$

with the initial state similar to (3)

$$x(t_0) = x_0, \quad x_0 \in X_0. \tag{6}$$

So we will use further the same notation $X(\cdot)$ for both trajectory tubes either for the control system (1)–(3) or for the differential inclusion (5)–(6).

It should be noted that the exact description of reachable sets $X(t)$ of a control system is a difficult problem even in the case of linear dynamics [1]. The estimation theory and related algorithms basing on ideas of construction outer and inner set-valued estimates of reachable sets have been developed in [3,4] for linear control systems.

In this paper the modified state estimation approaches which use the special quadratic structure of nonlinearity of studied control system and use also the advantages of ellipsoidal calculus [3,4] are presented.

The special type of nonlinearity chosen here for investigations is motivated by two reasons. First, we have found that it is very convenient in the theoretical analysis to correlate the nonlinear quadratic structure of system dynamics with the ellipsoidal assumptions on uncertain system data and therefore to extend the scope of ellipsoidal analysis for such nonlinear control problems. Second, there exist some applied models which may be described by considered nonlinear dynamical systems (e.g., [8]). Therefore researches in this direction may present interest for both the theory and applications.

2. Problem statement

Consider the nonlinear control system of the following type

$$\dot{x}(t) = A(t)x(t) + g(x(t)) + G(t)u(t), \ t_0 \leq t \leq T, \tag{7}$$

with unknown but bounded initial state

$$x(t_0) = x_0, \quad x_0 \in X_0, \tag{8}$$

and with control constraint

$$u(t) \in U_0, \quad t \in [t_0, T]. \tag{9}$$

Here matrices $A(t)$ and $G(t)$ (of dimensions $n \times n$ and $n \times m$, respectively) are assumed to be continuous on $t \in [t_0, T]$, X_0 and U_0 are compact and convex. The nonlinear n-vector function $g(x)$ in (7) is assumed to be of quadratic type

$$g(x) = (g_1(x), \ldots, g_n(x)), \ g_i(x) = x' B_i x, \ i = 1, \ldots, n, \tag{10}$$

where B_i is a constant $n \times n$ - matrix $(i = 1, \ldots, n)$.

Consider the following differential inclusion [7] related to (7)–(9) (with $P(t) = G(t)U_0$)

$$\dot{x}(t) \in A(t)x(t) + g(x(t)) + P(t), \ t \in [t_0, T], \ x(t_0) = x_0 \in X_0. \tag{11}$$

We introduce here the following additional notations. Denote as $B(a, r)$ the ball in R^n, $B(a, r) = \{ \ x \in R^n : \| \ x - a \ \| \leq r\}$, I is the identity $n \times n$-matrix. Denote by $E(a, Q)$ the ellipsoid in R^n,

$$E(a, Q) = \{ \ x \in R^n : \ (x - a)' Q^{-1} (x - a) \leq 1\}$$

with a center $a \in R^n$ and a symmetric positive definite $n \times n$–matrix Q. Denote by $\mathrm{hausd}(A, B)$ the Hausdorff distance for compact sets $A, B \subseteq R^n$. Assume now that $P(t) = E(a, Q)$ and $X_0 = E(a_0, Q_0)$ in (11), matrices B_i $(i = 1, ..., n)$ in (10) are symmetric and positive definite, $A(t) \equiv A$. We may assume that all trajectories of the system (11) belong to a bounded domain $D = \{x \in R^n : \| \ x \ \| \leq K\}$ where the existence of such constant $K > 0$ follows from classical theorems of the theory of differential equations and differential inclusions [7].

The problem of construction external and internal set-valued estimates of reachable sets $X(t)$ of the nonlinear system (7)–(9) (or, equivalently, of the inclusion (11)) was studied in [9,10]. The approach presented in [9,10] uses techniques of ellipsoidal calculus developed for linear control systems [3,4]. We note however that algorithms of estimating trajectory tubes of dynamical systems with quadratic nonlinearity described in [9, 10] are applicable only to small time intervals because of the necessity to guarantee the existence of the whole tube of solutions of the studied nonlinear system at the same time interval.

In this paper we consider the following uncertain system

$$\dot{x} \in Ax + h(\tilde{f}(x))d + E(\hat{a},\hat{Q}), \; x_0 \in X_0 = E(a_0,Q_0), \; t_0 \leq t \leq T, \qquad (12)$$

where $x \in R^n$, $\|x\| \leq K$, d is a given n-vector and a scalar function $\tilde{f}(x)$ has a form $\tilde{f}(x) = x'Bx$ with a symmetric and positive definite matrix B. We assume that a scalar nonnegative function $h(\cdot)$ is continuously differentiable and bounded. Denote by $L > 0$ the Lipschitz constant of the function $h(x)$ calculated for the domain $D = \{x \in R^n :\| x \|\leq K\}$.

Basing on results of ellipsoidal calculus [3,4] and using techniques of [9, 10] we present here a new state estimation algorithm which may be applied for producing external ellipsoidal estimate of reachable sets of system (12) at any finite time interval.

3. Main results

The approach discussed here is related to evolution equations of the funnel type that describe the dynamics of set–valued system states $X(t)$ of the differential inclusion (5)–(6). The basic assumptions on set–valued map $F(t,x)$ for the following result to be true may be found in [5,11].

Let us consider the equality which is called in the literature [5,11,12] as the funnel equation for set-valued function $X(t)$ $(t \in [t_0,T])$

$$\lim_{\sigma \to +0} \sigma^{-1} \text{hausd}(X(t+\sigma), \bigcup_{x \in X(t)} (x + \sigma F(t,x))) = 0, X(t_0) = X_0. \qquad (13)$$

Theorem 3.1 (Panasyuk, 1990). *The trajectory tube $X(t)$ of the system (5)–(6) is the unique set-valued solution to the evolution equation (13).*

In the case of the system (12) the funnel equation (13) of Theorem 3.1 takes the following form

$$\lim_{\sigma \to +0} \sigma^{-1} \text{hausd}(X(t+\sigma), \bigcup_{x \in X(t)} ((I+\sigma A)x + \sigma h(\tilde{f}(x))d + \sigma E(\hat{a},\hat{Q}))\ ) = 0, \quad X(t_0) = X_0, \quad t \in [t_0,T]. \qquad (14)$$

Note that the direct use of funnel equation (14) for finding trajectory tubes $X(t)$ is very difficult because it takes a huge amount of computations based on grid techniques [13]. Our goal here is to present another approach to avoid this complex grid technology.

The following new result presents an easy computational tool to find external estimates of $X(t)$ by step-by-step procedures and is the basis for further algorithm.

Theorem 3.2. *Let $X_0 = E(a, k^2 B^{-1})$ with $k \neq 0$. Then for all $\sigma > 0$ the following inclusion holds*

$$X(t_0 + \sigma, t_0, X_0) \subseteq E(a(\sigma), Q(\sigma)) + \sigma(rB(0,1) + E(\hat{a}, \hat{Q})) \\ +o(\sigma)B(0,1), \quad \lim_{\sigma \to +0} \sigma^{-1} o(\sigma) = 0, \tag{15}$$

where

$$a(\sigma) = (I + \sigma A)a + \sigma h(a'Ba + k^2) \cdot d, \tag{16}$$

$$Q(\sigma) = k^2 (I + \sigma A) B^{-1} (I + \sigma A)', \ r = 2Lk||a'B^{1/2}|| \cdot ||d||. \tag{17}$$

Proof. The proof of this result may be done along the lines of the proof of Theorem 3 [10] with some minor modifications due to the presence here of a nonlinear Lipschitz function h. □

Basing on this theorem we may formulate the following new scheme that gives the external estimate of trajectory tube $X(t)$ of the system (12) with given accuracy.

Algorithm 3.1. *Subdivide the time segment $[t_0, T]$ into subsegments $[t_i, t_{i+1}]$ where $t_i = t_0 + ih$ ($i = 1, \ldots, m$), $h = (T - t_0)/m$, $t_m = T$.*

(1) Given $X_0 = E(a, k_0^2 B^{-1})$ with $k_0 \neq 0$, define the ellipsoid $X_1 = E(a_1, Q_1)$ such that $E(a(\sigma), Q(\sigma)) + \sigma(rB(0,1) + E(\hat{a}, \hat{Q})) \subseteq E(a_1, Q_1) = X_1$ for $a(\sigma)$, $Q(\sigma)$, r defined in Theorem 3.2 with $\sigma = h$.

(2) Find the smallest constant k_1 such that $E(a_1, Q_1) \subseteq \tilde{X}_1 = E(a_1, k_1^2 B^{-1})$, and it is not difficult to prove that k_1^2 is the maximal eigenvalue of the matrix $B^{1/2} Q_1 B^{1/2}$.

(3) Consider the system on the next subsegment $[t_1, t_2]$ with $E(a_1, k_1^2 B^{-1})$ as the initial ellipsoid at instant t_1.

(4) Repeat steps 1–3 till the end of time interval.

At the end of the process we will get the external estimate $E(a(t), Q(t))$ of the tube $X(t)$ with accuracy tending to zero when $m \to \infty$.

4. Conclusion

The paper deals with the problems of state estimation for a dynamical control system described by differential inclusions with unknown but bounded initial state. Basing on results of ellipsoidal calculus developed for linear uncertain systems we present the modified state estimation approaches which use the special nonlinear structure of the control system and simplify calculations.

Acknowledgments

The research was supported by the Russian Foundation for Basic Research under Project RFBR 09-01-00223, by the Program 4576.2008.1 of State Support of Leading Scientific Schools and by the Program 29 of the Presidium of Russian Academy of Sciences.

References

1. A. B. Kurzhanski, *Control and Observation under Conditions of Uncertainty* (Nauka, Moscow, 1977).
2. N. N. Krasovskii and A. I. Subbotin, *Positional Differential Games* (Springer–Verlag, Berlin, 1988).
3. F. L. Chernousko, *State Estimation for Dynamic Systems* (CRC Press, Boca Raton, 1994).
4. A. B. Kurzhanski and I. Valyi, *Ellipsoidal Calculus for Estimation and Control* (Birkhauser, Boston, 1997).
5. A. B. Kurzhanski and T. F. Filippova, *On the Theory of Trajectory Tubes – a Mathematical Formalism for Uncertain Dynamics, Viability and Control*, in *Advances in Nonlinear Dynamics and Control: a Report from Russia*, ed. A.B. Kurzhanski, Progress in Systems and Control Theory, Vol. 17 (Birkhauser, Boston, 1993), pp. 122–188.
6. T. F. Filippova, *Math. Comput. Model. Dynam. Syst.* **11**, 2, 149 (2005).
7. A. F. Filippov, *Differential Equations with Discontinuous Right–hand Side* (Nauka, Moscow, 1985).
8. N. C. Apreutesei, *Appl. Math. Lett.* **22**, 7, 1062 (2009).
9. T. F. Filippova, State estimation in control problems under uncertainty and nonlinearity, in *Proc. 6th Vienna Conference on Mathematical Modelling (MATHMOD'2009)*, (Vienna, Austria, 2009).
10. T. F. Filippova and E. V. Berezina, *Lecture Notes in Comp. Sci., Springer* **4818**, 326 (2008).
11. A. I. Panasyuk, *J. Optimiz. Theory Appl.* **2**, 349 (1990).
12. P. R. Wolenski, *SIAM J. Contr. Optimiz.* **28**, 5, 1148 (1990).
13. A. L. Dontchev and F. Lempio, *SIAM Review* **34**, 263 (1992).

CONTROL PROBLEMS FOR A BODY MOVEMENT IN THE VISCOUS MEDIUM

D. S. ZAVALISHCHIN

Institute of Mathematics and Mechanics
Ural Division of Russian Academy of Sciences
Department of Optimal Control, 16 S. Kovalevskaya
Ekaterinburg, GSP–384, 620219, Russia
E-mail: zav@imm.uran.ru

The paper deals with the mathematical model of movement of a body in a viscous medium. The problem of control for a solid body moving in the viscous medium from initial position to the given one is considered. Movement occurs at Reynolds's greater numbers that generates effects of failure of the laminar boundary layer, caused by return difference of a gradient of pressure. Thus behind a body the vortex path is formed. The frequency of failure of whirlwinds is expressed in the form of the dimensionless parameter. Asymmetrical formation of whirlwinds leads to occurrence periodic cross-section to speed of power influences on a body. Oscillatory movements, especially as a result develop if the frequency of formation of whirlwinds comes nearer to own frequency of fluctuations of a body.

Keywords: Optimal power flow, control, Kármán trail, fluid dynamics.

1. Introduction

Autonomous vehicles intended for work in atypical environment has proved to form a great body of knowledge interesting from the viewpoint of challenging applications and being the source of new theoretical research. Particular emphasis is placed on mobile manipulation robots. Just this term is preferred in [1] intended for work in a viscous medium. Design of such vehicles is a complicated problem. The situation when one has to deal with rather limited energy supply of vehicles is natural and, sometimes, inevitable. Then, the following control problem is topical: to find the laws of the control forces and momentums behavior so as to move it from the initial position to a given one for minimum energy consumption. Such a problem is close to the ones of dynamic optimization considered by [1–4].

The totality of the problems solved in the present paper can be used in both the applied theory of dynamic optimization problems and design of perspective samples of new machines.

2. Problem statement

Throughout this paper we regard the term "medium" as fluid or gas. However, for sake of being intuitive we use the term "fluid". Hydrodynamic constraints listed below, being satisfied, give a possibility to analyze necessary conditions for optimality for the problems mentioned in the Introduction. It is assumed that an inertial system and, inside it, a right Cartesian coordinate system $Ox_1x_2x_3$ are chosen. Let $\mathbf{v}(t,x) = \mathbf{v}(t,x_1,x_2,x_3)$ be the velocity vector of fluid particle at the point $M(x_1,x_2,x_3)$ at the instant t, and v_1, v_2, and v_3 be its projections in the coordinate axes. The first two constraints are reduced to the following.

Claim 2.1. *The generalized Newton hypothesis (see [5]) is fulfilled*

$$P = -pE + \mu\Big(\frac{\partial \mathbf{v}}{\partial x} + \Big(\frac{\partial \mathbf{v}}{\partial x}\Big)^*\Big),$$

where P is the linear operator defined by the stress tensor, $p = p(t,x)$ denotes the scalar field of pressure, μ is the dynamic viscosity coefficient, E is the identity mapping, $\frac{\partial \mathbf{v}}{\partial x}$ is the Frechet derivative [6], and $\left(\frac{\partial \mathbf{v}}{\partial x}\right)^*$ is the conjugate operator.

Claim 2.2. *Fluid is incompressible.*

With account of the equation of continuity, this constraint is equivalent to zero velocity of the volume strain $\operatorname{div}\mathbf{v} = 0$. Let a body of bounded size with sufficiently smooth boundary S move in fluid. One of the fluid mechanics axioms is the sticking condition: at the body surface points the velocity vector of fluid particle is equal to the velocity vector of the corresponding body point. This condition and the claim 2.2 imply that in the case of translational motion of the body the following equality is fulfilled at its surface (see [5])

$$\Big(\frac{\partial \mathbf{v}}{\partial x}\Big)^* \mathbf{n} = 0,$$

where $\mathbf{n}$ is the unit vector of the outward normal to the surface S at the point x.

We need further the so-called moving coordinate system $O_c y_1 y_2 y_3$ with the body inertia center as the origin and the axes rigidly connected with the body. Then the Navier–Stokes equation is of the form

$$\frac{\partial \hat{\mathbf{v}}}{\partial t} = -\frac{\partial \hat{\mathbf{v}}}{\partial y}(\hat{\mathbf{v}} - \mathbf{V}) - \frac{1}{\rho}\Big(\frac{\partial \hat{p}}{\partial y}\Big)^* + \nu \operatorname{div}\frac{\partial \hat{\mathbf{v}}}{\partial y} + \mathbf{F}, \tag{1}$$

where $\mathbf{F}$ is the strength of the gravity field, ρ is the fluid density, $\nu = \mu/\rho$ is the kinematic viscosity coefficient.

Now, the above-mentioned boundary-value problem is reduced to finding the solution of a system of partial differential equations, namely, equation (1) plus the equation of continuity $\operatorname{div} \hat{\mathbf{v}} = 0$.

Suppose that the body has a symmetry axis. If the body moves in such a manner that this axis remains in a given plane (for example, in the plane Oxy), then, according to the statics theorems for an absolutely solid body, the totality of forces acting from fluid upon the body can be reduced to the resultant one called the hydrodynamic force. As usual the point of intersection of the symmetry axis and the line of the hydrodynamic force action is referred to as center of pressure. The hydrodynamic force is resolved into components parallel to the velocity vector $\mathbf{V}$ of the body inertia center and perpendicular to $\mathbf{V}$. The first component $\mathbf{D}$ is known as the drag force, and the second one $\mathbf{D}^l$ is called the lift force. In the considered case, such a presentation can be maintained for the magnitude of the stationary drag force $D = C_D \rho S V^2/2$. Analogously, the magnitude of the stationary lift force can be presented as $D^{\perp} = C_D^{\perp} \rho S V^2/2$. Here S is the area of the body projection onto the plane perpendicular to the velocity vector of the body inertia center.

In the framework of the listed constraints, the coefficient C_D is, following to [7], a function of the body shape, Reynolds number and, probably, the angle of attack between the velocity vector of the body inertia center and the symmetry axis, i.e., $C_D = C_D(\text{shape,Re}, \alpha)$. To determine the angle of attack, one can use the formula [4]

$$\alpha = -s \arccos |(\mathbf{e}, \mathbf{V}/V)|, \quad s = \operatorname{sign}((\mathbf{V}, \mathbf{e})(\mathbf{V}, \mathbf{e}^{\perp})),$$

where $\mathbf{e}$ is the directing vector of the body symmetry axis.

In this article the model of moving in the viscous medium of solid body for Reynolds's greater numbers is investigated. The increase in speed generates a separation of a boundary layer and occurrence of turbulent effects. In turn the last is the reason of occurrence of cross-section fluctuations operating on the body. Further attempt of modelling of movement of a body in such conditions is undertaken.

3. Mathematical model

In this section, we deal with a model of moving in a viscous medium of solid body in plane Oxy. The state of the body is described by the generalized coordinates x, y and φ. Let V be the vector of centroid velocity $\mathbf{V} = (\dot{x}; \dot{y})^T$,

F be the force acting along a body axis $\mathbf{F} = (F\cos\varphi; F\sin\varphi)^T$, E be the unit vector $\mathbf{E} = (\cos\varphi; \sin\varphi)^T$, D and $D^\perp$ are the drag force and lift force respectively

$$\mathbf{D} = (-D\cos(\varphi-\alpha); -D\sin(\varphi-\alpha))^T,$$
$$\mathbf{D}^\perp = (-D^\perp\sin(\varphi-\alpha); D^\perp\cos(\varphi-\alpha))^T,$$

U be the angular moment. Kinetic energy is equal to

$$T = \frac{1}{2}m(\dot{x}^2+\dot{y}^2) + \frac{1}{2}\frac{ml^2}{3}\dot{\varphi}^2.$$

Using the Lagrange equations

$$\frac{d}{dt}\left(\frac{\partial T}{\partial \dot{q}_i}\right) - \frac{\partial T}{\partial q_i} = Q_i$$

one can obtain body moving equations

$$\begin{aligned} m\ddot{x} &= Q_x \\ m\ddot{y} &= Q_y \\ \tfrac{1}{3}ml^2\ddot{\varphi} &= Q_\varphi \end{aligned} \tag{2}$$

The generalized forces corresponding to the generalized coordinates will be the following

$$\begin{aligned} Q_x &= -D\cos(\varphi-\alpha) - D^\perp\sin(\varphi-\alpha) + F\cos(\varphi) \\ Q_y &= -D\sin(\varphi-\alpha) + D^\perp\cos(\varphi-\alpha) + F\sin(\varphi) - mg \\ Q_\varphi &= U + M + F_T \end{aligned} \tag{3}$$

Here $F_T = h\sin(nt+\delta)$. The system of equations (2) and (3) describes body movement.

4. The case of Reynolds's greater numbers

The phase trajectory $y(x)$ of body movement is obtained by numerical integration of system (2) and (3) is represent on Fig. 1. It is visible that a body having overcome 60 metres deviates on a vertical axis on 3 metres. Control U – continuous line, velocities x – dot line, and y – dashed line, is represent on Fig. 2. At last angle φ – continuous line, and angle of attack α – dot line, (within 0,5 radians) is represent on Fig. 3. It should be noted that data for numerical experiment undertook from the book [8].

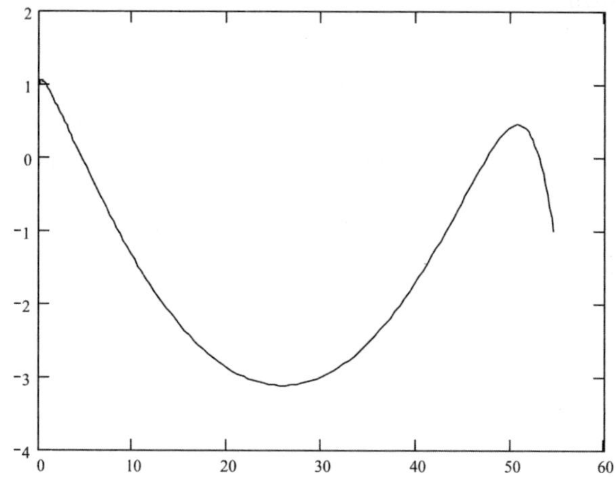

Fig. 1. The phase trajectory $y(x)$.

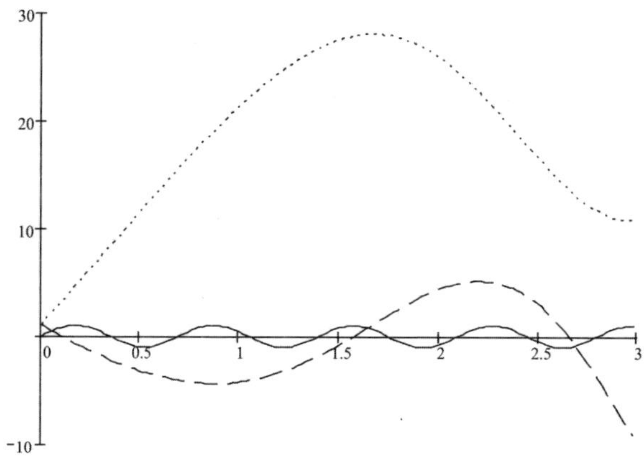

Fig. 2. Control U, velocities x and y.

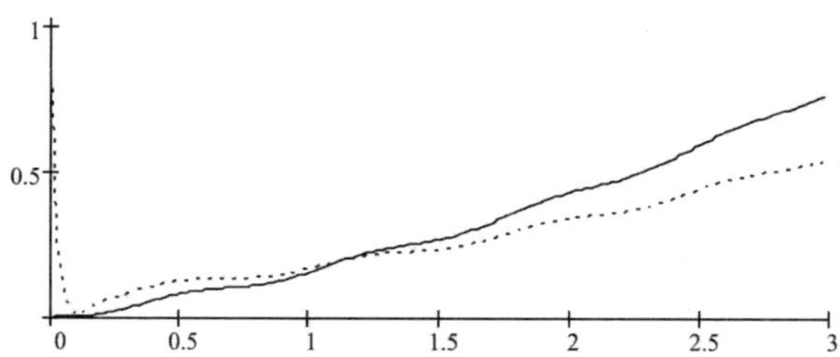

Fig. 3. Angles φ and α.

5. Conclusion

One of the classical open-flow problems in fluid mechanics concerns the flow around a circular cylinder, or more generally, a bluff body. At very low Reynolds numbers the streamlines of the resulting flow is perfectly symmetric as expected from potential theory. However as the Reynolds number is increased the flow becomes asymmetric and the so called Strouhal number relates the frequency of shedding to the velocity of the flow and a characteristic dimension of the body. It is defined as $S_t = f_{st}S/V$. In the equation f_{st} is the vortex shedding frequency (or the Strouhal frequency) of a body at rest. The Strouhal number for a cylinder is 0.2 over a wide range of flow velocities. The phenomenon of lock-in happens when the vortex shedding frequency becomes close to a natural frequency of vibration of the structure. When this happens large and damaging vibrations can vortex street occurs.

Analyzing results of numerical experiment it is possible to draw following conclusions. Adaptive control allows to smooth influence of the Kármán trail. Thus the border of its occurrence is probably removed. It would be the small contribution to struggle against turbulence.

Acknowledgments

Researches was supported by the Program no. P-04 of Presidium of the Russian Academy of Sciences "Fundamental Problems of Nonlinear Dynamics".

References

1. F. L. Chernous'ko, N. N. Bolotnik and V. G. Gradetskii, *Manipulation Robots: Dynamics, Control and Optimization* (Moscow, Nauka, 1989).
2. V. S. Beletskii, *Submarine Manipulation Vehicles Operated by Remote Control* (Leningard, Sudostroenie, 1973).
3. V. V. Avetisyan, L. D. Akulenko and N. N. Bolotnik, *Izv. Akad. Nauk SSSR. Tekhnicheskaya kibernetika* **3**, 100 (1987).
4. D. S. Zavalishchin and S. T. Zavalishchin, *Dynamic Optimizanion of Flow* (Moscow, Nauka, Physics and Math. Publishers, 2002).
5. N. A. Slezkin, *Viscous Incompressible Fluid Dynamics* (Moscow, GITTL, 1955).
6. L. Schwatz, *Théorie des Distributions, vol. I* (Paris, Hermann, 1950).
7. L. I. Sedov, *Solid Medium Mechanics, vol. 1* (Moscow, Nauka, 1973).
8. J. W. Daily and D. R. F. Harleman, *Fluid Dynamics* (Massachusetts, Wesley Publishing Co., 1966).

COOPERATIVE PATH PLANNING IN THE PRESENCE OF ADVERSARIAL BEHAVIOR

JOÃO BORGES DE SOUSA

Faculty of Engineering, Porto University
R. Dr. Roberto Frias, 4200-465 Porto, Portugal
E-mail: jtasso@fe.up.pt

JORGE ESTRELA DA SILVA

Institute of Engineering of Porto, Porto Polytechnic Institute
Rua Dr. António Bernardino de Almeida 431, 4200-072 Porto, Portugal
E-mail: jes@isep.ipp.pt

A collaborative control scenario is modeled in the framework of hybrid systems. The problem of reaching a given point under adversarial behaviour on a given time frame is modeled as a differential game for hybrid systems. System behaviour is discussed with the aid of Krasovskii's u-stable bridge.

Keywords: Collaborative control, differential games, hybrid systems.

1. Introduction

In previous work (see [1]) we discussed dynamic optimization for collaborative control problems with the help of a simple two-vehicle optimal path coordination control problem. This problem is representative of more general optimal coordination problems. The problem is modeled in the framework of hybrid systems. Here we extend that formulation and consider the case of optimization under adversarial behaviour. The adversarial behaviour models the effect of bounded disturbances.

In [1] we proved that, in the deterministic optimal control setting, it may be worthwhile for a vehicle to deviate from the optimal path (of isolated operation) to join other vehicle that will contribute to improved conditions of operation, resulting in a reduced overall cost to go. In the present case, it is assumed that the joint operation eliminates the effect of the adversarial action.

We consider the problem of a vehicle v_1 reaching a given target set $T := \{(t, x) : x \in S \subset \mathbb{R}^n, t \in [t_1, t_2]\}$ where S is a closed set. For the operation of v_1 in isolation, this is a classic differential game (see, for instance, [2]). However, the system dynamics becomes discontinuous and the problem gains a combinatorial flavour when collaboration with other vehicles is considered. This behaviour is modeled with a hybrid system. The discrete component of the system dynamics raises interesting questions: 1) if the initial condition for v_1 does not belong to the u-stable bridge corresponding to independent operation of v_1 it may happen that it is worthwhile for v_1 to meet other vehicles before reaching the target set; 2) if the initial condition for v_1 belongs to the u-stable bridge, then it may happen that we can impose stricter timing constraints for T. We extend Krasovskii's u-stable bridge in order to handle our hybrid systems formulation and answer these questions. An extended version of this paper was presented in [3].

2. System model

2.1. *Continuous dynamics*

The system is composed of N vehicles. Two types of vehicles are considered: *simple* and *jammer*. The *jammer* vehicles depart from their base and are subject to fuel constraints (this imposes fuel limitations on their operations). The problem of fuel constraints is assumed to be of no concern for the simple vehicles, which can depart from any position. The *simple* vehicles are subject to bounded disturbances which are modeled as adversarial actions. However, the adversarial action against a *simple* vehicle is eliminated when its position coincides with the position of a *jammer* vehicle. This joint operation is a case of collaboration. Notice that this "coincidence" of positions is a simplification; in practice, the *simple* vehicle would be required to be in a given neighbourhood of the *jammer* vehicle. In any case, this means that the dynamics of *simple* vehicles are a discontinuous function of their position relative to the *jammer* vehicles. The motion model for the *simple* vehicle i, when operating alone, is given by

$$\dot{x_i}(t) = f_i(t, x_i, u_i) + g_i(t, x_i, p_i) \tag{1}$$

where $x_i \in \mathbb{R}^n$ is the position of vehicle i, $u_i \in U_i$ is the respective control input, $p_i \in P_i$ is the adversarial input and U_i and P_i are closed sets. The considered model, assuming additive adversarial behaviour, is suitable to model several physical scenarios. Additionally, since the effects of the control and adversarial inputs are separable, the saddle point condition in a small game (Isaacs condition) is assured. The *jammer* vehicles are not affected by the adversarial behaviour (therefore the g_i term is droped for those vehicles).

The motion model for joint operation of a *simple* vehicle i with a *jammer* vehicle j is given by

$$\dot{x}_i(t) = f_{ij}(t, x_i, u_{ij}) \tag{2}$$

where $u_{ij} \in U_{ij}$ is the control input and $f_{ij}(t, x, u)$ and U_{ij} are defined such that $f_{ij}(t, x_i, U_{ij}) = f_i(t, x_i, U_i) \bigcap f_j(t, x_i, U_j)$. This means that the *jammer* vehicle will be capable of replicating the motion of the *simple* vehicle (we assume that $f_{ij}(t, x_i, U_{ij})$ is not empty).

The amount of available fuel of *jammer* vehicle i is modeled by the state variable $c_i \in \mathbb{R}$, with $w_i \leq 0$ and $c_i(0) = \theta_i$:

$$\dot{c}_i(t) = \begin{cases} w_i(x_i, u_i) & \text{if } c_i > 0 \\ 0 & \text{otherwise} \end{cases} \tag{3}$$

2.2. *Hybrid model*

A few operational constraints are considered. The vehicles are allowed to meet once and move together while the *jammer* vehicle has enough fuel to return to a base. If the *jammer* vehicle is not required to return to a base, the vehicles may travel together until the *jammer* vehicle runs out of fuel. It is possible to devise scenarios where this policy is not optimal. The advantage of this policy is that it allows us to reduce the dimension of the state space of the global system. It also simplifies the problem of coordination.

Consider the following two value functions: $V_i^f(t, x)$ and $V_i^b(t, x)$. $V_i^f(t, x)$ is a map of x to the minimum amount of fuel required by *jammer* vehicle v_i to reach x at time t after departing from its base. $V_i^b(t, x)$ is a map of x to the minimum amount of fuel required by *jammer* vehicle v_i to reach a destination base starting from x at time t. Those functions can be computed by dynamic optimization techniques for reachability analysis (see [4]).

Let $R_i(t) := \{x : \theta_i - V_i^b(t, x) - V_i^f(t, x) \geq 0\}$ be the set of points that can be reached by *jammer* vehicle v_i under the above mentioned operational constraints. Joint operation between *simple* vehicles and v_i may only occur in $R_i(t)$. This means that a *simple* vehicle will only benefit from reduced adversarial action when going through $R_i(t)$.

2.2.1. *Two vehicles scenario*

For $N = 2$ (one *simple* vehicle v_1 and one *jammer* v_2), we have three possible distinct modes:

Mode a**:** v_1 operating alone before meeting v_2 (which may never take place). This is the initial mode.

Mode b**:** v_1 moving together with v_2 in joint operation; this mode is optional. In this mode, the state variable corresponding to the fuel of v_2 is tracked along with the state of v_1. Given (2), this is enough to describe the system's continuous state.

Mode c**:** v_1 operating alone, after meeting with v_2; this mode happens after and only after mode b. In this mode, like on mode a, only the state of v_1 is tracked.

If it is decided that v_1 should meet v_2 (transition from mode a to mode b), the meeting point where the system will switch from mode a to mode b must be defined. From the perspective of v_1, all that really matters in what concerns v_2 is: 1) the instant t_{ab} when the meeting takes place; 2) the point where the meeting takes place, which must be inside $R_2(t_{ab})$; and 3) the amount of the fuel remaining in the fuel tank of v_2 at the meeting position, given by $\theta_2 - V_2^f(t_{ab}, x)$. When the vehicles reach the meeting position, the system switches to mode b. v_1 may decide to abandon v_2 still inside $R_2(t)$ (i.e., with positive fuel slack); on the other hand, as soon as v_1 leaves $R_2(t)$, v_2 must head back to its returning base. The former case corresponds to a controlled transition from b to c; the later corresponds to an autonomous transition from b to c.

2.2.2. *General case*

In a scenario with multiple *jammer* vehicles, there are $3^{(N-1)}$ possible modes of operation if all possible interactions between the *simple* vehicle and the remaining vehicles are considered. For instance, for $N = 5$, we could have, at a certain time, v_2 already used, v_3 and v_4 being used and v_5 still not used. We denote the set of all possible discrete modes as Q. The full continuous state for each discrete mode is $x_v, v \in Q$. The hybrid state is defined by the tuple (x_v, v). For $N = 2$, we have $Q := \{a, b, c\}$, $x_a = x_c = x$ and $x_b = (x, c_2)$.

3. Solution approach

For simplicity, a two-vehicle scenario is considered. Our approach is inspired by [5] and [6]. Krasovskii's u-stable bridge (SB) allows us to study the solution of this kind of problems and it may be also used to develop extremal aiming feedback strategies. The SB, $W(t)$, consists of the set of states the system may enter at time t while still ensuring reachability of the target set. If the initial condition of v_1 does not belong to the SB then there are no guarantees that v_1 reaches the target set. The extended SB corresponding to the hybrid system may be computed by noting that $W(t) := \{x : V(t,x,a) \neq \infty\}$, where $V(t,x,v)$ is the minimum time to reach S starting from continuous state x and mode v at time t. For each v, $V(t,x,v)$ is the solution of the Hamilton–Jacobi–Isaacs (HJI) equation with $V(t,S,v) = 0$, $t \in [t_1,t_2]$, and considering the system dynamics for mode v. Additionally, the following conditions must be enforced:

$$x_b(t) \in C_2(t) : V(t,x_b,b) > V(t,x,c) \Rightarrow V(t,x_b,b) = V(t,x,c) \quad (4)$$

$$x(t) \in R_2(t) : V(t,x,a) > V(t,x_b,b) \Rightarrow V(t,x,a) = V(t,x_b,b) \quad (5)$$

with $C_2(t) := \{(x,c_2) : V_2^b(t,x) \leq c_2 \leq \theta_2 - V_2^f(t,x)\}$. By comparing the extended SB to that of the independent operation one notices that, for some intervals of t, the set of admissible states is enlarged.

We illustrate the results by means of a one-dimensional scenario, with $f_1(t,x,u) = 20u$, $g_1(t,x,p) = 19p$, $|u| \leq 1$, $|p| \leq 1$, $-10 \leq f_{12}(t,x,u) \leq 10$, $c_2(0) = 1$ and $w_2(x,u) = -1$; the initial and final base for v_2 is $x_{20} = 10$. The SB for $T := \{(0,0)\}$ is illustrated on Fig. 1. The right diagram of Figure 1 corresponds to mode b; the bounds at $x = 5$ and $x = 15$ are due to the amount of initial fuel of v_2 and the requirement of v_2 returning to base; the fuel variable is not represented, therefore this diagram is only a projection of the complete SB.

Regarding the questions posed in the introduction, consider $t = -10.5$. Without collaboration from *jammer* vehicles, v_1 could only reach the origin in less than 10.5 time units (t.u.) when starting from $x \in [0, 10.5)$. With collaboration, that range is enlarged to $[0, 15)$. On other perspective, without collaboration, when starting from $x = 15$, v_1 could only reach the origin in 15 (t.u.) ; with collaboration, that time can be shortened to 10.5 (t.u.). It can be seen that for $x \geq 10$ the gain of collaboration is 4.5 (t.u.).

If v_1 starts on region I, it will have to coordinate with v_2 (on region III) in order to assure reachability of the target. When switching from mode b to mode c, v_1 must be inside region II. This is because the SB on mode c is composed only of region II.

4. Conclusion

The described approach allows a systematic qualitative and quantitative determination of whether cooperation is advantageous or not. Determination of optimal feedback laws or extremal aiming strategies is not discussed

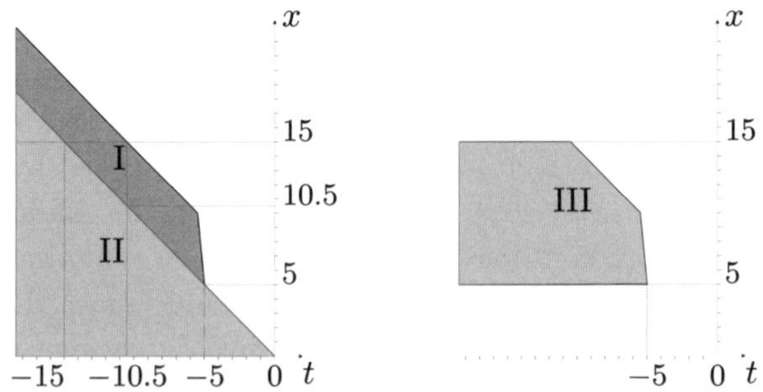

Fig. 1. The u-stable bridge for joint operation of two vehicles is the reunion of I and II. Region II is the stable bridge corresponding to independent operation of v_1. Region III corresponds to operation on mode b.

here due to space consideration. The u-stable bridge is computed through the solution of a set of HJI equations, coupled by the discrete logic of the hybrid system. The main limitation of the approach is the usual requirements for the computation of the HJI equation. Additionally, the concept of joint operation defined here (see (2)) may prove non-trivial for more complex systems, especially when one considers heterogeneous vehicles.

References

1. J. B. de Sousa, J. E. da Silva and F. L. Pereira, New problems of optimal path coordination for multi-vehicle systems, in *Proceedings of the 10th European Control Conference*, (Budapest, Hungary, 2009).
2. M. Bardi, T. E. S. Raghavan and T. Parthasarathy (eds.), *Stochastic and differential games: theory and numerical methods*, Annals of the International Society of Dynamic Games, Vol. 4 (Birkhäuser Verlag, Basel, Switzerland, 1999).
3. J. B. de Sousa and J. E. da Silva, Cooperative path planning in the presence of adversarial behavior, in *4th International Scientific Conference on Physics and Control*, (Catania, Italy, 2009).
4. A. B. Kurzhanskii and P. Varaiya, *Journal of Optimization Theory & Applications* **108**, 227 (2001).
5. J. Sethian and A. Vladimirsky, Ordered upwind methods for hybrid control, in *Proceedings of the hybrid systems workshop*, (Springer-Verlag, 2002) pp. 393–406.
6. H. Zhang and M. R. James, *SIAM Journal of Control and Optimization* **48**, 722 (2006).

PART J

Physics and control in fusion plasma devices

CONFINEMENT REGIME IDENTIFICATION AT JET VIA AN INTERPRETABLE FUZZY LOGIC CLASSIFIER

G. VAGLIASINDI * and P. ARENA

Dipart. di Ing. Elettrica, Elettronica e dei Sistemi
Università degli Studi di Catania, Catania, 95125, Italy
*E-mail: * gvaglia@diees.unict.it*
parena@diees.unict.it

A. MURARI

Consorzio RFX, Associazione EURATOM ENEA per la Fusione
Padova, 35127, Italy
E-mail: murari@igi.pd.cnr.it

and JET-EFDA Contributors†

JET-EFDA, Culham Science Centre, OX14 3DB, Abingdon, UK

In this paper a data driven methodology to automatically derive an interpretable Fuzzy Logic Classifier (FLC) has been applied to the problem of confinement regime identification in the Joint European Torus. The approach has been developed explicitly to handle the complexities of the inference process in Magnetic Confinement Nuclear Fusion (MCNF). The resulting FLC on the one hand attains very good performance in terms of generalization and classification, on the other hand provides a series of rules which can be easily interpreted and contributing to a very good first, intuitive understanding of the physics involved.

Keywords: Tokamak, confinement regime, knowledge from data.

1. Introduction

In magnetically confinement nuclear fusion, the data analysis process presents some unique challenges which have no direct counterpart in any other field of "Big Physics" research. The task of attaining a global and coherent view of the plasma state is a real challenge for a series of aspects of the involved physics. Since fusion plasmas are open systems far from equilibrium, it is very difficult to formulate satisfactory models starting from basic physics principles. Moreover the nonlinear interactions among

†See the Appendix of F. Romanelli *et al.*, Proceedings of the 22nd IAEA Fusion Energy Conference 2008, Geneva, Switzerland.

many variables not only increase the difficulty of the interpretation but also pose some limits to the degree of control of the experiments which can be performed.

For these reasons, in the last years some efforts have been devoted to the development of statistical and soft computing methods to support the analysis and interpretation of the experiments [1]. On the other hand, as summarized in [2] and [3], till nowadays the various developed automatic analysis techniques have been limited to exploratory aspects.

In this paper, a methodology [9] which, starting from raw data, proceeds up to the point of providing an intuitive interpretation of physics involved in a purely automatic way is applied to a specific nuclear fusion problem, i.e. magnetic confinement regime classification. The following criteria, proposed in [7] and [8], were followed to achieve an interpretable fuzzy system: 1. the fuzzy partition should be readable, i.e. it should be possible to interpret the fuzzy set as linguistic labels, the fuzzy sets must be distinguishable and they should be in moderate number; 2. the set of rules should be as small as possible, preserving the performance; 3. the rules should be incomplete, i.e. the rules premises should not exceed the limit of 7±2 conditions.

The FLC so produced provides a series of explicit rules which turn out to be very useful for the interpretation of the physics involved.

2. Confinement regime at Jet

Tokamak configuration can, indeed, be operated in different confinement regimes, which can be significantly different not only in terms of performance but also of physics interpretation and control requirements. The High Confinement regime, the so called H mode [5], is a particularly relevant example. Its performance can increase of more than a factor of two compared to the L-mode (Low confinement), but it is affected by edge instabilities. These require particular measures to avoid disruptions and could be very dangerous for the integrity of the entire device in the next generation of machines like ITER. On the other hand, the details of the plasma evolution from the L-mode to the H-mode state have not been fully understood yet, neither from the dynamics aspects, nor from power requirements to trigger the transition.In the perspective of ITER, in which accessing the H-mode is essential but disruptions can have very harmful consequences, it is becoming urgent to develop models to better interpret the H-mode physics and to identify the regimes in real time.

The database used for the analysis is described in [6]. Since the aim of the work is, also, to derive an intuitive understanding of the confinement transition in tokamak plasmas, the time slices of the various discharges have been divided in three subsets and analyzed separately. The three subsets include data around the transition from L to H (300 ms before and after), around the transition back from H to L (300 ms before and after), and away from the transitions in steady state L and H mode phases (500 ms in L mode, between 1200 ms and 700 ms before the L→H transition and intervals of 500 ms in H mode, between 1200 ms and 700 ms before the H→L transition).

3. Construction of the interpretable system

Since the objective of this work is to produce an interpretable fuzzy system, a limited number of input variables is, therefore, mandatory to have readable fuzzy sets and rule base (see [7] and [8]). This can be achieved trough a feature selection step to select among the variables in the database the most relevant for the description of the problem. The instrument selected to perform this step is Classification and Regression Trees [4]. It is a non-parametric statistical method, which uses a decision tree to solve classification and regression problems using both categorical and continuous variables.

The three subsets described in the previous section have been provided to CART in order to build three different trees. According to the variable ranking provided by CART the signals reported in Table 1 have been evaluated as the most relevant. In addition to these physical quantities, it has been considered important to test also the influence of geometrical

Table 1. The four most relevant variables for the three different subsets. The signals are sorted in descendent order of importance.

L→H		**H→L**		**Steady State**	
W_{mhd}	Magnetohydro-dynamic energy	β_N	β normalized over diamagnetic energy	β_N	β normalized over diamagnetic energy
β_N	β normalized over diamagnetic energy	B_T	Toroidal magnetic field	B_T	Toroidal magnetic field
T_e	Electron temperature	$FDWDT$	Time derivative of diamagnetic energy	*Lid4*	Outer interferometry channel
B_{T80}	Axial toroidal magnetic field at ψ=0.8	q_{95}	Safety factor at ψ=0.95	$FDWDT$	Time derivative of diamagnetic energy

parameters that can account for the position/shape of the plasma inside the vacuum vessel. Several tests have been performed using various quantities and the ones which have produced the best results are the radial and vertical position of the X point.

Once the most relevant variables have been selected, a new tree is produced providing only the selected variables as predictors. The output from the CART represents the input for the automatic FLC construction which comprises three steps:

- *Extraction of the crisp rules from the classification tree.*
  The tree is parsed automatically to determine a rule for each terminal node. For each of them, the corresponding branch is scanned up to the root and a specific rule is devised for each intermediate node, on the basis of the inequality used by CART at each node to perform the split.
- *Determination of the membership functions from the set of crisp rules obtained in the previous step.*
  In order to satisfy the first criteria for interpretability, i.e. the readability of fuzzy partitions, the number of fuzzy membership functions was limited to three trapezoidal function since they provides enough flexibility to cover each variable domain by dividing it in three different regions. The selection of the parameters identifying the trapezoidal function is described in [9].
- *Formulation of the fuzzy rules on the basis of the classification tree crisp rules and the membership functions.*
  Each rule provided by the tree has already a form similar to a fuzzy rule apart from the fact that the antecedent is composed of inequalities using crisp values. We have then to translate the inequalities of the tree rules for the terminal nodes into inequalities based on the membership functions defined in the previous section. This is achieved selecting the fuzzy membership function which best approximates the crisp inequalities produced during the tree construction.

A detailed description of the above mentioned steps is available in [9].

4. Evaluation of the performance and discussion

Since the total number of rules is affected by the number of terminal nodes of the related tree, the complexity of the produced FIS also depends on nodes retained in the CART trees. On the other hand some of the terminal nodes discriminate a very limited number of samples; therefore the corresponding rules may introduce an excessive increase in complexity compared

Table 2. Performance on full test set and specific test sets.

	Full Test			Specific Test		
	(%)	Thr	Rules	(%)	Thr	Rules
L → H	89.70	0.52	22	93.58	0.58	7
H → L	84.31	0.5	42	87.45	0.48	11
SteadyState	90.13	0.44	4	96.14	0.56	3

to the additional discrimination capability they provide. Therefore, an investigation of the number of terminal nodes, and consequently the number of rules, which give the best results in terms of correct classification of samples has been performed. At the same time, an investigation of the optimal threshold value of the output, which maximizes the classification performance, has also been carried out.

The three FIS have been tested on the full test data set and on a subset of it, in particular to data taken from the specific interval used for training them. Table 2 summaries the results, showing the maximum percentage of success for the various developed FIS, together with the number of nodes taken into account to build the FIS and the threshold value which best discriminate between L and H mode samples. With regard to the maximum percentages of success achievable by the various FIS, it can be noticed that while SS and LH are comparable, the classifier tuned on the HL exhibits significantly lower performance. This can be due to a greater uncertainty in the H→L transition times contained in the database and, therefore, in a more uncertain classification of the samples in the neighborhood of the H→L transition. This confirms the long suspected fact that the H→L is less defined and more difficult to pin point with the measurements available. The consequent uncertainties can lead to both erroneous learning during the training phase and a wrong estimation of the results during the testing phase.

The automatic procedure, in addition to classifying the plasma confinement regime with a very high rate of success, can also provide an intuitive interpretation of the plasma behavior at the transition. Qualitative indications about the plasma dynamics can indeed be obtained by the rules devised in an automatic way by the proposed methodology. The rules produced for the various cases have been analyzed and considered quite realistic since they summaries properly a significant amount of expert knowledge in the field.

5. Conclusion

An automatic learning system for fuzzy inference system based on crisp classification trees has been applied to a nuclear fusion problem. The method achieves a good compromise between accuracy and interpretability. It is, indeed, able to reach about 90% of accuracy on the whole data set and up to the 96% when tested on specific regions of the data set. The derived FIS, on the other hand, presents a high grade of interpretability. According to [7], in order for a set of rules to be interpretable it should satisfy the following requirements: the fuzzy partition should be readable, the set of rules should be as small as possible and the rules should be incomplete. All of these requirements are satisfied by the proposed approach. Indeed, the fuzzy partition is simple since the number of fuzzy set is limited to three and readable since each fuzzy set is associated to a linguistic label. According to Table 2, the best performance is usually achieved when the number of rules is small and, even when the best performing network has an high number of rules, the reduction in accuracy, when a lower number of rules is used, is limited to just a few %. Finally, most of the rules involve just a few of the input variables leading to an incomplete rule set.

Acknowledgments

The work was partially supported by the project "Realtime visual feature extraction from plasma experiments for real time control," funded by ENEA-EURATOM, 2008.

References

1. A. Rizzo and G. Xibilia, *IEEE Trans. Contr. Syst. Technol.*, **10**, 421 (2002).
2. A. Murari *et al.*, *Burning Plasma Diagnostics: An International Conference*, **988**, 457 (2008).
3. J. Vega *et al.*, *Fus. Eng. Des.*, **83**, 132 (2008).
4. L. Breiman *et al.*, *Classification and Regression Trees* (Chapman & Hall , New York, 1984).
5. F. Wagner *et al.*, *Physical Review Letters*, **49**, 1408 (1982).
6. A.J. Meakins, *A Study of the L-H Transition in Tokamak Fusion Experiments*, PhD Thesis, Imperial College London, (2008).
7. S. Guillame, *IEEE Trans. on Fuzzy Syst.*, **9**, 426 (2001).
8. S.M. Zhou and J.Q. Gan, *Fuzzy Sets and Systems*, **159**, 3091 (2008).
9. G. Vagliasindi *et al.*, CART data analysis to attain interpretability in a Fuzzy Logic Classifier, in *Proc. of. Int.l Joint Conference on Neural Networks (IJCNN09)*, (Atlanta, U.S.A., 2009).
10. L.A. Zadeh, *Infor. Control*, **8**, 338 (1965).

INSPECTION OF DISRUPTIVE BEHAVIOURS AT JET USING GENERATIVE TOPOGRAPHIC MAPPING

G. A. RATTA* and J. A. VEGA

Asociación EURATOM/CIEMAT para Fusión
Avda. Complutense, 22. 28040
*E-mail: *giuseppe.ratta@ciemat.es*
www.ciemat.es

A. MURARI

Associazione EURATOM-ENEA per la Fusione
Consorzio RFX, Padova, Italy

G. VAGLIASINDI

Dipartimento di Ingegneria Elettrica Elettronica
e dei Sistemi-Universit degli Studi di Catania
Consorzio RFX, Padova, Italy

Tokamaks are the most promising configuration of magnetic confinement fusion devices. However, a physical phenomenon that leads the plasma out to its operational bounds, called disruption, remains unavoidable. Disruptions cause the abrupt termination of the discharge and in addition to affecting the execution of the research program, they can constitute a risk for the structural integrity of the machine.

In this article two important aspects that can facilitate the better understanding of the phenomenon are presented. First, the selection of the physical parameters and their main characteristics related to disruptions are reviewed. Second, the application of Generative Topographic Mapping (GTM) to visualize and compare disruptive and non disruptive experiments at different times is shown. The resulting maps are aimed to evidence the evolution of the phenomenon, since it is unrecognizable till it can be distinguished. The identification of the instant when precursors of disruptions can be noticed is highly relevant in nuclear fusion since it determines the time margin the control systems have to apply mitigation or avoidance actions.

Keywords: Nuclear fusion, JET, GTM, disruptions, feature extraction.

1. Introduction

Tokamaks are the most promising configuration of magnetic confinement fusion devices. Presently, the biggest and most important machine of this kind is the Joint European Torus (JET), located in Culham (UK). A dangerous physical phenomenon often occuring in tokamak operation is called disruption: the plasma (aheated and ionized gas) confinement is suddenly lost and in tens of milliseconds its energy content is transferred to the first wall. As a consequence of the subsequent plasma abrupt current quench, large eddy currents can be induced in the vacuum vessel and surrounding structures creating forces potentially capable of producing severe damage to the device.

To measure the physical quantities of interest inside the vacuum vessel, advanced sensors systems are attached to the device. Those systems transform the acquired quantities into electrical signals. This information can be used to detect unusual instabilities or disruptions precursors to notice in advance the phenomenon and consequently to apply control or mitigation actions to reduce the possible damages. In this article two important aspects that can facilitate the better understanding of the phenomenon are presented. First, the selection of the physical measurements [1–3] and their main characteristics related to disruptions are reviewed [4]. Second, the application of Generative Topographic Mapping (GTM) [5] to visualize and compare the evolution of disruptive and non disruptive experiments is detailed. This unsupervised method can be considered as visual proof of the evolution of the phenomenon at different time periods before its occurrence.

2. Feature extraction

2.1. *Introduction*

This section is devoted to detail the feature extraction procedure, aimed to provide the adequate input characteristics to the GTM algorithm. The procedure consist in two general tasks, detailed in Secs. 2.2 and 2.3. The first one explains the implementation of CART [1] to choose the set of plasma measurements that carries the disruptive related information. The second subsection is focused on the adequate processing of the selected signals to obtain the feature vectors.

2.2. *Selection of the plasma measurements*

The selection of the most informative physical quantities is fundamental to properly identify a disruptive activity. To this end, decision trees, as the Classification and Regression Trees (CART) [1] have been employed. They consist on tree shaped diagrams that represent a classification system or predictive model.

The algorithm traverses the database attempting for each input variable to find the value that splits the dataset into the two preconceived groups of discharges. This value is called 'split value'. Several criteria are available for determining the splits, as is detailed in [1].

The obtained tree has the more important variables towards the root and the ones with less explanatory power towards the terminal nodes. Starting with 35 signals (each one sampled at 2 kHz), this method allowed selecting the collection of the 13 most relevant ones. These signals are:

1. Plasma current. 2. Poloidal beta. 3. Poloidal beta time derivative. 4. Mode lock amplitude. 5. Safety factor at 95% of minor radius. 6. Safety factor at 95% of minor radius time derivative. 7. Total input power. 8. Plasma internal inductance. 9. Plasma internal inductance time derivative. 10. Plasma vertical position. 11.Plasma density. 12. Stored diamagnetic energy time derivative. 13. Net power (total input power minus total radiated power).

Three previous studies [2–4] agree that a condensed number of waveforms (between ten and thirteen, and mostly the same ones) is enough to describe the phenomenon without a significant loss of information. This first step is crucial to decrease the complexity of the problem by dividing almost by 3 the amount of waveforms to be taken into account.

2.3. *Creation of feature vectors*

Still, the reduction in the selection of measurements has not been enough. To attain good results it is necessary to minimize redundant or useless data and to highlight the disruptive-related information. The 13 selected plasma parameters present amplitudes which differ by several orders of magnitude. To assign similar weights to all the signals they have been normalized according to the formula:

$$\text{Normalized signal} = \frac{\text{Signal - Min}}{\text{Max - Min}} \tag{1}$$

where Min and Max, respectively, represent the minimum and maximum values of each signal in the dataset. On the other hand, it is possible to recognize by a visual inspection that disruptions are rather linked to higher frequency components in the signals. Consequently, in a previous study [4], two additional procedures have been followed. The first one consists in splitting every signal in time windows of 30 milliseconds. By this way, the analysis can be performed on the 30 ms time portions of the signal instead on the whole waveform. The second procedure is based on compressing the information of the 30 ms time windows of shot in a single feature vector.

The feature extraction procedure to achieve this second step can be summarized as follows:

Let's consider as an example the specific time window of experiment [−60 ms, −30 ms] before the disruption. The steps to attain the feature $\text{vector}_{[-60\text{ ms},-30\text{ ms}]}$ are:

(1) FFT of each [−60 ms, −30 ms] time window of signal (applied independently to each one of the 13 signals). The offset component of each one of the 13 obtained spectra are discarded. Only their positive parts are retained.
(2) The standard deviation of each one of the 13 retained spectra are calculated. As result, one value per [−60 ms, −30 ms] of signal is attained.
(3) Finally, the 13 values are concatenated, creating the feature vector.

Summarizing, the initial set of 35 signals per experiment, with a total of 525 samples per 30 ms of shot (0.5 samples per ms $\times$ 30 ms $\times$ 35 signals) have been condensed to feature vectors of 13 values per 30 ms of discharge. The dimensionality reduction is considerable (from 525 data to 13 features per time window). For each shot, this feature extraction procedure has been applied to the time windows from [−30 ms, 0] to [−360 ms, −330 ms] before the disruption.

3. Generative topographic maps for a visual identification of the phenomenon

To provide an estimation of how different the behaviour of a disruptive and a non disruptive experiment are, the feature vector collections of the entire database, for different time windows before the disruption, were input to the GTM algorithm [5]. The purpose of GTM is to find a configuration of data points in a low-dimensional space such that the proximity between objects in the full-dimensional space is represented with a high level of reliability by the distances between points in the low-dimensional space. This implies that the objects that are close together in the high dimensional space became points also placed closely in a bi-dimensional space. The GTM algorithm is based on the self-organized maps [6] but it provides some advantages. It uses a cost function (using the well known Expectation Maximization algorithm [7]) and provides convergence guarantees [5].

For this application, and due to the fact that there are two different sets of discharges (disruptives and non-disruptives), it would be expected to visualize two clearly different groupings in the data. As feature vectors are closer in time to the disruption, a more clear distinction between clusters would have to be shown. On the contrary much in advance of the disruption, the distinction is expected to completely disappear.

A reference time is required to compare disruptive and non disruptive shots. Consequently, a 'disruptive equivalent' time, for non disruptive discharges, has been calculated. It has been determined as 7 seconds after the plasma X-point is created, the statistically most probable time for a discharge to disrupt at JET. Then, all the time windows to be compared are linked either to the disruption time or the 'disruptive equivalent' time for shots that end safely.

The feature vectors belonging to the different time windows of the database are provided to the GTM. Then, a map per each time window under analysis, for the 220 disruptive and the 220 non disruptive shots, has been developed. Three of them have been plotted in Fig. 1. There, the grey circles represent the low dimensional mapping of the non disruptive discharges and the black squares symbolize the disruptive ones. It should be noticed that in the top graph the phenomenon is evident and therefore the bi-dimensional representation of the experiments shows two clear groups of data. Also, it can be appreciated that the safe discharges are more similar themselves at those times than the disruptive ones. This issue can be explained in physical terms, because the operation during safe experiments is usually restricted to several well-known parameters. Besides, many types of disruptions and not only one exist and the signals near these events present a wide range of behaviours.

The transition between clear separation and the overlapping of the points in the selected feature space corresponds to the interval $[-210, -180]$. Finally, far away in time from the phenomenon, the serials of disruptive and

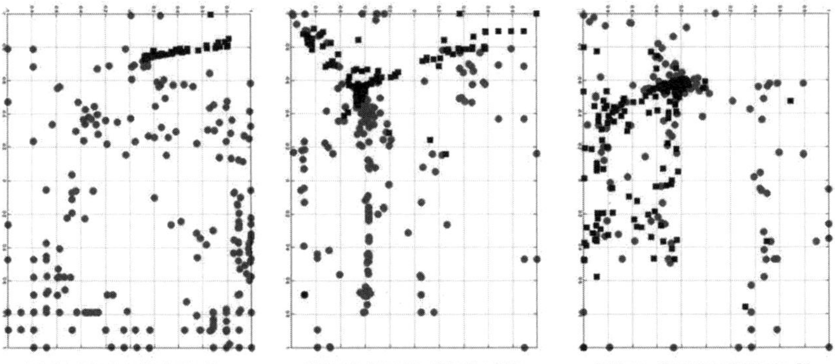

Fig. 1. 2D mapping of three different time windows before the disruption. Each map (in arbitrary units) represents the feature vectors of the 220 non disruptive (grey circles) and 220 disruptive (black squares) shots.

non disruptive shots are completely mixed, meaning that at those times even disruptive experiments behave as safe and therefore no clear distinction can be done between them.

4. Discussion

The results obtained with the GTM support the initial assumption that the closer in time the shots to the disruptions are the higher the differences between the physical parameters of the experiments are. Through this technique, an extremely high dimensional and complex phenomenon has been summarized in bidimensional plots where a simple visual inspection helps to understand the times when the behaviour of the disruptives discharges becomes more evident.

In spite of the dimensional reduction, this algorithm has the advantage of preserving the relative distances of the input data. To this end it can be also applied to any other high dimensional physical phenomena for the best comprehension of their evolution through a simple visual inspection.

References

1. L. Breiman, J.H. Friedman, R.A. Olshen and C.J. Stone. *Classification and Regression Trees* (CA:Wadsworth Inc.1993, New York, Chapman and Hall,1984).
2. A. Murari, G. Vagliasindi, P. Arena, L. Fortuna, O.Barana, M. Johnson. *Prototype of an adaptive disruption predictor for JET based on fuzzy logic and regression trees* (Nucl. Fusion 48 035010, 2008).
3. B. Cannas, F. Cau, A. Fanni, P. Sonato, M.K. Zedda. *Automated Disruption Classification at JET: Comparison of Different Pattern Recognition Techniques* (Nucl. Fusion 46 699-708, 2006).
4. G. A. Rattá, J. Vega, A. Murari, M. Johnson. *Feature extraction for improved disruption prediction analysis at JET* (Review of Scientific Instruments, Vol.79, Issue 10, 2008).
5. C. Bishop, M. Svensen, C. Williams. *The Generative Topographic Mapping* (Neural Computation 10, No. 1, pp. 215234, 1998).
6. T. Kohonen. *The self organizing maps* (Neurocomputing.21. pp.16, 1998).
7. W.L.Martinez and A.R.Martinez. *Exploratory data Analysis with Matlab* (Chapman and Hall, 2004).

MAXIMIZING RADIOFREQUENCY HEATING ON FTU VIA EXTREMUM SEEKING: PARAMETER SELECTION AND TUNING

D. CARNEVALE*, A. ASTOLFI† and L. ZACCARIAN‡

Dipartimento di Informatica, Sistemi e Produzione
Università di Roma, Tor Vergata
00133 Roma, Italy
E-mail: * *carnevale@disp.uniroma2.it*
† *astolfi@disp.uniroma2.it*
‡ *zack@disp.uniroma2.it*

L. BONCAGNI§, C. CENTIOLI¶, S. PODDA‖ and V. VITALE**

ENEA, Centro Ricerche Frascati
00044 Frascati (Roma), Italy
E-mail: § *boncagni@frascati.enea.it*
¶ *centioli@frascati.enea.it*
‖ *podda@frascati.enea.it*
** *vitale@frascati.enea.it*

In this paper we illustrate the use of a novel extremum seeking scheme recently proposed in [1] to minimize the percentage of reflected power on the Frascati Tokamak Upgrade (FTU) experimental facility during radiofrequency heating. The paper contains an explanation of how the parameters of the extremum seeking scheme should be selected to induce desirable closed-loop performance. The effectiveness of the tuning procedure will be shown via numerical simulations.

Keywords: Nonlinear control systems, optimization, Tokamak plasmas.

1. Introduction

Since the early 1950s the "extremum seeking" control has been introduced to minimize/maximize unknown functions at the output of dynamical systems (see [2] and [3]). In [4], for the first time, local stability properties of an extremum seeking feedback scheme for general nonlinear systems has been formally proved, motivating further interesting results (see [5], [6], [7]). Recently, in [8] an extremum controller slightly different from the one in [4] has been shown, under slightly stronger conditions, to formally guarantee non-local (semiglobal practical) stability properties.

An application that recently benefited from the use of extremum seeking techniques [9,9–11] is that of control of Tokamak plasmas, where radiofrequency heating seems to be the most promising and definitely the most widely experimented one. The reason why extremum seeking is needed in these applications is that there is an important coupling between the antenna and the plasma Scrape Off Layer (SOL), namely the outer surface of the plasma within the Tokamak vacuum vessel. If the coupling is poor, then reflected waves can damage the antenna and typically cause undesired safety shutdowns. Moreover, the effectiveness of the radiofrequency heating is evidently proportional to the coupling between antenna and plasma, because optimized coupling causes maximum absorbed power, therefore temperature increase.

In this paper we discuss about parameters tuning of the new extremum seeking scheme proposed in [1]. The choice depending on the plant and noise properties is discussed and shown via simulation examples The paper is organized as follows. In Sec. 2 the control scheme is recalled and some general ideas on the parameters tunings are outlined. Section 3 illustrates by two examples the extremum seeking construction and new phenomena related to the parameter selection. Conclusions are given in 4.

2. The control scheme

In this section we recall the control scheme proposed in [1], whose aim is to find a reference signal for a dynamical system such that an unknown function of its output is minimized. Due to space constraint, we please the reader to refer to [1] for all the details related to the control scheme in Fig. 1 that we consider in the sequel. The unknown map is $g(\cdot)$, with input y and d, the output of the first order linear dynamical system and the disturbance signal, respectively. The parameter $\varepsilon > 0$ sets the convergence speed of y to $\sigma\theta$, where $\delta > 0$ is the static gain of the linear plant. The noises ν_1 and ν_2 affect the measurements which are filtered by two SISO systems $F(s)$. The output of a unit saturation is fed with the signal $k_2 z_1(t) z_2(t)$ and is integrated and multiplied by k_1, yielding the plant reference $\theta(t)$, with positive scalars k_1 and k_2. Note that the saturation block in the feedback loop limits $\dot{\theta}$ below k_1. This is an appealing property for "risky" plants or when rapidly changing signals may excite high frequency dynamics. Moreover, this approach allows to meet rate saturation constraints of the actuators.

Note that the first order plant that we are considering can be the approximation of a higher order asymptotically stable system. The output difference between the real plant and its approximation may be enclosed in the signal $d(t)$. The main Theorems in [1] state the effectiveness of the control scheme to find the reference signal θ that minimizes $g(\cdot)$.

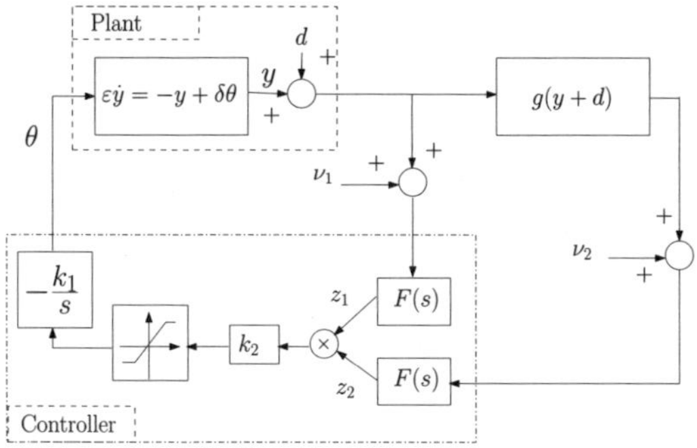

Fig. 1. The dynamic extremum seeking scheme.

The filter

$$F(s) = \frac{s}{(\tau_1 s + 1)^n}, \tag{1}$$

with $n \geq 2$, is considered in place of the ideal derivative (see [1]). The aim of this filter is twofold and is a key ingredient of the scheme: it has to estimate the time derivatives of the input and the output of $g(\cdot)$, to resemble the filters in the ideal case, and it has to filter out the measurements noises ν_1 and ν_2. To match these requirements, it is necessary that a "frequency" separation between (at least some components of) the signal $d(t)$ and the measurement noises exits. These considerations lead to choosing the filter parameters so that $F(s)$ approximates a time derivative action in the frequency range of (the useful components of) $d(t)$, *i.e.* when $\omega << 1/\tau_1$, and it is a low-pass filter at higher frequencies so as to filter out the measurement noise by tuning the value of $n \geq 2$ and resulting in a sharp band-pass filter.

Note that unlike the classical extremum seeking scheme, the "probing" signal $d(t)$ does not need to be sinusoidal but simply needs to satisfy the persistence of excitation condition.

3. Tuning the extremum seeking parameters

In this section, we show by means of two different examples how the control parameters k_1, k_2, τ_1, and n can be chosen to drive y close to $y^\star$. First note that k_1 multiplies the filter outputs $z_1(t)$ and $z_2(t)$ after the saturation, so it is associated with the large signal behavior and also corresponds to the maximum time derivatives of $\theta(t)$. Conversely, k_2 multiplies the signals before the saturation, so it is associated with the small signal behavior and its effect is negligible for large signals where the saturation is active. k_2 can then be seen as the square root of the static gain of $F(s)$. Increasing values of the integer parameter n increase the steepness of the Bode diagram for $\omega > 1/\tau_1$, resulting in a stronger low-pass action.

In this first example we consider the unknown function $g(y) = (y-4)^2$, $y(0) = 0$, $\varepsilon = 0.01$, $\delta = -2$, $d(t) = 0.05\sin(\sqrt{2}t) + 0.02\sin(30\sqrt{3}t)$ and no measurement noise, $\nu_1 = 0$ and $\nu_2 = 0$.
Since no noise is affecting the system, we may select the filter to resemble a time derivative as much as possible, so we set $\tau_1 = 10^{-4}$. With this choice, the filter is able to approximate the derivative of the two terms of $d(t)$. The simulation results are shown in Fig. 2 for different values of $k_1 = \{1, 2, 10\}$ and fixed $k_2 = 1$. It is clear the following role of the two gains k_1 and k_2:

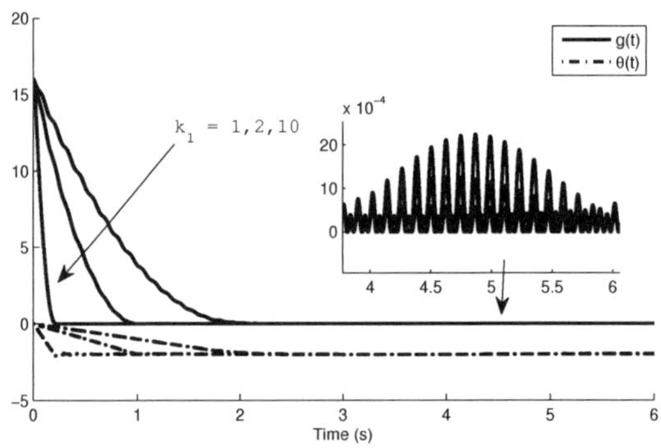

Fig. 2. First example: simulation results. $g(y(t))$ (solid) and $\theta(t)$ (dash-dotted) $k_1 = \{1, 2, 10\}$ and $k_2 = 1$.

k_1 strongly increases the convergence rate of y to $y^\star$ and $|\dot{\theta}| \leq k_1$, whereas k_2 acts more like a "magnifier" to converge to the minimum when z_1 and z_2 are small. Generally, the greater ε is, the smaller k_1 should be.

It is also interesting to analyze the case with $\tau_1 = 0.05$ depicted in Fig. 3, with $k_1 = k_2 = 1$. In this case the filter does not perform a sufficient approximation of $\dot{d}(t)$, then the feedback system induces oscillations which themselves are interpreted by the controller as the "probing signal". Those oscillations have lower frequency than those of d, and the filter is able to perform a slightly better approximation of their derivative. Therefore, the system starts to converge towards the minimum. This is an interesting property of this approach: any signal which is feed into the filter as long as its time derivative can be approximated with sufficient precision, can be regarded as an eligible "probing" signal.

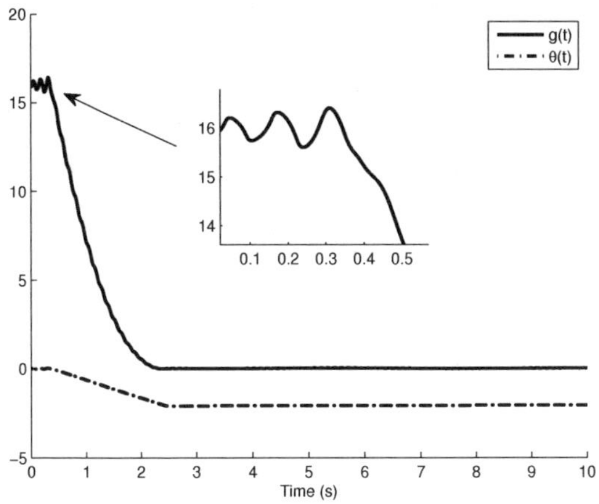

Fig. 3. First example: simulation result with $\tau_1 = 0.05$ and $k_1 = k_2 = 1$.

In the second example we consider a measurement noise given by

$$\nu_1(t) = w_1(t) + 0.05\sin(60t), \tag{2}$$

$$\nu_2(t) = w_2(t) + 0.05\sin(150t), \tag{3}$$

where w_1 and w_2 are band-limited white Gaussian noises with zero mean and power $2e-5$. Note that the second component of $d(t) = 0.05\sin(\sqrt{2}t) + 0.02\sin(30\sqrt{3}t)$ has almost the same frequency of the sinusoidal component of the noise ν_1. However, we may select $\tau_1 = 0.01$ obtaining a good approximation of the derivative of the first component of $d(t)$ and filtering out what remains. Finally, we try to exploit the self excitation property selecting $\tau_1 = 0.05$, $k_1 = 1$, $n = 6$ with $\varepsilon = 0.01$. The results are shown in Fig. 4.

4. Conclusions

In this paper we conveyed how the controller parameters of the new extremum seeking scheme proposed in [1] can be selected to induce desirable closed-loop performance. It has been shown how the filter can be chosen depending on the property of the disturbance affecting the nonlinearity input signal and of the noise affecting the measurements. The relation between

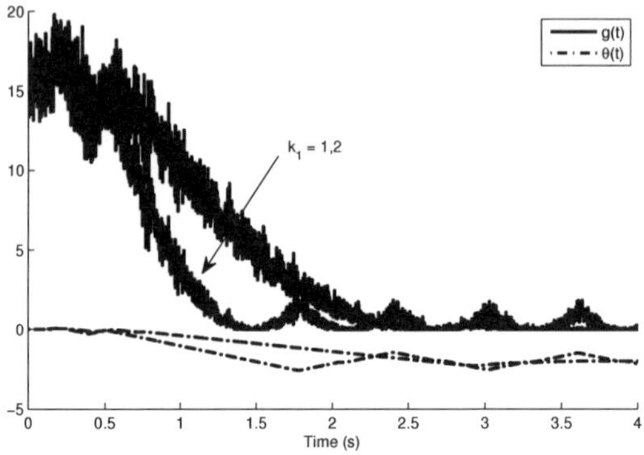

Fig. 4. Second example: decreasing k_1 from 2 to 1 to get closer to the minimum dealing with small $1/\tau_1$ and $n = 6$.

the controller gains k_1, k_2 and the filter parameter τ_1 has been discussed, highlighting the property of self excitation which may lead to improved convergence even when the approximated derivative of the signal $d(t)$ can hardly be evaluated.

References

1. D. Carnevale, L. Zaccarian, A. Astolfi and S. Podda, Extremum seeking without external dithering and its application to plasma rf heating on ftu, in *Proc. 4tth IEEE Conf. Decision and Control, Cancun, Mexico*, 2008.
2. C. Drapper and Y. Li, *ASME* **160**, 1 (1951).
3. I. Morosanov, *Automation and Remote Control* **18**, 1077 (1957).
4. M.Krstić and H.H.Wang, *Automatica* **36**, 595 (2000).
5. K. Ariyur and M. Krstic, *Real-Time Optimization by Extremum-Seeking Control* (Wiley-Interscience, 2003).
6. M. Guay, D. Dochain and M. Perrier, *Automatica* **40**, 881 (2004).
7. K. Peterson and A. Stefanopoulou, *Automatica* **40**, 1063 (2004).
8. Y. Tan, D. Nešić and I. Mareels, *Automatica* **42**, 889 (2006).
9. L. Zaccarian, C. Centioli, F. Iannone, M. Panella, L. Pangione, S. Podda and V. Vitale, *Fusion Engineering and Design* **74**, 543 (2005).
10. V. Vitale, A 10khz feedback control system for plasma shaping on FTU, in *11th IEEE NPSS Real Time Conference*, June 1999.
11. C. Centioli, F. Iannone, G. Mazza, M. Panella, L. Pangione, S. Podda, A. Tuccillo, V. Vitale and L. Zaccarian, *Control Engineering Practice* (2008, to appear).

EQUINOX: A REAL-TIME EQUILIBRIUM CODE AND ITS VALIDATION AT JET

D. MAZON

Association EURATOM-CEA, CEA Cadarache
DSM/IRFM 13108 St Paul Lez, Durance Cedex, France

J. BLUM, C. BOULBE and B. FAUGERAS

Laboratoire J-A Dieudonn (UMR 66 21)
Universit de Nice Sophia-Antipolis
CNRS Parc Valrose 06108 Nice Cedex 02, France

M. BARUZZO and A. BOBOC

Association EURATOM/UKAEA
Culham Science Centre Abingdon Oxon OX14
3DB, United Kingdom

S. BREMOND

Association EURATOM-CEA, CEA Cadarache
DSM/IRFM 13108 St Paul Lez, Durance Cedex, France

M. BRIX, P. DE VRIES, S. SHARAPOV and L. ZABEO

Association EURATOM/UKAEA
Culham Science Centre Abingdon Oxon OX14
3DB, United Kingdom

and JET-EFDA Contributors*

The real-time reconstruction of the plasma magnetic equilibrium in a Tokamak is a key point to access high performance regimes. Indeed, the shape of the plasma current density profile is a direct output of the reconstruction and has a leading effect for reaching a steady-state high performance regime of operation. We have seen in particular that non monotonic current density profiles can trigger enhanced particles and heat confinement. On top of this, the current density profile has a resistive diffusion time and any variation of the current drive systems takes some time to be efficient. The challenge is thus to develop methods and algorithms that reconstruct the magnetic equilibrium in the perspective to use these outputs for feedback control purposes.

*See the Appendix of F. Romanelli *et al.*, Proceedings of the 22nd IAEA Fusion Energy Conference 2008, Geneva, Switzerland.

1. Introduction

In present days tokamaks only the shape of the plasma boundary is routinely identifiable in real-time in less than few milliseconds using a set of magnetic and diamagnetic coils spread around the vessel. This information is mainly used for controlling the plasma shapes in real-time during a plasma discharge using coils current in a feedback control loop. In JET the so-called XLOC code is used routinely for plasma shape control [1]. But with this algorithm it is not possible to compute the internal magnetic flux configuration which is needed if we want to analyze the phenomenon occurring in the interior of the plasma. In this case the only way to get access to the current density profile is to use off-line codes that can compute accurately the profile but with no possibility to act in real time on it. This is rather a strong limitation because we know from the analysis performed that the shape of the current density profile is one of the key element to enhance the plasma performance [2] and insure stability but also performance [3,4].

2. Mathematical formulation of the plasma equilibrium

The problem of plasma equilibrium in a Tokamak is a free boundary problem in which the plasma boundary is defined as the last closed magnetic flux surface. Inside the plasma, the equilibrium equation in an axisymmetric configuration is called the Grad-Shafranov equation [5,6]. This equation is derived from the combination of the magnetostatic Maxwell's equations which are satisfied in the whole of space in presence of a magnetic field and the equilibrium of the plasma itself which occurs when the kinetic pressure is equal to the Lorentz force of the magnetic pressure. The expression of the Grad-Shafranov equation in a cylindrical coordinates system $(r,\ z,\ \phi)$ where $r = 0$ is the major axis of the torus reads:

$$-\Delta^* \psi = rp'(\psi) + \frac{1}{\mu_0 r}(ff')(\psi) \tag{1}$$

with

$$\Delta^* = \frac{\partial}{\partial r}\left(\frac{1}{\mu_0 r}\frac{\partial}{\partial r}\right) + \frac{\partial}{\partial z}\left(\frac{1}{\mu_0 z}\frac{\partial}{\partial z}\right) \tag{2}$$

Where μ_0 is the magnetic permeability of the vacuum, $\psi(r, z)$ the poloidal flux and f the diamagnetic function. Assuming that Dirichlet boundary conditions, h, are given on Γ which is the poloidal cross section of the vacuum vessel, the final equations governing the behavior of $\psi(r, z)$ inside the vacuum vessel become:

$$\begin{cases} -\Delta^*\psi = \left[\frac{r}{R_0}A(\bar{\psi}) + \frac{R_0}{r}B(\bar{\psi})\right] \\ \psi = h \quad \texttt{on } \Gamma \end{cases} \tag{3}$$

Defining

$$A(\bar{\psi}) = R_0 p'(\bar{\psi}) \tag{4}$$

and

$$B(\bar{\psi}) = \frac{1}{\mu_0 R_0}(ff')(\bar{\psi}) \tag{5}$$

the entire problem is thus resumed to identify in real time the plasma current i.e. the non linear functions A and B (function of the normalized flux in the previous equation).

3. The Equinox code

In order to meet the real-time requirements, a new version of the code called Equinox has been design and implemented in C++ using a finite element method and a non linear fixed point algorithm associated to a least square optimization procedure. The code relies on tokamak specific software like XLOC providing flux values on the first wall of the vacuum vessel. By means of least-square minimization of the difference between measurements and the simulated ones the code identifies the source term of the non linear Grad-Shafranov equation. The experimental measurements that enable the identification are the magnetics on the vacuum vessel, the interferometric and polarimetric measurements on several chords and the motional Stark effect measurements. For the magnetic measurements the flux loops give the poloidal flux on particular nodes Mi such that $\psi(M_i) = h_i$ on Γ. Thanks to an interpolation (performed by XLOC at JET) between the points M_i these measurements provide the Dirichlet boundary condition h. The problem is thus resumed to find a solution that minimizes the cost function defined as:

$$J(A, B, n_e) = J_0 + K_1 J_1 + K_2 J_2 + J_\varepsilon \tag{6}$$

with

$$J_0 = \sum_i \left(\frac{1}{r}\frac{\partial\psi}{\partial n}(N_i) - g_i\right)^2 \tag{7}$$

$$J_1 = \sum_i \left(\int_{C_i} \frac{n_e}{r}\frac{\partial\psi}{\partial n}dl - \alpha_i\right)^2 \tag{8}$$

$$J_2 = \sum_i \left(\int_{C_i} n_e dl - \beta_i \right)^2 \tag{9}$$

where g_i, α_i and β_i are respectively the measurements of the magnetic poloidal field, the Faraday rotation and the line integrated density along the chords C_i. The weighting parameters K_1 to K_2 enable to give more or less importance to the corresponding experimental measurements [8]. As the inverse problem of the determination of A and B is an ill posed one a Tikhonov regularization term [9] J_ε constrains the functions A, B and n_e (n_e being the plasma density) to be smooth enough and its expression is given by the following expression:

$$J_\varepsilon = \varepsilon_1 \int_0^1 [A''(x)]^2 \, dx + \varepsilon_2 \int_0^1 [B''(x)]^2 \, dx + \varepsilon_3 \int_0^1 [n_e''(x)]^2 \, dx \tag{10}$$

where ε_1, ε_2 and ε_3 are the regularizing parameters. Equation (3) is solved using a finite element method [10]. A careful implementation leads to execution time less than 60ms per iteration on a 2GHz PC, complemented with excellent robustness. The unknown functions A, B, n_e are approximated by decomposition in a reduced basis. The basis can be made of different types of functions (polynomials, B-splines, wavelets etc) [11]. In our case we choose B-splines. The Picard type (fixed point) algorithm is then used to solve iteratively the inverse and direct problem.

4. Equinox validation

The validation of the Equinox code has been performed starting from a database of about 130 pulses, well representative of the JET discharges with different shape and triangularity of the plasma boundary and with global parameter varying in the whole JET interval. For some pulses clear MHD signatures have been identified and help in particular at the validation of the current density profile. The validation of the Equinox version constrained by magnetic measurements has been done mainly using the results of the well assessed EFIT equilibrium code [12]. The shape of the plasma is perfectly reproduced. Global quantities like the internal inductance li showed some differences the medium error value being around 0.1beeing of the order of the error bars on the results. To fully assess the Equinox reconstruction we have used PROTEUS [13] that solves the direct problem of the Grad Shafranov equation. The idea is to compute the flux mapping starting from a given and known current density profile. In that particular case a monotonic current density profile was chosen, the equilibrium has been

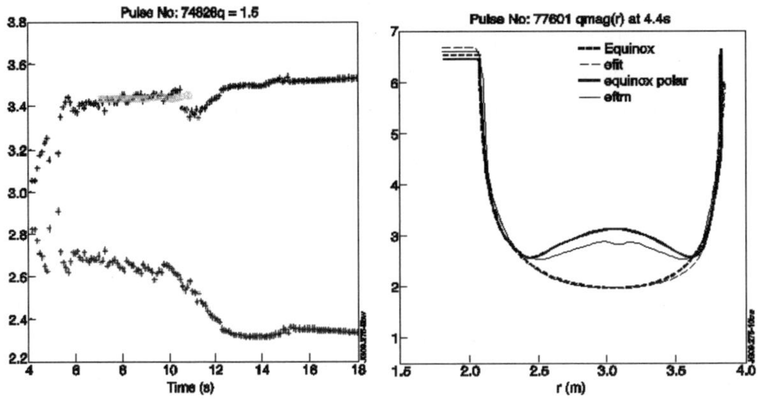

(a) (b)

Fig. 1. (a) Time traces of the location of the $q = 1.5$ surface as found by Equinox (low field side and high field side) and in pale grey location of that mode deducted from Fourier analysis of Magnetic and ECE data. (b) Profiles comparison between EFIT constraint by magnetic measurements (efit) by MSE (eftm) and Equinox constraint by magnetic measurements (equinox), by MSE (equinox polar) for shot 77601 at t=44.4s.

reconstructed by PROTEUS who computed also the boundary conditions requested by Equinox. The outputs of Equinox are then compared with the ones coming from PROTEUS. A very good agreement is found. In the case of the Equinox constrained with polarimetry measurement the same results were obtained for the validation of the plasma shape and position, which were not modified by the inclusion of internal measurements. More interesting was to validate the obtained current density profile. The first strategy was to use clear MHD signatures of some shots of the database for checking the location of the corresponding mode. An example can be seen in Fig. 1 where the location of the $q = 1.5$ mode is given by Equinox (high and low field side) and also identified (low field) from Fourier analysis of the magnetic measurements and electron temperature (Electron Cyclotron Emission) during a shorter period of time. The agreement is almost perfect. The second strategy, see Fig. 2 for example, was for the other shots of the database to compare the Equinox reconstruction with some reconstruction using EFIT constraint by MSE. In dotted lines are represented for the same shot at the same time the profiles obtained with magnetic only. Here again the agreement is very good.

Acknowledgments

This work, supported by the European Communities under the contract of Association between EURATOM and CEA, was carried out within the framework of the European Fusion Development Agreement. The views and opinions expressed herein do not necessarily reflect those of the European Commission.

References

1. D.P O'Brien, Nuc. Fus, Vol 33 (3) p331 (1993).
2. X. Litaudon, Nuclear Fusion, Vol.43, No.7, July 2003, p.565-572.
3. D..Mazon, PPCF, Vol.45, No.7, July 2003, p.L47-L54.
4. D. Moreau, Nuclear Fusion, Vol.48, No.10, October 2008, p.106001.
5. H.Grad, 2nd U.N conference on the peaceful uses of Atomic energy, Geneva 1958 Vol 31 pp 190-197.
6. V. Shafranov, Soviet Physics JETP 6 1013 (1958).
7. H.Grad Physical Review Letter 24 pp 1337-1340 (1970).
8. J. Blum, Nuclear Fusion 30 1475 (1990).
9. A. Tikhonov, "Solutions of ill-posed problems" Winston, Washington D.D (1977).
10. P. Ciarlet, "The finite element method for elliptic problems", North Holland, (1980).
11. J. Blum, IMA Volumes in Mathematics, Part 1, Biegler, Coleman, Conn and Santosa, 1997, Vol 92 pp 17-36.
12. L. Lao, Nuclear Fusion 30 1035 (1990).
13. Albanese R, - 12th Conference on the Numerical Simulation of Plasmas, S. Francisco, Sept. 1987.

PLASMA SCENARIOS AND MAGNETIC CONTROL IN FAST

G. ARTASERSE[1,*], F. MAVIGLIA[1], R. ALBANESE[1], G. AMBROSINO[1], G. CALABRÒ[2], V. COCILOVO[2], F. CRISANTI[2], A. CUCCHIARO[2], M. MATTEI[1], G. MAZZITELLI[2], A. PIRONTI[1], A. PIZZUTO[2], G. RAMOGIDA[2], C. RITA[2] and F. ZONCA[2]

[1] *Associazione Euratom-ENEA-CREATE sulla Fusione*
Via Claudio 21, 80125 Napoli, Italy

[2] *Associazione Euratom-ENEA sulla Fusione*
Via Fermi 45, 00044 Frascati, Italy
*E-mail: * giovanni.artaserse@unirc.it*

The Fusion Advanced Studies Torus (FAST) conceptual has been proposed as possible European ITER Satellite facility with the aim of preparing ITER operation scenarios and helping DEMO design and R&D. Insights into ITER regimes of operation in Deuterium plasmas can be obtained from investigations of nonlinear dynamics that are relevant for the understanding of alpha particle behaviours in burning plasmas by using fast ions accelerated by heating and current drive systems. In this paper the plasma scenarios that can be studied on FAST are presented, with emphasis on the aspect of its flexibility in terms of both performance and physics that can be investigated. Plasma position and shape control studies are also discussed.

Keywords: Tokamak, plasma equilibrium codes, SVD, PID.

1. Introduction

FAST has been proposed to help preparation of ITER scenarios [1], [2] and the development of new expertise for DEMO design and R&D in an integrated fashion, simultaneously addressing many aspects of non linear dynamics that are relevant for the understanding of alpha particle behaviours in burning plasmas and their interaction with plasma turbulence and turbulent transport, exploiting advanced regimes with long pulse duration with respect to the current diffusion time and up to full non-inductive current driven (NICD), testing technical innovative solutions for the first wall/divertor directly relevant for ITER and DEMO, and providing a test bed for ITER and DEMO diagnostics as well as an ideal framework for model and numerical code benchmarks, verification and validation in ITER and DEMO relevant plasma conditions. FAST equilibrium configurations have been designed in order to reproduce those of ITER with scaled plasma current, but still suitable to fulfil plasma conditions for studying operation problems, plasma wall interaction and burning plasma physics issues in an integrated framework. The paper is arranged as follows: the FAST load assembly and equilibrium configurations and the plasma scenarios are presented in Sec. 2, the design of the main parameters of the magnetic control system of FAST is described in Sec. 3.

2. FAST load assembly and equilibrium configurations

FAST has been designed to achieve a plasma behaviour sufficiently close to ITER together with a significant flexibility in the operation space [2]. The maximum plasma current, I_P, ranges from 2 MA in the full NICD scenario to 8 MA in the high performance H-mode scenario, while the pulse duration can extend up to 160 s ($\sim$ 40 resistive times τ_{res}) in the longest AT (Advanced Tokamak) scenario at 3 MA/3.5 T. This features have been accomplished in a compact (major radius $R = 1.82$ m, minor radius $a = 0.64$ m, triangularity $= 0.4$), cost effective design with a complete set of plasma diagnostics and a full choice of auxiliary heating systems (ICRH - Ion Cyclotron Resonance Heating, LH - Lower Hybrid Heating , ECRH - Electron Cyclotron Resonance Heating and Negative Ion Neutral Beam Injection - NNBI). This design permits at the same time and in integrated framework, non linear dynamics effects in the fast particles behaviours [1] typical of the burning plasma physics, plasma-wall interaction under ITER relevant power load and ITER relevant operational issues, including the AT regimes up to fully NICD scenarios. FAST is based on a Load Assembly shown in Fig. 1(a) and consisting in a vacuum vessel (VV) with its internal

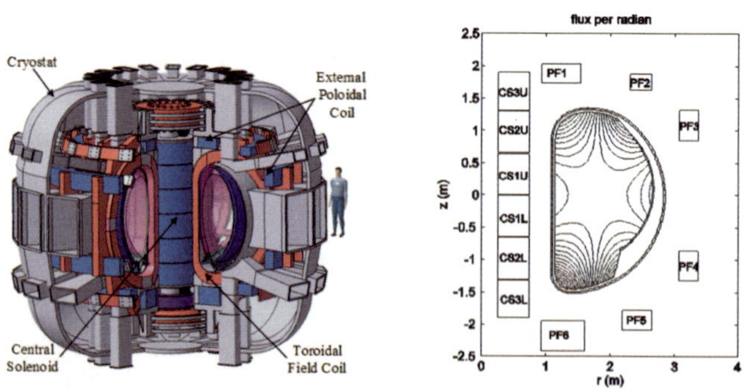

Fig. 1. (a) The load assembly of FAST. (b) Poloidal Field Coils system and field null region during the plasma breakdown.

components (First Wall, Divertor, Stabilizing Shell), 18 Toroidal Field Coils (TFC) designed to keep the whole toroidal magnet structure in wedged configuration, 6 External Poloidal Field Coils (PFC) and a Central Solenoid (CS) vertically segmented in 6 coils to increase plasma shaping capability, manufacture easiness and cooling efficiency. Copper has been chosen as material for all the coils (TFC, PFC and CS) to limited the machine cost. The adiabatic heating during the plasma pulse have been minimized keeping all the coils at cryogenic temperatures (30 K). The VV is a single shell 25 to 40 mm thick, made of Inconel 625 steel to minimise the flux consumption during the plasma start-up and to guarantee the stiffness requested to withstand the loads arising from plasma disruptions. The access to the VV is provided through vertical, oblique and equatorial ports in each 20 degree module. A passive stabilizing shell, consisting of 26 mm thick toroidally segmented copper plates, has been envisaged on the outboard, inside the VV, to make easier the vertical control of the plasma by slowing down the growth rate of the vertical instability around 13 s^{-1}. A set of ferromagnetic inserts, located in front of TFC inside the VV on the outboard between the stabilizing shell and the VV itself, have been designed to keep the Toroidal Field Ripple (TFR) below 0.3% on the plasma separatrix. The plasma facing components in FAST should withstand very high thermal fluxes relevant to ITER and DEMO Plasma Wall Interactions regimes (for more details [3]). FAST could offer a full choice of auxiliary heating systems: 30 MW of ICRH able to accelerate plasma ions in the range $0.7 \div 0.8$ MeV, 4 MW of ECRH for MHD control, heating, current drive at low density and 6 MW of LH for the current drive and profile control. Moreover an additional 10 MW NNBI system could be used to accelerate plasma ions in the extreme scenarios [1]. The poloidal magnetic field system has been optimized, regards to locations and sizes of the coils (Fig. 1(b)), to minimize the stored magnetic energy and the copper temperature rise while maximizing the available magnetic flux swing and the extension of the field null during the plasma break-down (a very large central hexapolar region as shown in Fig. 1(b)). FAST has been designed as a very flexible device able to reproduce, with scaled plasma current, the three main ITER equilibrium configurations: standard H-mode with broad pressure profile, hybrid mode with narrower pressure profile and AT scenario with peaked pressure profile. FAST could then work in a dimensionless parameter range close to ITER, with similar equilibrium profiles, dominant electron heating (with $T_e \sim 10$ keV) and plasma performance in the fusion parameter space with $Q \geq 1$. An outline of the possible plasma configurations analysed for FAST is given below. All these plasma configurations have the same geometrical features, as shown in Fig. 1(b) (for further details [3]). Sufficient flexibility is maintained to allow different plasma shapes, triangularity scan and strike point sweeping. The duration of the plasma flat-top is significantly greater than the resistive diffusion time in all these scenarios, with a ratio $\Delta t_{\text{flat-top}}/\tau_{res}$ about half than in ITER. The reference H-mode scenario ($I_P = 6.5$ MA, $B_T = 7.5$ T) has been designed to be used in the extensive integrated studies and is characterized by a high density ($n = 2\times10^{20}$ m^{-3}), high current plasma ($I_p = 6.5$ MA) with a X-point equilibrium that can be sustained as long as the magnetic flux sweeping allows (13 s of high current flat-top i.e. about two resistive decay time, in a discharge lasting around 22 s and assuming 32 MA$/m^2$ as maximum current density in the PFC). The time evolution (see [3]) is characterized by a rise of the plasma current with a circular equilibrium up to $I_p = 2$ MA in 1.5 s after the break-down, an increase of the elongation while the current keep raising for the next 3 s when the final X-point shape configuration is achieved with $I_p = 4.5$ MA and then a further increase of the current until the current achieves its target value $I_p = 6.5$ MA at $t = 7$ s. The full additional heating, applied at $t = 7.5$ s, causes a large increase of the internal kinetic energy (and then of β_N) on a time scale (about 1 s) longer than the plasma energy confinement time: during this increase the plasma boundary shall be preserved by using a technique like the Extreme Shape Controller (XSC) adopted in JET [4]. The extreme H-mode scenario ($I_P = 8$ MA, $B_T = 8.5$ T with a safety factor $q_{95} \sim 2.6$) corresponds to the highest achievable performance in terms of Q, by assuming the use of the additional NNBI system. In this transient scenario ($\tau_{\text{flat-top}} = 2$ s with $\tau_E \sim 0.7$ s and $\tau_{res} \sim 5$ s) a large species coupling (and then $T_e = T_i$) is foreseen to happen due to the high plasma density close to the Greenwald limit. Several Hybrid and AT scenarios can be achieved by FAST, with quite different features: Hybrid ($I_P = 5$ MA, $B_T = 7.5$ T, able to reach $Q \sim 1$ with $\beta_N \sim 2$ and $n/n_GW = 0.8$), AT (large $B_T = 6$ T with moderate $\beta_N \sim 2$ and plasma current $I_P = 3$ MA driven by LH for 22% and bootstrap for 38%), AT2 ($I_P = 3$ MA, lower $B_T = 3.5$ T with $\beta_N \sim 3.2$ greater than the MHD stability), full NICD ($B_T = 3.5$ T with very large $\beta_N \sim 3.4$, density $n = 1 \times 10^{20}$ m^{-3} and fully non inductively driven plasma current $I_p = 2$ MA consisting of 60% $\div$ 70% bootstrap and 30% $\div$ 40% LH driven fractions). In all these configurations the plasma shape is the same as the reference H-mode. The plasma discharge is sustained in the AT scenarios as long as poloidal flux is available to drive the residual inductive current ($t_{\text{flat-top}} \sim 60$ s in AT and $t_{\text{flat-top}} \sim 160$ s in AT2 assuming a plasma residual resistivity about $60 \div 100$ mV) while in the NICD scenario the duration is constrained by the TFC adiabatic heating ($t_{\text{flat-top}} \sim 160$ s). In all these cases the discharge lasts quite longer than the resistive time (from 20 up to 40 times).

3. Plasma position and shape control

The control of the plasma position and shape is a crucial issue as in every compact, elongated and high performance tokamak as FAST. The capability of the PFC system, as presently designed, to provide an effective vertical stabilization of the plasma has been investigated using the CREATE_NL response model [5], assuming axisymmetric deformable plasma described by few global parameters. The plasma chamber has been schematized as a Inconel 625 vessel, 25 mm thick, with a resistivity equals to 1.29 $\mu\Omega$m at operating temperature: the resulting torus resistance is 62.6 $\mu\Omega$, neglecting the $3D$ effects of the ports. A stabilizing copper shell inside the vacuum vessel (Fig. 1(b)) has been designed, optimizing its thickness (26 mm) and location to provide a slowing down of the growth rate of the vertical instability around 13 s^{-1}. To avoid flux shielding during plasma breakdown the shell has been toroidally segmented, providing the up-down connection by the poloidal path around the ports, so the net total toroidal current flowing in it is zero. Preliminary analyses have been performed to study the control of the plasma current, shape and position during the flat-top of the reference H-mode plasma scenario. The structure of the proposed controller [3] consists in a feedback loop which controls the derivative of the vertical position and a slower multivariable feedback loop, which controls the plasma current, shape and position. The closed loop system guarantees that, in the presence of a disturbance, the plasma vertical velocity goes to zero, while the plasma vertical position of the current centroid is not recovered. The current and shape controller structure uses as controlled variables, besides plasma current, six linear combinations of 39 gaps (between the plasma separatrix and the plasma facing components), strike points and X-point descriptors, obtained using a SVD (Singular Value Decomposition) approach based on PID (Proportional Integrative Derivative) control scheme. Consequently the controller also requires the measures of the PF coil currents. The power supply system has been modelled, in a conservative approach as a pure time delay of 10 ms: under this assumption the vertical stabilization controller and the power supplies voltage limits have been designed so as to guarantee a settling time for the plasma velocity of about 800 ms. As far as the current/shape disturbances rejection (recovery of the gaps within 1 cm), the controller has been evaluated simulating the system response to a 1 cm plasma Vertical Displacement Event (VDE), to a 100 kA step in the plasma current and to a minor disruption characterized by a 20% fall in internal inductance and poloidal beta. In all these cases, the recovery is guaranteed by the current/shape controller as presently designed, with a settling time less than 2 s. The maximum power required for this stabilization is about 14 MW, in the range of the capabilities of the designed PFC system, with the most demanding disturbances, VDE and minor disruption.

4. Conclusion

The Fusion Advanced Studies Torus (FAST) has been designed to provide a European ITER Satellite facility able to explore Fast Particle Physics, to investigate ITER relevant Plasma Operations issues, to study the physics and test the technologies required to deal with large heat loads on ITER and DEMO plasma facing components, to investigate long lasting AT regimes up to fully non inductive scenarios, to validate numerical simulation codes predictions of ITER fusion and burning plasma performance. FAST will be able thus to support the preparation of ITER operation scenarios by using fast ions accelerated by heating and current drive systems, working with deuterium plasmas in a dimensionless parameter range close to that of ITER. The FAST flexibility in terms of both performance and physics that can be investigated is emphasized by the variety of plasma scenarios that can be studied, from the extreme high performance H-mode to the full NICD scenario. The feasibility of a proper plasma position and shape control with the current Poloidal Field system design has been also introduced, showing the possibility of guaranteeing a wide stability region and of rejecting undesired shape modification.

References

1. Pizzuto on behalf of the Italian Association, "The Fusion Advanced Studies Torus (FAST): A Proposal for an ITER Satellite Facility in Support of the Development of Fusion Energy", Proc. 22nd IAEA Fusion Energy Conference, Geneva, Switzerland, October 13 - 18, 2008; submitted to Nucl. Fusion.
2. The FAST Team, "FAST Conceptual Study", Technical Report ENEA/FPN-FAST-RT-07/001, Frascati, Italy, 2008.
3. G. Calabrò, et al., proceedings of 4th International Scientific Conference on Physics and Control - PhysCon2009.
4. R. Albanese, et al., Fus. Eng. Des. 74 (2005) 627-632.
5. R. Albanese and F. Villone, Nucl. Fusion 38 (1998) 723.

PERFORMING REAL-TIME PLASMA EQUILIBRIUM OF THE FTU TOKAMAK IN AN RTAI VIRTUAL MACHINE

Y. SADEGHI[1,2,*], L. BONCAGNI[2], G. CALABRO[2], F. CRISANTI[2], G. RAMOGIDA[2], E. VITALE[2] and L.ZACCARIAN[1]

[1]*Department of Computer Science, Systems and Production*
University of Rome "Tor Vergata"

[2]*EURATOM - ENEA Fusion Association*
Frascati Research Centre, Division of Fusion Physics
*E-mail: *ysadeghi@disp.uniroma2.it*

An important topic in plasma equilibrium study [1] and control in a tokamak is to find out and reconstruct the magnetic iso-flux surfaces by using plasma boundary condition. This can be done by using the multi-polar moments method which results from the homogeneous solution of the Grad-Shafranov equation. The equilibrium code ODIN [2] is based on the above described technique and is used to reconstruct the magnetic flux and the equilibrium in the Frascati Tokamak Upgrade (FTU) experiment. The real-time reconstruction of the magnetic field map is important to compute quantities necessary to control the plasma. In this paper we discuss the real-time implementation of ODIN in an RTAI virtual machine and as a result of the real-time implementation, we will show the time evolution of the reconstructed magnetic iso-flux surfaces.

Keywords: Plasma equilibrium/control, Grad-Shafranov equation, iso-flux, RTAI, ODIN code.

1. Introduction

In other works [2], [3] the relations between the multi-polar moments and the magnetic flux have been discussed. The goal of our research is to provide a procedure that implements in real-time the existing equilibrium code ODIN, which is currently working (off-line) at FTU. In this paper we will first give a brief explanation of the plasma current/position control scheme and of the Grad-Shafranov equation, then we will show the proposed real-time ODIN algorithm implementation. This implementation amounts to pre-computing a number of constant parameters before the experimental pulse (shot) and then evaluating key quantities in real-time, based on the magnetic measurements available from the plant and acquired by the real-time acquisition boards during the experiment [4]. The developed real-time algorithm will be shown to be effective by running it in real-time on an RTAI virtual machine with characteristics of Pentium II@1.5 GHz using a Linux

Fig. 1. (a) Outer view of the torus. (b) Inner view of the torus.

Kernel 2.4.18 patched with RTAI 24.1.10, whose operation exactly coincides with the operating conditions of the real-time control system during the FTU experiments (see [5] for a specification of the underlying architecture). Finally, we will present time evolution of the magnetic flux reconstruction as a result.

2. The plasma boundary

One of the available methods for plasma boundary estimation is based on equilibrium reconstruction. Equilibrium codes, such as ODIN and EFIT calculate the distributions of flux and toroidal current density over the plasma and the surrounding vacuum region that best fits the external magnetic measurements (in a least squares sense), and that, at the same time, satisfy the Magneto Hydro Dynamics (MHD) equilibrium equation (the solution of the Grad-Shafranov equation).

3. The plasma position and current feedback system

To maintain the plasma column in the center of the vacuum vessel and to control the movement of the plasma in the horizontal and vertical directions in a tokamak, a set of poloidal field coils which is placed symmetrically with respect to the tokamak equatorial plane is compulsory. In the FTU control system, two SISO PID controllers are used to control the plasma position and current. Four sets of windings named respectively T, H, V, and F coils [6], [7] generate the poloidal magnetic fields required for the plasma position/current control in FTU. The T winding regulates the plasma current, the V coil generates a pre-programmed poloidal field which is able to regulate the plasma column position during normal operation while the F coil generates a poloidal field to compensate for the horizontal plasma displacements during the experiment. The H poloidal winding performs, instead, a slight control of the vertical plasma displacement. The currents (I_T, I_V, I_F and I_H) flowing in the above-mentioned windings represent the actuating signals of the feedback system.

4. Solution of the Grad-Shafranov equation

The main equation that describes the plasma force balance is the Grad-Shafranov equation (GSE), which is one of the most famous equations arising from MHD. The solution of the equation is used to reconstruct the magnetic equilibrium (the operator Δ^* is: $R\frac{\partial}{\partial R}[\frac{1}{R}\frac{\partial}{\partial R}] + \frac{\partial^2}{\partial Z^2}$).

$$\Delta^*\psi = -\mu_0 R^2 p'(\psi) - {\mu_0}^2 f(\psi) f'(\psi) \tag{1}$$

This equation is a second order partial differential equation, where the function $p(\psi)$ is the plasma pressure and $f(\psi)$ is the poloidal current density in an axial-symmetric torus. These functions are arbitrary and must be determined from considerations other than the theoretical force balance. The GSE can be analytically solved in two cases: inside ($\Delta^*\psi = 0$) and outside ($\Delta^*\psi = 2\pi\mu_0 R J_\phi$) the plasma, where J_ϕ is the current density source [3]. According to [2] the solution of the GSE in toroidal coordinates correspond to:

$$\psi(\theta,\tilde{\omega}) = \frac{1}{\sqrt{(ch\theta - cos\tilde{\omega})}} \sum_{m=0}^{\infty} \{[M^i_m(\theta) f_m + M^e_m(\theta) g_m] ch(\theta)\} cos(m\tilde{\omega}) \tag{2}$$

(see [2] for the explicit expression of the internal and external multi-polar moments M^i_m , M^e_m which are evaluated as an integral of the current density $J_\phi(\theta_0, \tilde{\omega}_0)$ that flows inside the torus with toroidal coordinates $\theta, \tilde{\omega}$). The antisymmetric part (*sine*) can be described by similar equations. Moreover:

- *Outside the plasma, M^i_m, M^e_m are constant and depend on the conditions at the boundary of the domain.*
- *Inside the plasma, $M^i_m(\theta)$, $M^e_m(\theta)$ depend on the current distribution inside the radial coordinate.*

The total plasma current is proportional to the sum of all the internal multi-polar moments [8], according to; $I_p = -\sqrt{2}(2\pi\mu_0 R_0)^{-1} \sum_{m=0}^{\infty} M^i_m$ (3).

It is possible to estimate the plasma boundary under the approximation that the multi-polar expansion is constant at all the points of such a surface [8]. In addition, this approximation is strictly accurate only in the case of a circular plasma boundary coinciding with a constant θ coordinate. The

ODIN code computes the distribution of the magnetic flux over the plasma and the surrounding vacuum region (see [2], [4] for details about the ODIN structure).

5. The real-time environment

RTAI stands for Real-Time Application Interface. RTAI provides deterministic and preemptive performance in addition to permitting the use of standard Linux drivers, applications and functions. RTAI's performance is very competitive with some of the best commercial real-time Operating Systems such as VxWorks, QNX etc. We run the ODIN implementation using the magnetic probes data stored in the FTU database, using the feedback simulator on a Virtual Machine [4], [5] with characteristics of a Pentium II@1.5 GHz using the Linux Kernel 2.4.18 patched with RTAI 24.1.10.

6. Structure of real-time ODIN

The algorithm starts from reading a set of constants and known data which are evaluated off-line. In the real-time implementation we use an arbitrary and tabled guess for ψ (for example a conic surface centered in R_0 and monotonically decreasing or increasing along θ and constant along $\tilde{\omega}$). In addition, at the contact point ψ, the guess is forced to zero.
Next, the moments are calculated and a least squares approximation which fits the magnetic measurements is performed. The moments are then adjusted based on this approximation and a new guess of ψ is computed until the difference between two subsequent iterations is less than a specified tolerance: $(|\psi - \psi'| < \varepsilon)$ over the whole mesh. Finally important quantities associated with the equilibrium are computed, such as the poloidal beta β_p and the internal inductance $l_i/2$. The above algorithm is repeated at each data sample.

7. Results on RTAI virtual machine

The experimental data were analyzed for shot ♯30145 with the following conditions: plasma current 501 kA, toroidal magnetic field 6.0 T, average plasma density 0.79×10^{20} m^{-3}, elongation 1.042, tolerance $\varepsilon = 10^{-5}$. The magnetic flux estimated in real-time by our algorithm are shown in Figs. 2 and 3. Figure 2(a) is the initial guess of ψ (a cone, as described before) and the final estimate of ψ available in real-time, referring to the experimental time $t = 0.6$ s. Figures from 2(b), 3(a) and 3(b) show the equilibrium

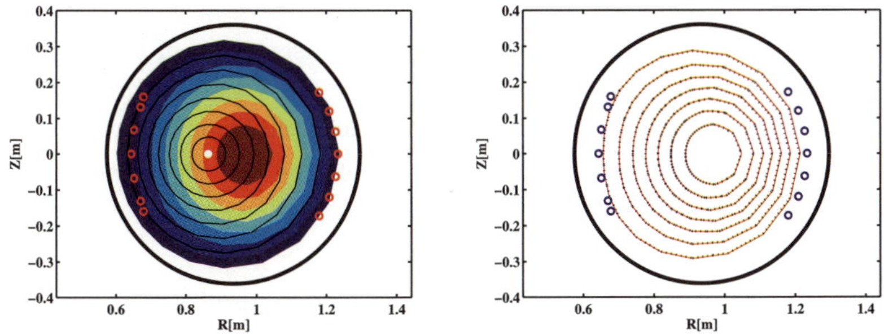

(a) (b)

Fig. 2. (a) Initial guess of the magnetic flux ψ and final estimate of ψ in real-time at time $t = 0.6$ s. (b) Real-time estimate of the magnetic flux at times $t = 0.6$ s (red) and $t = 1.0$ s (black dot).

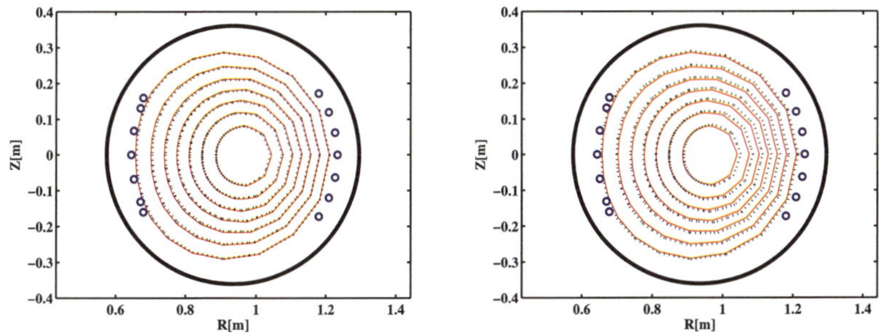

(a) (b)

Fig. 3. (a) Real-time estimate of the magnetic flux at times $t = 1.0$ s (red) and $t = 1.5$ s (black dot). (b) Real-time estimate of the magnetic flux at times $t = 1.5$ s (red) and $t = 1.8$ s (black dot).

reconstructions performed in real-time throughout the experimental time. In particular, to appreciate the evolution of the reconstruction with flowing time, each figure represents the last snapshot of the previous one and the first snapshot of the next one. The selected time instants are $t = 0.6$ s, $t = 1.0$ s, $t = 1.5$ s and $t = 1.8$ s. The gradual drift of the reconstructed flux along time can then be appreciated by looking at the sequence of the four figures. For each of the snapshots in Figs. 2 and 3, the estimation algorithm converges in 4 iterations.

8. Conclusion

In this paper we reported on the real-time reconstruction of the magnetic flux at FTU on an RTAI virtual machine using the ODIN algorithm. This method allows us to carry out fast real-time equilibrium reconstruction using magnetic probe data on above-mentioned virtual machine. The following goals were achieved within this context:

(1) *real-time testing of the code using the magnetic probes data stored in the FTU database;*
(2) *reconstruction of the magnetic flux using multi-polar moments with few iterations (4 or 5).*

Acknowledgment

The authors would like to thank the ENEA Frascati Research Center for the support in carrying out this research. This research was supported in part by ENEA-Euratom and MIUR under PRIN and FIRB projects.

References

1. V. D. Shafranov, "Equilibrium of a plasma toroid in a magnetic field," Sov. Phys. JETP, Engl. Transl., 37(10), 775, 1960.
2. F. Alladio, F. Crisanti,"Analysis of MHD equilibria by toroidal multi-polar expansions," Nuclear Fusion 26, 1143, 1986.
3. B. J. Braams, "The interpretation of tokamak magnetic diagnostics," Plasma Phys. Controlled Fusion 33,715, 1991.
4. Y.Sadeghi, L.Zaccarian, C.D'Epifanio, G.Ramogida, L.Boncagni, V.Vitale and F.Crisanti, "Real-time reconstruction of the magnetic flux in FTU using multi-polar current moments", ICOPS/SOFE 2009, 36th International conference on plasma science and 23rd symposium on fusion engineering, San Diego, California USA, May 31 - June 5, 2009.
5. L. Boncagni, C. Centioli, L. Fiasca, F. Iannone, M. Panella, V. Vitale, and L. Zaccarian, "Introducing a virtualization technology for the FTU plasma control system," In Proc. of the 18th topical meeting on the technology of fusion energy TOFE, San Francisco, USA, September 2008.
6. F. Crisanti and M. Santinelli, "Active Plasma Position and Current Feedback in the FTU Tokamak Machine", Proc. 16th Symp. Fusion Engineering, London, UK, Sept. 3-7, 1990.
7. F. Crisanti, C. Neri, and M. Santinelli, "FTU Plasma Position and Current Feedback", presented at 16th Symp. Fusion Engineering, London, United Kingdom, Sept. 3-7, 1990.
8. G.N. Deshko, T.G. Kilovataya, Yu.K. Kuznetsov, V.N. Pyatov, V.N. Yasin, Fusion 23, 1309, 1983.

MODEL-BASED PLASMA CONTROL IN TOKAMAKS

D. MOREAU*, D. MAZON* and JET-EFDA CONTRIBUTORS†

JET-EFDA, Culham Science Centre
OX14 3DB, Abingdon, UK

Y. ADACHI and Y. TAKASE

The University of Tokyo, Graduate School of Frontier Sciences
277-8561, Kashiwa, Japan

Y. SAKAMOTO, S. IDE and T. SUZUKI

JAEA, Fusion Research and Development Directorate
311-0193, Naka, Japan

The basic elements of an integrated model-based control strategy for extrapolating present-day advanced tokamak scenarios to steady state operation are described. Taking advantage of the large ratio between the time scales involved in the magnetic and thermal diffusion processes, the model identification procedure makes use of a multiple time scale approximation. The methodology is generic and can be applied to any device, with different sets of heating and current drive actuators, controlled variables and/or parameter profiles. It has been applied to experimental data from JET and JT-60U, and satisfactory models have been obtained. First closed loop experiments were performed on JET with three H&CD actuators to control the safety factor profile. Simultaneous real-time control of the q-profile and toroidal velocity profile on JT-60U has been simulated.

Keywords: Tokamak, plasma control, modeling.

1. Introduction

The design of a steady state fusion reactor relies on the development of advanced tokamak operation scenarios in which a high performance magneto-thermal plasma state is achieved and controlled in real time [1,2]. The multiple magnetic and kinetic parameter profiles that define the non-linear plasma state (safety factor, plasma density, velocity, pressure, etc), and will need to be regulated, are known to be strongly coupled. The heating and

*Perm. address: CEA, IRFM, CEA-Cadarache, 13108 Saint-Paul-lez-Durance, France.
†See the Appendix of F. Romanelli *et al.*, Proceedings of the 22nd IAEA Fusion Energy Conference 2008, Geneva, Switzerland.

current drive (H&CD) control actuators are generally quite constrained and their number is limited. An integrated controller must therefore be designed to regulate the global plasma state through a minimization algorithm, rather than each plasma parameter or profile accurately and separately [3]. An identification technique has been developed to find an appropriate plasma response model from the analysis of experimental data. The state variables appear naturally to be the variations of a magnetic variable, μ, such as the internal poloidal magnetic flux, Ψ, or inverse safety factor, ι, and some fluid/kinetic variables, ρ, such as the plasma toroidal velocity, V_{tor}, pressure, p, (or temperature, T) with respect to their reference values (their values in the reference state). After projection onto radial basis functions, a lumped-parameter version of the state space model is then derived, which reads:

$$\partial\mu/\partial t = A_{11}\mu(t) + A_{12}\rho(t) + B_{11}n(t) + B_{\mu V}V_{ext}(t) \qquad (1)$$

$$\varepsilon\partial\rho/\partial t = A_{21}\mu(t) + A_{22}\rho(t) + B_{21}P(t) + B_{22}n(t) \qquad (2)$$

with inputs P(t) = [P1(t), P2(t), P3(t), etc ...], the heating and current drive input powers, V_{ext}, the plasma surface loop voltage, and $n(t)$, the plasma density. The small parameter, ε, represents the typical ratio of the thermal and resistive diffusion time scales. The model order can therefore be further reduced by using the theory of singularly perturbed systems [4]. We shall therefore seek two models of reduced orders, a slow model

$$\partial\mu/\partial t = A_s\mu + B_s u_s; \quad \rho_s = C_s\mu + D_s u_s \qquad (3)$$

and a fast model,

$$\partial\rho_f/\partial t = A_f\rho_f + B_f u_f \qquad (4)$$

Here ρ_s and ρ_f are the slow and fast components, respectively, of the kinetic variables ($\rho = \rho_s + \rho_f$), and u_s and u_f are the slow and fast components, respectively, of the input vector ($u = u_s + u_f$). Two examples will be considered in this paper. The first example refers to the control of the safety factor profile (representative of the current density profile) on JET and the second example will be dedicated to the identification, from some JT-60U experimental data, of a two-time-scale (magnetic/kinetic) state space model describing the coupled dynamics of the safety factor and toroidal rotation profiles in a non-inductive, high-bootstrap-current scenario.

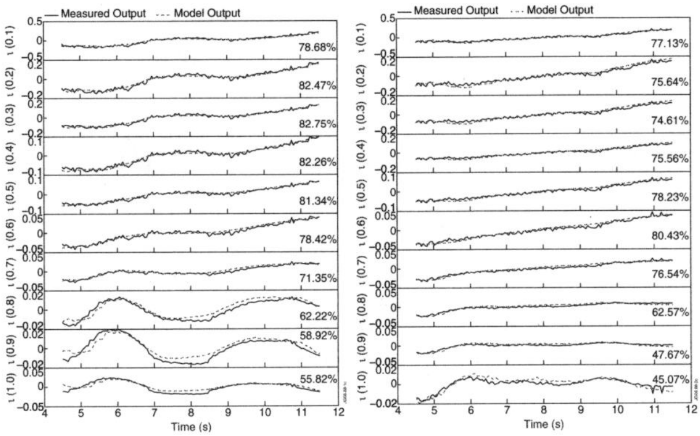

(a) (b)

Fig. 1. (a) Coefficients of the $\iota(x)$ profile at knots $x = 0.1, 0.2, 0.3, 1$. The figure shows a comparison between the experimental data and the model output. JET pulse ♯67840: modulations of V_{ext} with constant H&CD powers. (b) Coefficients of the $\iota(x)$ profile at knots x=0.1,0.2,0.3,... 1. Figure shows a comparison between the experimental data and the model output. JET pulse ♯67874: modulations of the NBI power with constant LH and ICRH powers, and constant request on V_{ext}.

2. Identification of a slow model for the control of the current profile

In tokamaks, a non-dimensional parameter that characterizes the current density profile and the helicity of the magnetic field lines on a given toroidal flux surface is defined as $\iota(x) = d\Psi(x)/d\Phi(x)$, where $\Psi(x)$ and $\Phi(x)$ represent the poloidal and toroidal magnetic fluxes, respectively, and x is a normalized radial variable ($0 \leq x \leq 1$). Its inverse, $q(x) = 1/\iota(x)$, is called the safety factor. In order to identify the response of $\iota(x)$ to variations of the control actuators around a given reference equilibrium state on JET, a number of specific open-loop experiments were performed at 3 Teslas, with a plasma current around 1.5 MA and an average plasma density of about 3.5×10^{19} m^{-3}. The available actuators were modulated randomly around a given set of input values that define our reference state. The selected actuators consisted of: (i) neutral beam injection (NBI), (ii) ion cyclotron resonance heating (ICRH), (iii) lower hybrid (LH) current drive, (iv) surface loop voltage, V_{ext}. Comparing the experimental $\iota(x)$(x) data with predictions using the measured inputs and the identified model shows good agreement (Figs. 1(a)–1(b)).

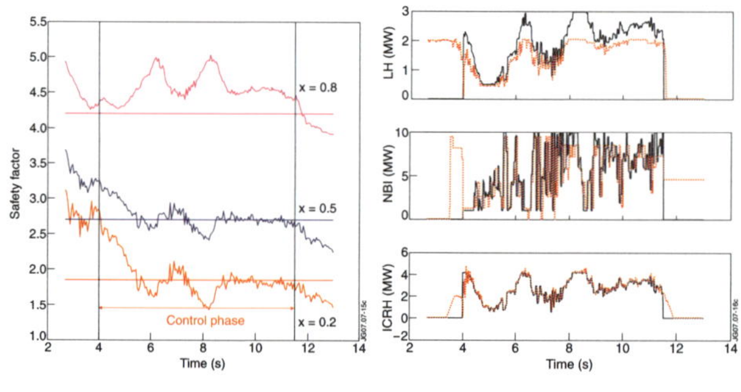

(a) (b)

Fig. 2. (a) Control of the q-profile at three normalized radii, $x = 0.2$ (red), $x = 0.5$ (blue), and $x = 0.8$ (magenta) using the three H&CD actuators (Pulse ♯70395). During the control phase V_{ext} is requested to be constant (32mV/rad). Target values are represented by horizontal lines. (b) Requested (full traces) and delivered (dotted traces) LH, NBI and ICRH powers for JET pulse ♯70395.

Altogether, the slow model thus identified was found to be sufficiently accurate for some closed-loop control experiments to be attempted. Figs. 3–4 illustrate such an experiment where the controlled variables were $q(x)$ at

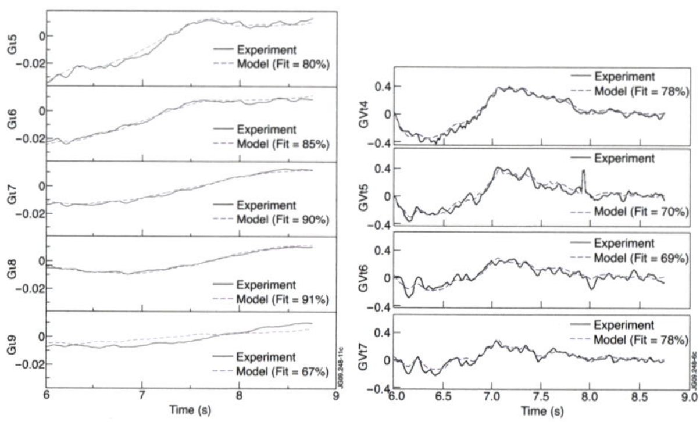

(a) (b)

Fig. 3. (a) Comparison between the experimental data (JT-60U pulse ♯45862) and the slow model output. Coefficients of the $\iota(x)$ profile at knots $x = 0.5, 0.6, 0.7, 0.8, 0.9$. (b) Comparison between the experimental data (JT-60U pulse ♯45862) and the fast model output. Coefficients of the $V_{tor}(x)$ profile at knots $x = 0.4, 0.5, 0.6, 0.7$.

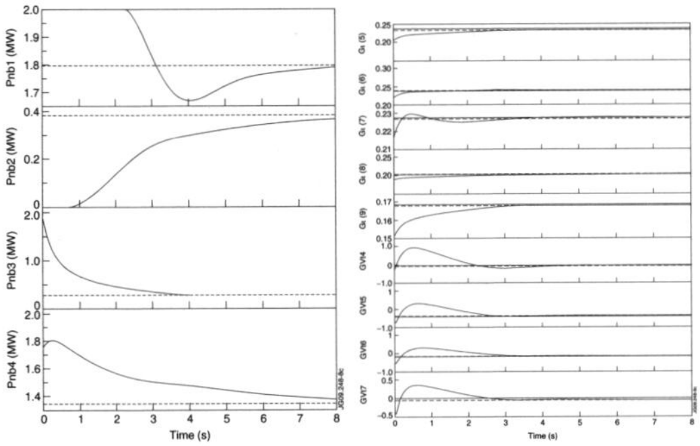

(a) (b)

Fig. 4. (a) Evolution of the four actuator powers in a closed-loop simulation showing the simultaneous control of the safety factor and the toroidal velocity profiles on JT-60U. The time origin refers to the start of the control phase. (b) Evolution of $\iota(x)$ at $x = 0.5$–0.9 (5 upper traces) and $V_t(x)$ at $x = 0.4$–0.7 (4 lower traces) in the closed-loop simulation showing the simultaneous control of the safety factor and the toroidal velocity profiles on JT-60U (see also Fig. 2(b)). The requested target values are represented by the horizontal lines.

$x = 0.2$, 0.5 and 0.8. The corresponding target values were 1.85, 2.7 and 4.2, respectively. The controller was active between $t = 4$ s and $t = 11.5$ s, and the requested value of the surface loop voltage was 32 mV/rad during the control phase. The initial behavior of the controller is dominated by a transient in the boundary flux control which causes large oscillations of the loop voltage and, as a consequence, of the H&CD powers. Control becomes really effective and successful when the boundary flux has finally tracked the requested waveform, and the loop voltage has settled to the desired value. Figure 2(b) shows a comparison between the requested actuator powers, and the delivered ones. The requested target for the q-profile was not reached exactly at $x = 0.8$ because the LH power could not exceed 2 MW while 3 MW were requested. At constant loop voltage, more LH power would have driven more current, with a larger off-axis component, decreasing q in the outer region and reducing the error around $x = 0.8$.

3. Two-time-scale model for magnetic and kinetic control

The same methodology has been applied to JT-60U data to identify a two-time-scale model for the simultaneous control of magnetic and kinetic profiles. A series of high-bootstrap-current advanced tokamak discharges were analyzed. The reference plasma state was characterized by a magnetic field of 3.7 T, a plasma current of 0.9 MA at zero loop voltage (i.e. fully non-inductively driven), and a central plasma density of 3×10^{19} m^{-3}. The selected actuators consisted of four groups of neutral beam injectors corresponding to : (i) on-axis perpendicular injection, (ii) off-axis perpendicular injection, (iii) on-axis co-current tangential injection, (iv) off-axis co-current tangential injection. The response to changes in the line-averaged density was also identified because it plays an important role in the model.

The comparison between the measured data and the model simulation for the dynamics of the inverse safety factor, the toroidal plasma velocity and the ion temperature shows the good potential of the technique. An example is displayed here on Figs. 3(a)–3(b). The results of typical closed-loop simulations based on the identified two-time-scale model is also displayed on Figs. 4(a)–4(b).

4. Conclusion

In conclusion, it is shown that the technique described can be applied to different devices, for simple as well as more comprehensive controls, and with different sets of actuators and sensors. Experiments on other pulsed and steady-state tokamaks would also be beneficial to possibly validate and improve this methodology. They could provide a broad basis for developing integrated profile control and reactor relevant steady-state scenarios in ITER.

Acknowledgments

This work, supported by the European Communities under the contract of Association between EURATOM and CEA, was carried out within the framework of the European Fusion Development Agreement. The views and opinions expressed herein do not necessarily reflect those of the European Commission. Part of this work was also carried out at the Takase-Ejiri Laboratory, University of Tokyo, while the first author was a visiting professor.

References

1. C. Gormezano *et al.*, *Nucl. Fusion*, **47**, S285 (2007).
2. Y. Gribov *et al.*, *Nucl. Fusion*, **47**, S285 (2007).
3. D. Moreau *et al.*, *Nucl. Fusion*, **48**, 106001 (2008).
4. P. V. Kokotovitch, H. K. Khalil and J. O'Reilly, *Singular Perturbation Methods in Control: Analysis and Design* (Academic Press , London, 1986).

PART K

Modeling and optimization of beam and plasma dynamics

PARAMETRIC OPTIMIZATION FOR TOKAMAK PLASMA CONTROL SYSTEM

S. ZAVADSKY* and A. OVSYANNIKOV

Faculty of Applied Mathematics and Control Processes
Saint-Petersburg State University
Saint-Petersburg, Russia
*E-mail: * sergzav@pochtamt.ru*

N. SAKAMOTO

University of Nagoya, Nagoya, Japan

In order to design the control system for plasma current, shape and position the structural parametric optimization of transient processes is suggested. Optimization approach to plasma dynamic is based on the consideration of transient processes of the full-sized control object that is closed by a regulator of a decreased dimension. It is suggested to use an integral performance criterion as a functional that allows optimizing the transient process, perturbed at the initial point set and the set of external disturbances. In the framework of this approach the optimization of plasma dynamics of the ITER tokamak is given.

Keywords: Tokamak, plasma, control system, optimization.

1. Introduction

The problem of control plasma in nuclear reactor tokamak occupies the leading place in controlled thermonuclear fusion. The main task in this area is a plasma feedback control system design. Problems of analysis and synthesis of stabilizing regulators of current, position and plasma shape in tokamak are of great importance.

The mathematical model of ITER tokamak plasma control system is a very complex object which includes various subsystems and differential equations that define plasma behavior. The structural schema of ITER control system links external disturbances, plasma state equations, filters system, vertical controller, current and shape controller, power system and set of diagnostic signals. Mentioned schema can be represented using the structural diagram in Fig. 1. Matrices of these subsystems are known with constant components.

The plasma state equations are done based on the linearization of differential equations that define plasma behavior in deviation of equilibrium

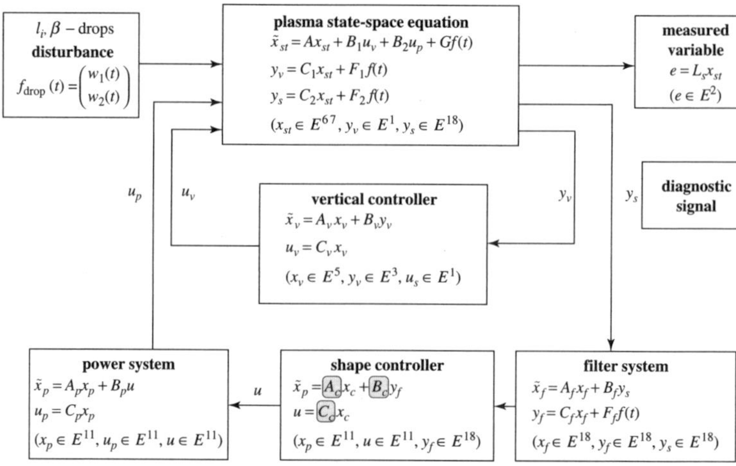

Fig. 1. Structural model of ITER plasma control system.

position, where $x_{st} \in E^{67}$ is the state space vector, $u_p \in E^{11}$ is the control voltages vector, $y_s \in E^{18}$ is the diagnostic signals vector, $e \in E^7$ is the measurement variables vector. Measurement vector includes deviation of plasma current and six checked clearances between plasma and tokamak chamber, which signed as $g_1, ...g_6$ and called *gaps*. The function $f(t)$ is external plasma disturbance, which is called l_i, β – *drops* disturbance and defined as known function. Power and filters system are given and defined by construction features of tokamak. To plasma shape stabilize the control object is closed with a shape controller of a decreased dimension with the following structure:

$$\begin{aligned} \dot{x}_c &= A_c\, x_c + B_c\, y_f \\ u &= C_c\, x_c \end{aligned} \tag{1}$$

where vectors $u \in E^{11}$, $y_f \in E^{18}$ are the control voltages and the diagnostic signals of the tokamak control system respectively, matrices A_c, B_c, C_c are the constant matrices of the controller, which must be obtained. The control object closure by the obtained regulator is done in accordance with the scheme in Fig. 1.

By the "regulator synthesis" we mean such a choice of component of controller matrices that gives us a stable closed object and sufficient quality of stabilization. The stabilization quality performance base on the numerical characteristics such as the integral of squared gaps

$$I_{gaps} = \int_0^T \sum_{i=1...6} g_i^2(t)\ dt \tag{2}$$

and the settling time

$$t_{\text{settling}} = \min_{\tilde{t}}\{\tilde{t} : \max_{i=1...6} |g_i(t)| \leq 0.01, \quad \forall t \geq \tilde{t}\}, \tag{3}$$

Also, we need take into account the nonlinear amplitude constraints on control voltage signals. These constraints have various numerical values and complicate the design of the controller, and allow no simple raise in the magnitude of control voltage signals. Note that the controller design is analyzed by means of the linear model, however this controller is tested on the nonlinear model, and has to possess appropriate characteristics.

2. Parametric optimization method

In order to design the control system for plasma current, shape and position the structural parametric optimization of transient processes is suggested. In the framework of this approach, the optimization of transient processes of the full-sized control object that is closed by a regulator of a decreased dimension is conducted. It is suggested to use an integral performance criterion as a functional that allows optimizing the transient process, perturbed at the initial point set and the set of external disturbances.

As a rough approximation for optimization the shape controller can be obtained, for example, using the reduction procedure and the LQG-optimal synthesis, see [1-6] for more details. So, let us investigate the equations of the control object presented in Fig. 1 with constantly applied perturbation, which is closed by the regulator of decreased dimension (1) and the another subsystem of the structural diagram. Further, let the elements of matrices A_c, B_c, C_c of the dynamic shape controller will be taken as parameters that are to be optimized and combined into a vector of parameters p

$$p = \{p_k\} \longleftrightarrow \{A_c,\ B_c,\ C_c\}\,. \tag{4}$$

We combine the structural diagram of the control system shown in Fig. 1 into a system of linear differential equations with disturbances in the following form:

$$\begin{aligned}
&\dot{x} = P(p)\,x + N(p)\,f(t),\\
&x(0) = x_0\\
&f(t) = f(d_\beta, t_\beta, d_l, t_l,\ t),\\
&e = L\,x(t,\,x_0,\,p),\\
&u = \ K(p)\,x(t,\,x_0,\,p),
\end{aligned} \tag{5}$$

where $x \in E^{112}$, $x = \{x_{st},\, x_v,\, x_p,\, x_f,\, x_c\}$ is the extended state space vector, $e \in E^7$ is the measurement variables vector, $u \in E^{11}$ is the control voltages vector, P, N, L, K are the transformed constant matrices, $f(t)$ is the l_i, β – *drops* disturbance, $p = \{p_k\}$ is a vector of parameters. Note, that the matrices A_c, B_c, C_c of a designed regulator (1) will be taken as parameters that are to be optimized.

It is suggested to use the following integral performance criterion that allows optimizing the transient processes perturbed by the initial point and external disturbance

$$\begin{aligned}
I(p) = \int_0^T \{e^*(t)\,Q\,e(t)\, + \,u^*(t)\,R\,u(t)\}\;dt&\\
+\; e^*(T)\,Q_1 e(T) \to \min,&
\end{aligned} \tag{6}$$

where Q, R, Q_1 are symmetrical weight matrices. The minimization algorithm of this functional by parameters $p = \{p_k\}$ is suggested bellow. Let entered additional differential equation

$$\begin{aligned}
&\tfrac{d\psi}{dt} = -P^*\psi + 2\,(L^*Q\,L + K^*R\,K)^*x(t),\\
&\psi(T) = -2\,L^*Q_1\,L\,x(T).
\end{aligned} \tag{7}$$

Then, using vector $\psi(t)$ we obtain a representation for the gradient of the functional

$$\begin{aligned}
\tfrac{\partial I(p)}{\partial p_k} = -\Bigg(\int_0^T \psi^*(t)\,\left(\tfrac{\partial P}{\partial p_k} x(t)\, + \,\tfrac{\partial N}{\partial p_k} f(t)\right)&\\
-2x(t)^*\,\tfrac{\partial K^*}{\partial p_k} R\,K(p)\,x(t)\;dt\Bigg)&
\end{aligned} \tag{8}$$

Based on the analytical expressions (4)–(8) a gradient optimization method for the functional (6) with respect to the parameters $p = \{p_k\}$ is implemented for C++ and MatLab environments.

3. Optimization results

The optimization results are considered based on transient processes of checked clearances – *gaps,* which are the members of the measurement

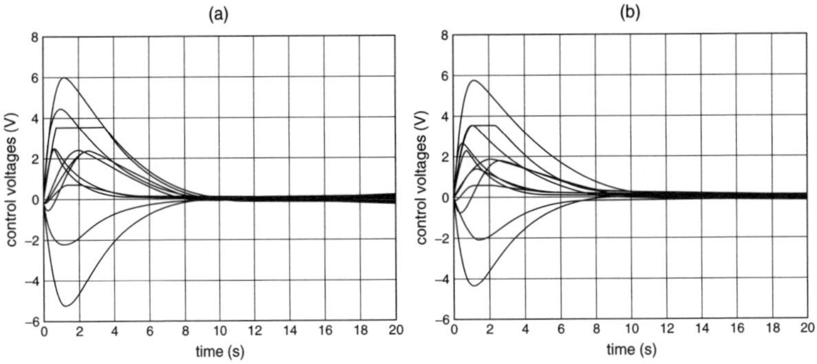

Fig. 2. Control voltage signals $u_1, \ldots, u_{11}$ corresponding to (a) initial controller and (b) optimized controller.

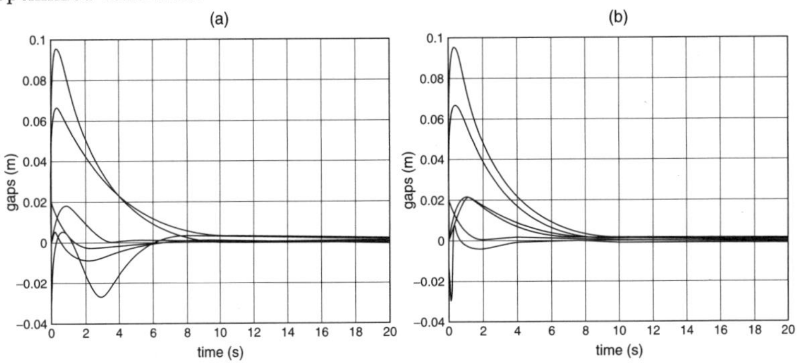

Fig. 3. Transient processes of gaps $g_1, \ldots, g_6$ corresponding to (a) initial controller and (b) optimized controller.

variables vector $e \in E^7$ for the control object (5). The modeling of transient processes of the control object closed with the optimized and initial controller is shown. This initial controller is obtained using the approaches described in [4,6] and already possesses proper characteristics. The graphical results of optimization are presented in Figs. 2 and 3. The Fig. 2 shows control voltage signals $u_1, ...u_{11}$ for control object closed with initial and optimized controllers and Fig. 3 shows transient processes of gaps $g_1, ..., g_6$ for control object closed with initial and optimized controllers of the nonlinear model with voltage limitations.The Fig. 2 shows that by using equation for u_i we can choose maximums of control voltages according to their limitations, then raise them and obtain better performance for measurement variables e_i. This is illustrated in Fig. 3. Also, we consider optimization numerical characteristics the integral of squared gaps and the settling time mentioned above which are presented in Table 1.

Table 1. Settling time and integral of squared gaps for control object closed with initial and optimized controllers.

Control object closed with initial controller	Control object closed with optimized controller
$I_{\text{gaps}} = 0.027$	$I_{\text{gaps}} = 0.024$
$t_{\text{settling}} = 6.373\,\text{sec.}$	$t_{\text{settling}} = 5.634\,\text{sec.}$

4. Conclusion

This work is dedicated to questions concerning synthesis and optimization of tokamak plasma control system. The structural model of ITER plasma control system is discussed and the structural parametric optimization method is suggested. Results of the computations are obtained and discussed. Numerical characteristics such as the integral of squared gaps and the settling time are presented for the both initial and the optimized controllers. For the optimized controller the squared gaps and settling time are 11% and 12% lower, correspondingly.

References

1. D. A. Ovsyannikov, E. I. Veremey, A. P. Zhabko et al. *Mathematical methods of plasma vertical stabilization in modern tokamaks* // Nuclear Fusion. 2006. Vol 46. P. 652-657.
2. G. J. McArdle, V. A. Belyakov, D. A. Ovsyannikov, E. I. Veremey. *The MAST plasma control system* // Proc. 20th Intern. Symp. on Fusion Technology SOFT'98. Marseille, France. 1998. P. 541–544.
3. D. A. Ovsyannikov, A. P. Zhabko, E. I. Veremey et al. *Plasma current and shape stabilization with control power reducing* // Proc. of the 5th Intern. workshop: "Beam Dynamics & Optimization". St.Petersburg, 1998. P. 103–111.
4. D. A. Ovsyannikov, E. I. Veremey, A. P. Zhabko et al. *Mathematical methods of tokamak plasma shape control* // Proc. of the 3th Intern. workshop: "Beam Dynamics & Optimization". St.Petersburg, 1996. P. 218–229.
5. B. A. Misenov, D. A. Ovsyannikov, A. D. Ovsyannikov et al. *Analysis and Synthesis of Plasma Stabilization Systems in Tokamaks* // Control Application of Optimization: Preprints of the Eleventh IFAC Intern. Workshop. St.Petersburg, 2000. P. 249–254.
6. E. I. Veremey, N. A. Zhabko. *Plasma current and shape controllers design for ITER-FEAT tokamak* // Book of abstracts of Workshop on Computational Physics Dedicated to the Memory of Stanislav Merkuriev. St.Petersburg, 2003. P. 52.

OPTIMIZATION OF THE INITIAL CONDITIONS OF THE PLASMA DISCHARGE IN THE ITER TOKAMAK

MARIA MIZINTSEVA*, ALEXANDER OVSYANNIKOV† and EVGENY SUHOV‡

Faculty of Applied Mathematics and Control Processes
Saint-Petersburg State University, Russia
*E-mail: * lqi87@yandex.ru*
† ovs74@mail.ru
‡ suhovevgeny@mail.ru

Mathematical model and optimization at the initial stage of discharge in the ITER tokamak. Results and computer realization.

Keywords: Control, modeling, optimization, tokamak, ITER, plasma.

1. Introduction

The electro-magnetic system of the tokamak has an extremely complicated structure, numerous plasma parameters are to be controlled on every stage of discharge so the mathematical modeling of this system demands a lot of effort. In this paper the modeling and optimization on the initial stage of the discharge will be considered.

2. Mathematical model

The whole system of the tokamak to some extend can be represented by a set of conducting contours.

The electromagnetic system of the ITER tokamak to some extend can be described by 11 Kirchhoff equations:

$$\frac{d(LI)}{dt} + RI = U \tag{1}$$

where L stands for the matrix of inductivities, R is a diagonal matrix of conductivities, I is a vector of currents, U is a vector of voltages which are none zero for coils PF2-PF5. In this case U is a vector of time-dependent parameters and represents itself programmed control. The dimension of the system is a rational balance between the accuracy of the model and the volume of computations.

3. Limitations and conditions

To get the plasma discharge in the ITER tokamak started it is necessary to reach a certain state of the electro-magnetic field in the tokamak vessel that puts respective terms and limitations on the major parameters. Some of conditions are due to the physics of plasma which determine the configuration of the electro-magnetic field under which avalanche ionization of plasma is possible. Some of the contingencies are determined by the parameters of the device and materials it is made of. All of the conditions considered can be divided into two major groups: integral and terminal. The first group should be satisfied alongside the trajectories (i.e. solutions) of the system and the second group of conditions should be met at the very end of the beam of trajectories which corresponds to the moment of the breakdown.

3.1. *Terminal conditions*

Firstly let us have a look at the so called terminal constraints, which are conditions which should be delivered at the end point of the time-interval of modeling that is the starting point of the discharge. (1) By the moment of the breakdown the loop voltage on the contour, which passes the 0-point should be equal to 14,1 V:

$$U_{loop}(T) = 14,1V \tag{2}$$

$$U_{loop}(t) = 2\pi R_0 Z_0 E_0 = \sum \frac{dI_k L_k}{dt} \tag{3}$$

where (R_0, Z_0) corresponds to the 0-point center of the breakdown, E_0 voltage of the electric field in the 0-point, k - number of conducting contours, I_k - current in the contours, L_k its co-enductivity with the countour passing via the 0-point.

(2) There are constraints, which are determined by the physical processes during the discharge. In particular the magnitude of the magnetic induction in the breakdown area should be less than 2 mlT:

$$B(R, Z, T) \leq 2mlT \tag{4}$$

(3) There is a demand to maximize the magnetic flow reserve in the breakdown area by the time when the discharge starts with the given starting flow in the central solenoid of 90 W:

$$\psi(t_0) = 90W \tag{5}$$

This condition is due to the fact that the plasma discharge goes while the flow of the CS is rising.

3.2. *Integral conditions*

These are conditions that should be delivered from the start to the end of the modeling interval. (1) The following correlation should be maintained:

$$\frac{dU_{\text{loop}}(t)}{dt} > 0 \forall t \in [t_0, T] \tag{6}$$

That condition demands constant rise of the current which allows to avoid pre-breakdown.

(2) The alternating currents in the active coils are limited by amplitude by some maximum value:

$$|I_k(t)| \leq I_k^{\max} \forall t \in [t_0, T] \tag{7}$$

Where $I_k(t)$ stands for the current in the coil and is $I_k^{\max}$ a maximum magnitude for each coil.

(3) The magnitude of the magnetic fields in the coils limited from above that is because under strong magnetic fields the super-conducting material of the coils loses its qualities. Magnetic field reaches its maximum on the edge on the conductor, so the following correlation should be maintained: WHAT

(4) The maximum voltage at the active coil is also limited due to the design of the device. So alongside the trajectories the voltage is limited from above:

$$|U_k(t)| \leq U_k^{\max} \forall t \in [t_0, T] \tag{8}$$

where $U_k^{\max}$ is a maximum voltage for the k-coil.

4. Functional

To build an adequate and effective control system all of these conditions are to be considered. It can be done in a form of a functional an integral characteristic of a dynamic system. The most general form of it can be written down as:

$$J(u) = \iint \varphi(t, x_t) dx_t dt + \int g(x_T) dx_T \tag{9}$$

Usually conditions in any system can be put into a quite simple arithmetical form. Using this functional approach an estimation of the parameters of the system can be made. And thus the task of the creating of the optimal control can be replaced by minimizing of the proper functional of the dynamic system. Functional represents itself a function of many variables, which are control parameters. Its minimum can be found in particular using the method of gradient descend. So the task of optimization consists of three steps: writing down a functional, finding components of its gradient by control parameters, descend to minimum.

(1) Maximization of the poloidal magnetic flow in the 0-point by the moment of breakdown:

$$g_1(I(T)) = \begin{cases} \frac{1}{\alpha^*_\psi I(T)}, \alpha^*_\psi I(T) \neq 0 \\ 0, \alpha^*_\psi I(T) = 0 \end{cases} \tag{10}$$

Where T stands for the moment of breakdown, $\alpha^*_\psi I(t)$ - instant poloidal magnetic flow in the 0-point.

(2) Delivering the loop voltage U=14,1 V in the 0-point at the start of the discharge:

$$g_2(I(T), UK, R) = \alpha^*_\psi(L^-1U(T) - L^-1RI(T)) - 14,1)^2 \tag{11}$$

(3) Limitation of the magnitude of the poloidal field in the control points at the start of the discharge:

$$g_3(I(T)) = \sum_{i=1}^{5} g_3(I(T)) \tag{12}$$

where

$$g_3 = \begin{cases} (\alpha^*_{r_i} I(T))^2) + (\alpha^*_{z_i} I(T))^2) - 4 \times 10^-3, \\ \sqrt{(\alpha^*_{r_i} I(T))^2) + (\alpha^*_{z_i} I(T))^2)} > 2 \times 10^-3 \end{cases} \tag{13}$$

Where $\alpha^*_{r_i} I(T)$ and $(\alpha^*_{z_i} I(T))$ are respectively radial and vertical components of the poloidal magnetic field in the control points.

(4) Limitation of the maximum magnitude of the current in the coils:

$$\varphi_1(t, I, UK, tk, R) = \sum_{n=1}^{11} (\varphi_1 n(t, I, Uk, tk, r)) \tag{14}$$

where

$$\varphi_{1n} = \begin{cases} (I_n(t) - I_{\max n})^2, I_n(t) > I_{\max n} \\ (I_n(t) + I_{\max n})^2, , I_n(t) < I_{\max n} \end{cases} \tag{15}$$

(5) Limitation on the magnitude of the fields in the coils:

$$\varphi_2(t, I, UK, tk, R) = \sum_{n=1}^{11} (\varphi_{2n}(t, I, Uk, tk, r)) \tag{16}$$

where

$$\varphi_{2n} = \begin{cases} (\alpha^*_{r_i} I(T))^2) + (\alpha^*_{z_i} I(T))^2) - 4 \times 10^{-3}, \\ \sqrt{(\alpha^*_{r_i} I(T))^2) + (\alpha^*_{z_i} I(T))^2)} > B_{i_{\max}} \end{cases} \quad (17)$$

Here $B_{i_{\max}}$ - stands for the maximum magnetic field in each coil. The resulting functional can be obtained as a sum of the above.

5. Variation of the functional

Another way to describe a dynamical system is a variational approach. System 1 in variational terms can be written as follows:

$$\frac{d}{dt}\delta I = \frac{\partial f}{\partial I}\, \delta I + \Delta_{Uk} f + \Delta_{tk} f + \Delta_R f \quad (18)$$

With the initial condition:

$$\delta I(t_0) = \Delta I_0 \quad (19)$$

Using the extra $\psi(t)$ functions the functional can be presented as:

$$\delta J = \delta J + \int_{t_0}^{T} \psi^*(t)(\tfrac{d}{dt}\, \delta I - \tfrac{\partial f}{\partial I})\delta I - \tfrac{\partial f}{\partial Uk}\, \delta Uk - \tfrac{\partial f}{\partial tk}\delta tk - \tfrac{\partial f}{\partial R}\delta R)dt \quad (20)$$

Via the variation of the components the variation of the functional can be found as:

$$\delta J = \tfrac{\partial g}{\partial R})\, \delta R - \tfrac{\partial g}{\partial UK}\delta UK + \int_{t_0}^{T} -\psi^*(t)\tfrac{\partial f}{\partial Uk} + \tfrac{\partial \varphi}{\partial Uk}\, dt\delta Uk +$$
$$+ \int_{t_0}^{T} -\psi^*(t)\tfrac{\partial f}{\partial tk} + \tfrac{\partial \varphi}{\partial tk}\, dt\delta tk + \int_{t_0}^{T} -\psi^*(t)\tfrac{\partial f}{\partial R} + \tfrac{\partial \varphi}{\partial R}\, dt\delta R - \psi^*(t_0)\delta T(t_0) \quad (21)$$

The expression above for the variation of the functional actually contains the components of the gradient. After that one of the methods of directional descend can be applied.

6. Results and computer realisation

The mathematical model described above was realized in a program Discharge Initial Stage developed at the Faculty of Applied Mathematics and Control Processes. This program has the initial conditions (i.e. currents at t=0), resistances and voltages as model parameters. After those are entered the program executes modeling of the process. The output in this case are curves for such characteristics as loop voltage, magnetic flow etc. To execute the optimization scheme in addition optimization parameters should be given. Optimisation parameters represent themselves weight coefficients with which different parts of the functional are included into the resulting expression. Usually already after the 5th iteration a noticeable decrease of the value of the functional is observed, after a hundred of iterations that value is almost zero. As a result of optimization a set of values of control parameters is obtained, after what new better trajectories of the system are gained. Thus approach and its computer realization, described above, are very useful though not too complicated model of the initial stage of the discharge in the ITER tokamak.

References

1. Ovsyannikov D.A, Egorov N.V. *Mathematical modeling of the focusing systems for electron and ion beams*
2. Gluhih V.A., Belyakov V.A., Mineev A.B. *Physical and technical basis of the thermonuclear fusion*
3. Zavadsky S.V., Ovsyannikov D.A., Chung S.L. *Parametric optimization methods for the tokamak plasma control problem*

MULTI-OBJECTIVE OPTIMIZATION FOR BEAM LINES

A. A. CHERNYSHEV

Applied Mathematics and Control Processes, Saint-Petersburg State University
Saint-Petersburg, 198504, Russia
E-mail: andr.chernyshev@gmail.com
www.spbu.ru

In the present paper some problems of global optimization for focusing beam lines are discussed. The main features of this concept are described, and there are cited solutions for a variant of micro- and nanoprobe systems. For this purpose some analytical and numerical methods, and tools are realized and also discussed are some results of numerical experiments.

Keywords: Beam line, modeling, optimization, computing, nanoprobe.

1. Introduction

Today, big place in science and technology is occupied with focusing systems that form a beam on the target with size less than a micrometer (up to nanometers) and named micro- and nanoprobes. Their specifics is that their construction stage always includes a stage of theoretical research of possible choice of such systems. The thing is, that the calibration process of finished systems (the adjustment) doesn't lead to significant system improvements and therefore the need to upgrade the systems by adding new elements arises, thus increasing the installation price. The traditional process of designing and tuning of focusing systems such as micro- and nanoprobes in order to produce certain desired properties is not straightforward. So the process of searching for optimal accelerator parameters has to come by with a thorough research of structures suiting the experimenter physicist, and can be divided into the following steps:

✓ aligning the succession of structure optimality criteria based on the existing systems;

✓ selection and classification of optimization parameters and influences, i. e. constructing a vector of operating parameters and operating functions;

✓ creating a physical and mathematical model of the whole focusing system suiting all the possibilities and expectations of the experimenter physicists.

✓ linear model adjustment, which result is the local optimum set for selected criteria of the system;

✓ rejection of solutions not suiting the researcher's additional criteria, some of which could be admission satisfactions, realization possibilities and so on;

✓ including of additional effects in the scope of the linear model, such as fringing fields, self-charge etc;

✓ the account of nonlinear effects with the purpose of consecutive rejection of the local minima found at the previous stages.

As a result of performance of the given list of actions, the researcher receives a final set of acceptable decisions from which final customer carries out a choice based on additional, weakly-formalized criteria such as cost, technological realization restrictions of the given decisions.

The above told implies that the problem has to satisfy multiple criteria. At the same time, many of those criteria are contradicting (antagonistic), therefore antagonism account problem instantly appears, by means of the weight factors, indicators or any other.

The ideology described above is considered in the present paper on an example of a probe forming system (see [1]).

2. Physical backgrounds

Among the big family of beam lines, the special place is occupied by ion-optical systems, to which, in particular, micro- and nanoprobes concern. Under an ion-optical system we understand a system intended for transferring a beam from one part of the space to another (transportation), in which the basic attention is given to the formation of cross-section phase characteristics of the beam (focusing).

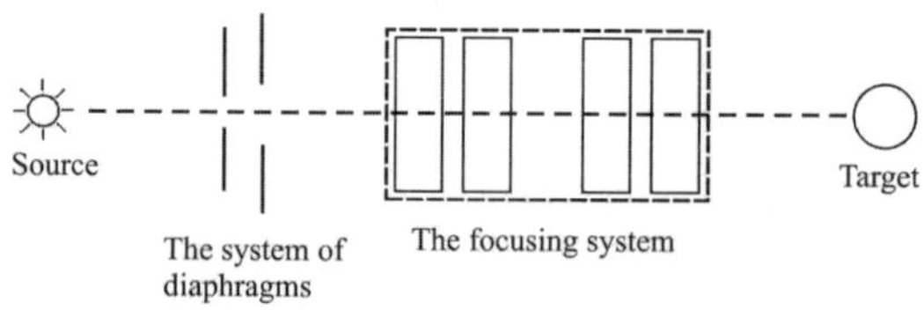

Fig. 1. An exemplary structure of probe forming system.

Figure 1 presents an example of a nanoprobe.

Distances between lenses, lense lengths, distance from objective to target ('working distance'), 'pre-distance' and fields created by lenses serve as the controls in systems of this type. At the same time, a number of parameters, such as, for example, lense lengths, remain fixed after the construction of the beam line or demand considerable expenses for customization, and the other part can be subsequently corrected via adjustment.

When designing the given kind of beam lines, it is necessary to consider the following complex of quality functionals and restrictions:

1. The criteria defining focusing characteristics.
2. Aperture restrictions.
3. Additional restrictions (for example, fixation of distance between lenses).
4. Luminosity of the beam line.

There are various approaches to the methodology of the formation of these functionals and the account of restrictions. The matter is, in some cases it is favourable to consider this or that restriction as a functional component, and in others – to include it in the list of restrictions in the form of equalities or inequalities.

Thus we are in the need to minimize a set of functionals at a time, which, in general form, could be represented with one using α_i and P_i weights, the selection of which defines the level of significance of a criterion:

$$I(\vec{B},\vec{U}) = \sum_{i=1}^{k} \alpha_i I_i^{2P_i}(\vec{B},\vec{U}),$$

where $\mathcal{B}$ is the control parameters set, $\mathcal{U}$ is the control functions set, $\vec{B} \in \mathcal{B}$, $\vec{U} \in \mathcal{U}$.

The selection of the control functions set defining the controlling field from a certain appropriate class allows us to introduce parameters. So instead of $[\mathcal{B},\mathcal{U}]$ we will be using the pair $[\mathcal{B},\mathcal{B}_u]$ where $\mathcal{B}_u$ is the set parameters describing the controlling field.

3. The problem definition

While describing fringing fields, a transition from functions to parameters can be made. Thus, the selection of a function approximately describing a given fringing field and its parametrical representation is made at the same time.

Consequently, after we have the appropriate quality functionals and restrictions, the following problem of nonlinear programming can be formulated:

find inf $I(\vec{B},\vec{B}_u)$,
limited by the equations: $h_i(\vec{B},\vec{B}_u) = 0,\ i = 1,\ldots,m$,
and inequalities: $g_i(\vec{B},\vec{B}_u) \geq 0,\ i = m+1,\ldots,p$.

Here $I(\vec{B}, \vec{B}_u)$, $h_i(\vec{B}, \vec{B}_u)$ and $g_i(\vec{B}, \vec{B}_u)$ could be either linear or nonlinear functions.

The author's method requires:

— the using of random search for getting initial approximations and the selection of areas suspicious to be enclosing the optimum;

— the further minimization of the said approximations using the sliding tolerance method (see [2]).

4. Global optimization

The solving of a nonlinear programming problem is a problem of global optimization. Usually, under global optimization, a search of one best decision giving a global minimum to the quality functional is understood. However, at the decision of real physical problems on beam lines designing, the given approach is impracticable.

In the present paper, under global optimization, we understand the consecutive process, allowing to receive a limited set of decisions optimum in a certain sense, their gradual rejection on a number of additional criteria, their research on satisfaction to admissions and realization possibility. The result of the given process is a small set of decisions on the basis of which the experimenter carries out a definitive choice of the decision for realization based on cost, reliability and convenience of the realization criteria.

The given process is mostly iterative. Varying different parameters that define the importance of some quality criteria, as well as entering of various criteria into the functional or their consideration in the form of restrictions and the use of personal experience in analysis of the results approaches the researcher to the required decision. Thus, the big role in the given process is played by the presence of the qualitative software and tools, allowing to carry out the necessary calculations and represent the received results in the form convenient for the analysis.

5. Solution to the task

Let us analyze the possible solutions for a so-called "russian quadruple" [3] system (see Fig. 2). For mathematical model details see [4,5]. The controlling parameters vector in this case is $\vec{B} = (s, \lambda, a, g; k_1, k_2)$. Let's then apply a number of limitations onto the controlling parameters as follows: $a = 150$, $g = 1$, $\lambda = 2$, $s = 0.5$.

Figure 3 shows graphs representing the load curves (solid line) and the working distance length restriction $g = 1$ (broken line). The corresponding

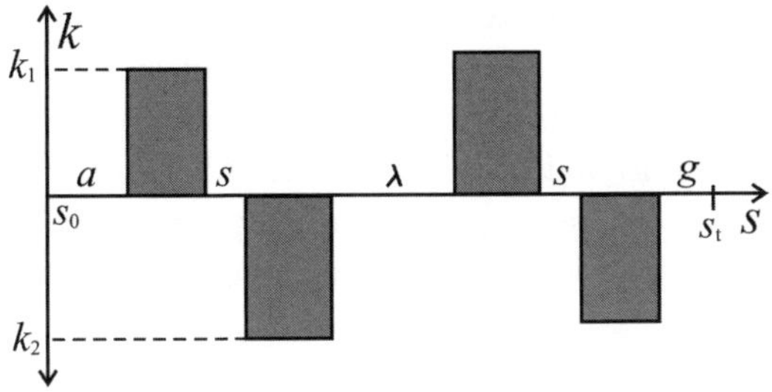

Fig. 2. Russian quadruplet focusing system.

intersection points are marked as follows: $K^1 = (0.54345, 0.69629)$, $K^2 = (1.37421, 1.00696)$, $K^3 = (2.23405, 1.07949)$.

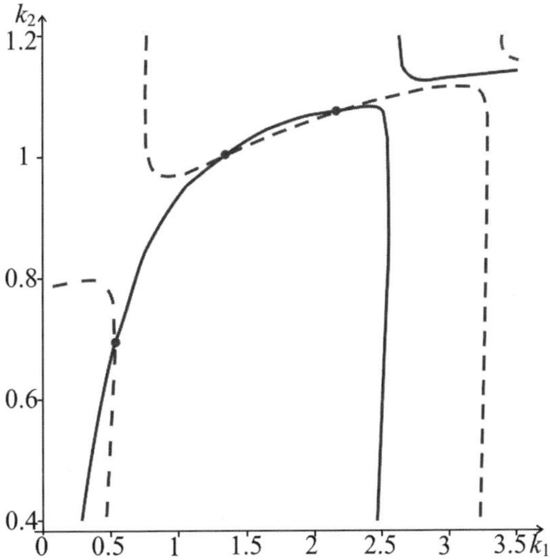

Fig. 3. Load curves and working distance length restriction $g = 1$.

The following beam compression factors correspond with the aforementioned points in the scope of our linear model: $r_{11}(K^1) =$ -0.0352203, $r_{11}(K^2) = 0.0061025$, $r_{11}(K^3) = 0.0087011$.

Regarding the compression factors, the most interesting is the point K^2. Note, however, the stability of this solution regarding the deviation of control fields from the values set.

For the solutions K^2 and K^3, a minor deviation from the optimal parameters (k_1, k_2) leads to a major break of load curves condition, which may also lead to the point-to-point focusing condition break.

Inside the neighborhood of the solutions K^1 and K^2 there is a stable enough conservation of the compression factor value. K^3 shows a major deviation of the compression factor while the certain minor parameter deviation takes place.

Therefore, at this stage we have the screening of the solution K^3. The two remaining solutions require further analysis for satisfying additional criteria such as fringing fields (see [6], for example) and nonlinear effects (see, for example, [5]), as well as influence of parameter deviation sensitivity of the load curves condition to the solution K^2.

6. Conclusion

In the present paper, the methodology of modeling and finding the optimal parameters for beam lines with high beam compression requirements, based on multi-objective analysis, is considered. The given methodology aims for solving the problem of taking into account many, and often contradicting, criteria of quality, consequently considering them. The linear model building and analysis, followed by the inclusion of additional limitations and criteria, leads to the shrinking of a set of solutions down to a finite fixed set, appropriate enough for the selection of a certain solution to be realized.

References

1. S. A. Lebed, *Double-mode probe forming systev for modern ion nanoprobe*, J. of Applied Rhysics, Vol.72, 2002, pp.92–95 (in Russian)
2. D. Himmelblau *Applied Nonlinear Programming* (1975).
3. S. N. Andrianov, A. D Dymnikov and G. M. Osetinsky, *Forming system for proton microbeams*, Instr. and Exper. Techn., Vol. 1, pp. 39–42 (1978) (in Russian).
4. S. Andrianov, N. Edamenko, A. Chernyshev and Yu. Tereshonkov. *Synthesis of Optimal Nanoprobe (Linear Approximation)*, Proc. of the EPAC'2008 (Genoa, Italy, 2008) pp. 2125–2127.
5. S. Andrianov, N. Edamenko and Yu. Tereshonkov *Synthesis of optimal nanoprobe (Nonlinear approximation)*, Proc. of the EPAC'2008 (Genoa, Italy, 2008) pp. 2972–2974.
6. Yu. Tereshonkov *Load Curves Distortion Induced by Fringe Fields Effects in the Ion Nanoprobe*, Proc. of the EPAC'2008 (Genoa, Italy, 2008) pp. 1514–1516.

MATHEMATICAL MODELING OF FRINGE FIELDS IN BEAM LINE CONTROL SYSTEMS

Yu. V. TERESHONKOV

Applied Mathematics and Control Processes, St. Petersburg State University
University avenue 35, Peterhof, Saint Petersburg, 198504, Russia
E-mail: yury.tereshonkov@gmail.com

Real magnet fields have a bell-shaped form, which describes a field distribution along the electrical axis of a steering element. This paper deals with methodology, which allows including fringe field effects during an initial stage of a modeling process. Also a wide spectrum of modeling functions for fringe field distributions is presented. It is known, that fringe field effects are intrinsic and unremovable effects and could heavily impact on the beam dynamic and corresponding beam characteristics. Mathematical and computer models for fringe field help to estimate their influence and include information of real magnet field into the designing model. Research is based on the matrix formalism for Lie algebraic tools. This approach gives large flexibility, because it could admit the usage of computer algebra methods and technologies. Computer algebra methods can be easily paralleling, which gives line benefit with increasing number of processors (cores). Besides, this approach can be extended for nonlinear aberrations without loss of generality.

Keywords: Fringe fields, beam optics, matrix formalism, Lie methods.

1. Introduction

Nanoprobes are very high-precision systems and sensitive to linear and nonlinear aberrations in accordance with experimental (see e. g. [1,2]) data and theoretical investigations (see, e. g. [5]). Moreover, the latest researches show that selection problem of *optimal* Focusing Probe Systems (FPS) variant (like nanoprobe) is not limited with one or two alternatives (see e. g. [5]).

In the present paper, one of the most significant factor, which affects on beam characteristics, is so-called *fringe field* of control elements (magnet lenses) (see e. g. [4,6]). According to the features of control elements it is required to differ the "in" and "out" fringe fields. Information about fringe fields could be obtained using experimental data from various sources or field maps, or as the result of numerical solution of Laplace's equation. All applied papers are attached to specified magnet lenses or some types of lenses. There are no mass measuring of real fringe fields due to high cost of similar measuring. So it is impossible to create some new system structures. One can create systems, which based on previous systems. In order to construct absolutely new system structure it is necessary to consider a set of approximation modeling functions for fringe fields. These functions help investigating of fringe field influence on beam characteristics.

The aim of the approximation process is to prepare a set of suitable functions in some appropriate class, in which the search of optimal configuration of control FPS will be performed. In other words, it is required to solve an "inverse problem": using the required final beam characteristics it is necessary to find a number of possible variants of steering fields.

2. Mathematical model of beam dynamics

Evolution operator (matrix propagator) is constructed as infinite dimensional up-triangular matrix, which consists of block matrices according to representation in Poincare-Witta basis(see e. g. [1]). With a glance of above mentioned assumptions matrix propagator is not depend on initial beam status if a space charge is neglected.

Particle motion equation in the neighborhood of optical axis in common case can be written as following:

$$\frac{d\mathbf{X}(s)}{ds} = \mathbf{F}(\mathbf{X}, s), \quad \mathbf{F}(\mathbf{0}, s) \equiv 0. \tag{1}$$

Using quite suitable assumption one can solve the initial value problem (1) in Poincare-Witta basis and obtain an infinite dimensional Taylor series as a solution. The corresponding equation and its solution in terms of matrix formalism can be written as following

$$\frac{d\mathbf{X}(s)}{ds} = \sum_{k=1}^{\infty} \mathbb{P}^{1k}(s)\mathbf{X}^{[k]}(s),\ \mathbf{X}(s) = \sum_{k=1}^{\infty} \mathbb{R}^{1k}(s|s_0)\mathbf{X}_0^{[k]},$$

where $\mathbf{X}^{[k]}(s)$ is the Kronecker k-th power for the phase vector $\mathbf{X}(s)$, $\mathbf{X}_0 = \mathbf{X}(s_0)$ is an initial phase vector, s_0 is an initial point. Here $\mathbb{P}^{1k}(s)$ are matrices with the entries equal to k-th derivative of the components of vector function $\mathbf{F}(\mathbf{X}(s), s)$. The matrices $\mathbb{R}^{1k}(s|s_0)$, $k \geq 2$ are called k-th order aberrations matrices, and they store the influence of all nonlinear effects up to k-th order. The whole matrix propagator can be presented *exactly* as production of partial matrix propagators using group property.

In common case it is possible to approximate steering field with piecewise constant, piecewise linear or more smoothness functions, for which the analytical matrix propagators are known or can be determined. Sometimes for some classes of steering fields one can find modeling functions (including composite), for which it is known the analytical solution (matrix propagator), and these functions are very closely to steering fields in norm.

In the present paper, analytical calculations are privileged due to its flexibility, lightness of parameters varying and multiple applicability in similar classes of tasks. It helps to make a decision about fundamental construction possibility of alike FPS with the glance of technological features of manufacturing and alignment for a quite wide class of facilities.

3. Particle motion equations

In linear approximation particle motion equations can be represented as following for quadruples and linear optical axis:

$$\begin{cases} x'' + k(s)x = 0, & x' = dx/ds, \\ y'' - k(s)y = 0, & y' = dy/ds, \end{cases} \tag{2}$$

where $k(s) = qG/(m_0 c\beta\gamma)$, c is a light speed, $\beta = |v|/c$, $\gamma = 1/\sqrt{1-\beta^2}$, $G = \partial B_x/\partial y|_{x=y=0} = \partial B_y/\partial x|_{x=y=0}$ is a gradient of magnet field, s is a length, which is measured along some reference orbit. Scalar equations (2) could be written in vector form $\dfrac{d\mathbf{X}(s)}{ds} = \mathbb{P}(s)\mathbf{X}(s)$, where $\mathbf{X}_0 = \mathbf{X}(s_0)$ is a initial vector. The aim of linear approximation (2) is to construct linear propagator $\mathbb{R}^{11}(s|s_0)$ (matrizant) for the whole system, with a glance of steering fields along the optical axis.

4. Control functions and parameters

In the present paper FPS structure allows representing control function $k(s)$ as piecewise smooth functions

$$k(s) = \begin{cases} 0, s \in [s_0, s_1), \Delta s_1 = s_1 - s_0, \\ k_2(s), s \in [s_1, s_2), \Delta s_2 = s_2 - s_1, \\ 0, s \in [s_2, s_3), \Delta s_3 = s_3 - s_2, \\ \ldots \\ k_n(s), s \in [s_{n-1}, s_n), \Delta s_n = s_n - s_{n-1}, \\ 0, s \in [s_n, s_{n+1}), \Delta s_{n+1} = s_n - s_{n+1}. \end{cases}$$

where $k_i(s)$ is a field of i-th control element, which can be also split on intervals. In mathematical model input and output fringe fields are control functions. Let us introduce the additional segmentation, which allows interpreting fringe fields as virtual control parameters, during modeling process for optimal solutions retrieval.

After similar segmentation the task of optimization of FPS can be formulated in terms of control functions and control parameters. The right part in (1) can be rewritten as $\mathbf{F}(\mathbf{X},\mathbf{U},\mathbf{B},s)$, where $\mathbf{U}(s) = (u_1(s),\ldots,u_n(s)) = (0,k_2,\ldots,k_n)$ is a vector of control functions and $\mathbf{B} = (\Delta s_1,\ldots,\Delta s_{n+1})$ is a vector of control parameters. Segmentation performs according to control elements location. Presentation of $k_i(s)$ as input and output parts of fringe field, and central part leads to increasing a set of control functions and parameters. In other words, $\mathbf{U}(s) = (k_2^{\text{in}}, k_2^{\text{out}},\ldots,k_n^{\text{in}}, k_n^{\text{out}})$, where $k_i^{\text{in}}, k_i^{\text{out}}$ are modeling functions for input and output fringe field parts of i-th control element correspondingly. The vector of control parameters in this case $\mathbf{B} = (\Delta s_1, k_{\max}^2, L_0^2,\ldots,k_{\max}^n, L_0^n, \Delta s_{n+1})$, where $k_{\max}^i, L_0^i$ is a maximum value of field gradient and length of central part of i-th control element. After introducing virtual modeling functions one can convert the control of functions and parameters to the whole set of control parameters, where parameters: $\Delta s_i, k_{\max}^{i+1}, L_0^{i+1}$ and $A_{i+1}^{\text{in}}, A_{i+1}^{\text{out}}$ are vectors of parameters describing $i+1$-th input and output fringe field modeling functions.

5. Fringe field forming problem

Some special case of fringe field are discussed, for example, in [4–6]. However it is not enough for thorough analysis and detailed modeling. Mathematical and computer models of fringe fields are not work out in detail.

All standard control elements generate magnet field, which is symmetric relative to the center of control element. Thereby, we consider only symmetric fringe fields (see e. g. [4]) relative to the center point $s_c = (s_2 - s_1)/2$ for each element. In the present paper is supposed, that fringe fields of nearby control elements (doublets, triplets) do not interact and the result steering field is determined as linear fields superposition.

Due to laws of electrodynamics and experimental data, control magnet field is a smoothness function, which can be presented in following form:

$$f(s) = f_0 \begin{cases} f_{\text{in}}(s), & s \in [s_0, s_1), \\ k_{\max}, & s \in [s_1, s_2), \\ f_{\text{out}}(s), & s \in [s_2, s_3]. \end{cases}$$

Functions $f_{\text{in}}(s)$ and $f_{\text{out}}(s)$ describe input and output fringe fields correspondingly. In order to make $f(s)$ smooth in positions of joint it is required to demand the additional conditions:

$$\begin{aligned} f_{\text{in}}(s_0) = f_{\text{out}}(s_3) = f'_{\text{in}}(s_0) = f'_{\text{out}}(s_1) = f'_{\text{in}}(s_2) \\ = f'_{\text{out}}(s_3) = 0,\ f_{\text{in}}(s_1) = f_{\text{out}}(s_2) = 1, \end{aligned} \tag{3}$$

where $f'(s) = df(s)/ds$. Then $f_{\text{in}}(s)$ is determined with the aids of different functions approximation. The output fringe field $f_{\text{out}}(s)$ can be automatically found using the symmetric reflection relative to s_c.

Instead of Eq. (3) it is possible to introduce the asymptotic analogues, which allow considering more extensive class of modeling functions:

$$\begin{aligned}
&\lim_{s\to+s_0} f_{\text{in}}(s) = \lim_{s\to+s_0} f'_{\text{in}}(s) = \lim_{s\to-s_1} f'_{\text{in}}(s) = 0,\\
&\lim_{s\to-s_1} f_{\text{in}}(s) = \lim_{s\to+s_2} f_{\text{out}}(s) = 1,\\
&\lim_{s\to-s_3} f_{\text{out}}(s) = \lim_{s\to+s_2} f'_{\text{out}}(s) = \lim_{s\to-s_3} f'_{\text{out}}(s) = 0.
\end{aligned}$$

6. Fringe fields simulation

On the initial stage of modeling it is easy to use the piecewise constant model for fringe fields approximation, which allows varying number of segmentation intervals. If it is necessary to consider some fringe field effects in detail, one can use piecewise linear approximation, bell-shaped function, Enge functions [4] and so on, which are better than piecewise constant model.

On the next stage it is required to find modeling functions, which can approximate fringe field on the whole interval or on its parts. It is useful to approximate fringe fields with modeling functions, for which it is known an analytical solutions (matrizants). Modeling functions can be composed and consists two or more parts. It is preferred to select modeling functions with free parameters in order to approximate real steering field more accurately with a glance of experimental data and find some optimal solutions for FPS.

7. Some function classes with analytical solution

Using the methodology from [3] let us show, for example, three classes of functions and some their entries, for which on can found the analytical solution (matrizants): $f_1(s) = \psi(s)e^{\varphi}(s)$, $f_2(s) = \varphi(s)\cos\psi(s)$, $f_3(s) = \psi^n\dot{\psi}^m$. Examples from the first class can be the following functions: $(a+bs)^{-4}, \alpha^2 + (a\sin\alpha s + b\cos\alpha s)^{-4}$, $1+2n-s^2+(e^{2s^2})/[H_n(s)]^4$, where $H_n(s)$ is a n degree Hermitian polynomial. On the For the second class one can find the following examples $n^2 s^{2n-2} - (n^2-1)/(4s^2)$, $1/2 - \cos^2\alpha s + 3/4\tan^2\alpha s$. Finally, third class consists of e. g. the following functions $n - n(n-1)\tan^2\alpha s$, $n(3-s^2) - n(n-1)(s-1/s)^2$, $(1+4s^2-4n^2)/(4s^2) - 2J_n(s)/\left(sJ_{n+1}(s)\right)$, where $J_n(s)$ is a n degree Bessel function.

8. Conclusion

The methodology for mathematical and computer modeling of fringe fields are presented in detail. Some parts of fringe fields can be presented as virtual control elements with the number of parameters. It allows transferring from control functions to the representation with control parameters only. Moreover, several examples of fringe fields modeling functions, which have analytical solutions are presented. For functions which do not have analytical propagators it is proposed to approximate them with piecewise constant, piecewise linear or even more smoothness functions in order to get analytical propagator in common case. The proposed approaches allows finding optimal working modes and structure of FPS with a glance of fringe field effects. Moreover, it is possible to investigate fringe fields influence on beam characteristics.

References

1. S. N. Andrianov, *Dynamical Modeling of Beam Particle Control Systems*, St. Petersburg State University, St. Petersburg, 2004. (in Russian).
2. S. N. Andrianov, A. D. Dymnikov and G. M. Osetinsky, Forming system for proton microbeams. *Instr. and Exper. Techn.*, Vol. 1, pp. 39–42, 1978. (in Russian).
3. T. A. Antone and A. A. AL-Maaitah, Analytical solutions to classes of linear oscillator equations with time varying frequencies, *J. of Mathematical Physics*, Vol. 33, Issue 10, pp. 3330–3339, 1992.
4. Martin Berz, Béla Erdélyi and Kyoko Makino, Fringe Field Effects in Small Rings of Large Acceptance. In *Phys. Rev. ST Accel. Beams*, Volume 3, 124001, 11 P, 2000.
5. Yu. Tereshonkov and S. Andrianov, Load Curves Distortion Induced by Fringe Fields Effects in the Ion Nanoprobe. In *Proc. EPAC'08*, Italy, Genova, pp. 1514–1516, 2008.
6. M. Venturini, D. Abell, and A. Dragt, Map Computation from Magnetic Field Data and Application to the LHC High-Gradient Quadrupoles. In *Proc. Int. Computational Accel. Phys. Conf.*, USA, Monterey, California, pp. 184–188, 1998.